An Introduction to Population Ecology

FRONTISPIECE: A memorial of a time of high natality and high mortality. Sir Ralph Pudsay with his three wives and twenty-five children, 1468. Bolton by Bowland, Yorks. F. A. Greenhill, *Incised Effigial Slabs*. London: Faber & Faber, 1976 (by kind permission).

G. EVELYN HUTCHINSON

An Introduction to Population Ecology

New Haven and London
Yale University Press

Published with assistance from
the Louis Effingham deForest Memorial Fund.

Fourth printing, 1979.

Designed by Thos. Whitridge
and set in Monophoto Bembo type by
Asco Trade Typesetting Ltd., Hong Kong.
Printed in the United States of America by
The Murray Printing Company, Westford, Mass.

Published in Great Britain, Europe, Africa, and
Asia (except Japan) by Yale University Press,
Ltd., London. Distributed in Australia and New
Zealand by Book & Film Services, Artarmon,
N.S.W., Australia; and in Japan by Harper & Row,
Publishers, Tokyo Office.

Library of Congress Cataloging in Publication Data

Hutchinson, George Evelyn, 1903–
An introduction to population ecology.

Includes indexes.
1. Ecology. 2. Biotic communities. I. Title
QH541.H87 574.5'24 77-11005
ISBN 0-300-02155-0

In Piam Memoriam

A. L. H.

Fratris Amici

Contents

Preface

THIS book is ultimately derived from a course of lectures that I gave for many years under the title Ecological Principles. Though primarily addressed to first year graduate students and senior undergraduates whose main interests were biological, the course was open to, and sometimes taken by, students from other disciplines, not all in the natural sciences.

After I had given the lectures to so diversified an audience, they seemed to turn into a written version that would appeal not only to beginners in ecology but also to an occasional historian, economist, or sociologist, and even more to those biologists in other areas of the science who may have wondered what the people who talked unceasingly about populations were trying to say.

It is impossible to write about population ecology without using some mathematics. In spite of the appearance of a few pages, the amount employed here is pared down to a minimum. All that one really needs to follow the argument is a memory of the simplest parts of high school algebra, a knowledge that dN/dt means the rate of change of N (here the size of the population, with time denoted by t), and enough faith to take a few results from linear algebra on trust. It also helps to have an idea of what integration is; this is explained in a very simpleminded way in an appendix, which was handed out to the class, with the title *Ratiocinator infantium*. Once over the early pages of the initial chapter, most of the mathematics can be skipped at first reading without losing the ultimate biological content of the work.

The subject matter of the book is developed along classical lines. This may appear old-fashioned to those who would prefer a more stochastic treatment or who put most of their trust in computer simulation. The approach, however, seems to have been appreciated by a surprisingly large number of people who have become distinguished ecologists. I cannot help thinking that what I have found interesting and useful may also interest and help others.

Even with the obvious restrictions that are imposed on the field, it is impossible, in a book intended as an introduction, to discuss everything that might appear related to the themes that are developed. For those who want more, there are now a number of compendious works. I would particularly recommend:

R. E. Ricklefs, *Ecology*. Newton, Mass.: Chiron Press, 1973, x, 861 pp.

R. Margalef, *Ecología*. Barcelona: Ediciones Omega, S.A., 1974, xv, 951 pp.

F. Schwerdtfeger, *Ökologie der Tiere*. Hamburg and Berlin: Paul Parey; vol. 1: *Autökologie*, 1963, 461 pp.; vol. 2: *Demökologie*, 1968, 448 pp.; vol. 3: *Synökologie* 1975, 449 pp.

Anyone seriously interested will need to be familiar with:

C. S. Elton, *Animal Ecology*: London. Sidgwick and Jackson, 1927, xvii, 207 pp., a short, quite unmathematical but deeply fundamental book, which has improved with reading over half a century. I would also suggest:

L. B. Slobodkin, *Growth and Regulation of Animal Populations*. New York: Holt, Rinehart, and Winston, 1961, viii, 184 pp.

Both these books, being small, are particularly recommended for desert-island reading.

I would also note, as a close relative of what I have written, though its title may not indicate this:

Alison Jolly, *The Evolution of Primate Behavior*. New York: Macmillan, 1972, xiii, 397 pp.

The best places to get an idea of what is being done at the moment are, of course, the recent numbers of journals such as *Ecology*, *Evolution*, and *The American Naturalist*. Two books, moreover, give a very up-to-date picture:

M. L. Cody and J. M. Diamond, editors, *Ecology and Evolution of Communities*. Cambridge, Mass.: Belknap Press of Harvard University Press, 1975, xiv, 545 pp.

R. M. May, editor, *Theoretical Ecology, Principles and Applications*. Philadelphia and Toronto: W. B. Saunders, 1976, viii, 317 pp.

I am deeply aware of the help of a great variety of friends. That birds figure largely in any work devoted to animal ecology is partly due to most species being diurnal and possessing color vision, but also to the genius of two of the outstanding investigators of the subject, David Lack (1910–73) and Robert Helmer MacArthur (1930–72), whose close friendship I enjoyed and whose memories I cherish. Many other students of birds have helped me. I particularly would acknowledge, with great affection, the influence of S. Dillon Ripley which extends far beyond the conventional limits of ornithology.

Every student who asked questions or contributed to the discussions that followed the lectures probably added something to the book. I thank them all.

The debt to my colleagues at Yale University is great; individual enumeration would be invidious as the number of people to be thanked is so great that some would be accidentally omitted. Four institutions in the university have, however, meant so much to me and my work that I want to mention them specifically. I first take this opportunity to express my deep appreciation, over more than the past three decades, to the fellowship of Saybrook College in Yale University. The historical and other ancillary material used here owes a great deal to friendships formed in that institution where Basil Duke Henning, Master of the College for many years, and his wife Alison created a tradition which I know their successors, Elisha and Elizabeth Atkins, will continue, of the loving exchange of learning, which is the greatest value that a university can afford. Of the present fellows I would particularly thank Richard S. Miller who has read the book in manuscript, Phoebe Ellsworth who helped in the hunt for the elusive Roswell Johnson, the reluctant inventor of the ecological niche, and who encouraged my belief that what I have written can be read by really literate social scientists, and Marjorie Garber who unwittingly led me, by a beautiful if circuitous path, to the final lines of the book. Second, the collections of the Peabody Museum were always available to provide classroom demonstrations of organisms mentioned in lectures. No one can have profited from this more than I did. At a time when such institutions do not take a very high place in the priorities of university administrators, I would emphasize the importance of a natural history museum with a really extensive collection in giving substance to names and ideas that might otherwise become rather meaningless abstractions. I would specifically like to thank Charles Remington for perennial help relating to insects and Eleanor Stickney for her kindness in the bird room. Third, the Kline Science Library and its staff, ever willing to help satisfy the most impossible requests and indeed often able to do so, demand my admiration as well as my gratitude. Last, the Beinecke Rare Book and Manuscript Library proved to contain a remarkable number of books and pamphlets bearing on the history of ideas about population. It is also a delightful place in which to work.

Both Eric L. Charnov of the University of Utah and Robert H. Whittaker of Cornell University have read the book in manuscript. I am deeply indebted to them for many suggestions.

I thank Messrs. Faber and Faber for permission to publish the frontispiece and the Guildhall Museum, London, for figures 3, 4, and 5. I am extremely grateful to Maxine

Watson for figures 66, 67, and 104 based on material now in press in her own papers, and James Porter for figure 134. Pamela Parker and Rudi Strickler both made available manuscripts of papers now in press. William B. Keller made the transcription of the epitaph on the memorial to Malthus in Bath Abbey. Mary M. Poulson helped in innumerable ways. For assistance in the final stages of prepaning the manuscript, I thank Anna Aschenbach. For the excellence of the typescript derived from a refractory longhand manuscript I thank Beverley Dooling and Alice Pickett, and I am specially grateful to Virginia Simon for the loving care lavished on the illustrations. At the Yale University Press, Lottie M. Newman has been a superb copyeditor and Jane Isay, as a perfect godmother, has watched over the growth of the book from its earliest infancy.

My gratitude to my wife increased daily as each new page was written.

G. E. H.

New Haven, Connecticut

All Hallow's Eve 1976

Chapter One

M. Verhulst

IN this first chapter we will look at some of the ways in which we can construct mathematical models of populations. For this purpose we shall need, in addition to the ordinary postulates of mathematics which we may assume to be satisfactory for our purposes, at least two additional biological postulates. These appear to follow from everyday experience, though the first of them was not generally believed of all organisms until the nineteenth century, while a glance at a newspaper often suggests that many people behave as if the second postulate were false.

The postulate of parenthood

This states that every living organism has arisen from at least one parent of like kind to itself; it is often called the principle of abiogenesis and expressed epigrammatically as *omne vivum ex vivo*. For anyone who believes in the initial terrestrial origin of life, the postulate is not universally valid; but since under present conditions spontaneous generation has never been observed, we can take it as true enough to use in our investigation of contemporary populations.

The use of this postulate limits our investigation to living beings, trees but not the telegraph poles that can be made from them, people but not the cars that they drive. The stochastic dynamics of destruction or death can be applied to sets of both nonliving and living objects, but the equivalent dynamics of birth only to living populations.

The postulate of an upper limit

The second postulate is that in a finite space there is an upper limit to the number of finite beings that can in some way occupy or utilize the space under consideration. This may be merely a geometrical limitation as in some sessile animals like barnacles; in such a case it is deducible from mathematics. Much more often it is due to the objects requiring a supply of energy at a certain rate to maintain their stability; obviously the space cannot contain more of such objects than utilize the energy input into the space. Comparable limits may be set by the rate of supply of nutrient materials such as water, carbon dioxide, phosphorus, nitrogen, or soluble iron. In the case of animals, more complicated situations, involving limitation by a food supply that itself depends on the rate of supply of radiation, water, or nutrients, are of course frequent. More subtle limitations, notably involving territoriality, are common among the more complex invertebrates and the vertebrates. Some of these types of limitations will be discussed in greater detail in later chapters.

Convention of continuity

In addition to the biological postulates of parenthood and of an upper limit, it is convenient for mathematical reasons initally to adopt the convention that the variation of a population of size N behaves as if N were a continuous variable, capable of taking any value, integral or fractional, between the possible lower and upper limits of the population. This permits the use of the infinitesimal calculus. The convention, though strictly untrue, is harmless when we are dealing with a sufficiently large population of organisms not having definite breeding, or dying, seasons, in which reproduction occurs at random among all members of the appropriate age class, and death occurs according to some statistically defined pattern not varying with time. When definite breeding seasons occur, or when mortality is much greater at some times of year than at others, finite difference equations have to be used. At first this seems an inelegant and inflexible approach, but as will later be apparent, it can lead to some remarkable results.

The logistic equation of Pierre-François Verhulst

The initial approach to an equation of population growth, ultimately due to Pierre-François Verhulst, of whom more hereafter, will be that of Lotka, as developed in his remarkable book *The Elements of Physical Biology*.[1] This book was written to provide for biology, or at least parts of biology, a basis comparable to that given by theoretical physics to experimental physics. Lotka probably knew a smaller proportion of the relevant biology of his time than a theoretical physicist usually knows of experimental physics. This gives parts of his book a curiously naïve character. In spite of this limitation it is a great work, one of the foundation stones of contemporary ecology.

Lotka generalizes the behavior of a population in the following equation:

$$\frac{dN}{dt} = f(N). \qquad (1.1)$$

This merely tells us that the rate at which the number of individuals, N, in the population changes with time depends in some way on the number present. This sounds very reasonable, but it could be untrue, in a sexually reproducing species, for all $N > 2$. Suppose a species with a restricted breeding season and the property that any individual that starts to breed inhibits chemically or behaviorally the breeding of all other individuals, except its own mate, in a small localized population. This could mean that only one pair could breed at a time and consequently the rate of growth of the population would be independent of N. Cases where a single dominant male inhibits the breeding activity of many other males are well known, so the discovery of a case of this sort is not quite out of the question; small populations of babblers of the genus *Turdoides* can indeed behave in this way.[2] This example at least indicates that we may not always find $f(N)$ to be a very simple function.

To go back to equation (1.1), it at first looks trite and uninteresting. Since, however, from Taylor's Theorem it may be expanded as a power series, we may write

$$\frac{dN}{dt} = a + bN + cN^2 + dN^3 + . \qquad (1.2)$$

This looks more interesting than equation (1.1). From our postulate of parenthood, if $N = 0$, $dN/dt = 0$, so $a = 0$, and we may write

$$\frac{dN}{dt} = bN + cN^2 + dN^3 + . \qquad (1.3)$$

Since we are looking for a suitable equation we may regard (1.3) as a set of equations in which the coefficients have any rational value including zero. We then invoke the principle of parsimony or simplicity, often called Ockham's razor, after the fourteenth-century English Franciscan friar William of Ockham,[3] which tells us that it is vain to do by more what can be done by less. In this case we begin by considering just one term, bN,

1. A. J. Lotka, *Elements of Physical Biology*. Baltimore, Williams and Wilkins, 1925, xxx, 460 pp. Reprinted in 1956 as *Elements of Mathematical Biology*. New York, Dover.

2. For the reproduction of babblers see A. J. Gaston, Brood parasitism by the pied crested cuckoo *Clamator jacobinus*. *J. Anim. Ecol.*, 45:331–48, 1976.

3. William of Ockham, Ockam, or Occam, was born at Ockham in Surrey, England, late in the thirteenth century. He studied at Merton College, Oxford, which at that time was a center of scientific thought. At some time while at Oxford he became a Franciscan. He taught in the university until 1323. His later life was largely taken up in controversy with Pope John

before going on to $cN \cdot N$ or two or more combinations of terms. Our first attempt gives

$$\frac{dN}{dt} = bN \tag{1.4}$$

or

$$N = e^{bt}$$

which is what Malthus, and indeed his predecessors, thought in principle, and correctly, about the inherent growth of populations, but which here must be rejected since there is no upper bound. This is true of all equations using just one term of the expansion.

In equation (1.4), b, which represents the unrestricted rate of increase per individual, birthrate minus death rate, in the kind of ideal population that we are considering, is often called the Malthusian parameter.

XXII and his successors, at first over the concept of evangelical poverty and later on the question as to whether the Emperor could depose the Pope. After captivity at Avignon, from which he escaped, he spent the later years of his life in Munich, under the protection of Ludwig IV of Bavaria. He died late in the 1340s, probably as an excommunicated heretic; he was buried in the Franciscan Church in Munich, which was pulled down early in the last century.

The famous passage *quia frustra fit per plura quod potest fieri per pauciora*, "because it is vain to do by more what can be done by fewer," occurs casually, almost as if it were a universally accepted opinion, as may well have been the case, in a demonstration that substance itself is quantity, *ipsamet substantia est tunc quantitas*, in the *Tractatus quam gloriosus de sacramento altaris, et in primis de puncti, lineae, superficiei, corporis, quantitatis, qualitatis et substantiae distinctione venerabilis Inceptoris Guilhelmi Ockham Anglici* . . . (The very celebrated tract on the sacrament of the altar and particularly on the distinction of the point, line, surface, solid, quantity, quality, and substance, of the venerable Inceptor William Ockham the Englishman . . .). An inceptor was someone who had not yet achieved a higher degree; on a copy of his lost tombstone he appears as doctor. The book, apparently based on lectures given at Oxford, seeks to establish a foundation theory of a geometrical kind as the basis for the exposition of the doctrine of transubstantiation in the Holy Eucharist. The *De Sacramento Altaris* is an extremely difficult book for anyone not fully familiar with the language of medieval philosophy. The most accessible edition, which, however, seems at times not to meet the most exacting standards of modern scholarship, is one edited by T. Bruce Birch (Burlington, Iowa, Lutheran Literary Board, 1930, xlvii, 576 pp.). This edition provides, opposite the Latin, a literal English translation which lovingly preserves all the obscurities of the original; the passage quoted may be found on on p. 104 (Latin), p. 105 (English).

Ockham had an important influence on Luther and has been hailed as the forerunner of many later philosophers from Locke to Bertrand Russell. Modern writers have discussed the simplicity postulate in a variety of ways. The approaches used by Harold Jeffreys in *Scientific Inference* (Cambridge University Press, 1931, 1937) are of particular scientific interest.

Jeffreys argued more or less as follows. Given any set of data the total number of laws that can explain the data is infinite. Jeffreys restricts his treatment to quantitative laws, expressible as differential equations that can be rationalized to have integral exponents and coefficients. There is a denumerable infinity of such laws. If we consider any set of data, the sum of the prior probabilities of all the laws must be unity, unless we assume that full explanation of the data is forever impossible. The probabilities cannot be infinitely small, and evidently also cannot be finite and equal. Jeffreys supposes that they are finite and unequal and form a convergent series summing to unity. The differential equations representing different laws can each be given a number by summing the order, degree, and a term representing the numerical coefficients. In the kind of equation that we use in biology, in which the coefficients such as r or K can have any of a wide range of values, representing boundary conditions of a sort, it is probably best to use just the irreducible number of such terms. Thus

$\frac{dN}{dt} = bN$ would have a complexity of 3.

$\frac{dN}{dt} = bN - cN^2$ would have a complexity of 5.

$\frac{dN}{dt} = bN - cN^2 + dN^3$

would have a complexity of 7.

Jeffreys rejects the idea that simple laws are initially chosen for their convenience, because this mode of choice would give no confidence that the law would be correct if a value of N previously untested was considered in the equation. He therefore maintains that the simple law is in some sense actually more likely to be true than any randomly chosen law of greater complexity. This view leads to certain difficulties, though it does describe the attitude of the investigator trying one of a hitherto unknown body of laws. This aspect of simplicity does not seem to have been much considered in recent years. Most readers will probably feel that whatever other reasons we may have for adopting Ockham's razor, it does give the best entry into further complexities. As Lagrange said, "Seek simplicity but distrust it."

In later discussions we shall use r rather than b for the Malthusian parameter or unrestricted rate of increase, since r has been so widely used that it has become part of compound nouns, notably r-selection. For the present, while we are constructing our equation, b will be retained.

Still following William of Ockham, we now try the two simplest terms together, taking

$$\frac{dN}{dt} = bN + cN^2 \qquad (1.5)$$

which after a moment's thought, writing $c = -b/K$, becomes

$$\frac{dN}{dt} = bN\frac{(K - N)}{K}$$

or as we shall now write

$$\frac{dN}{dt} = rN\frac{(K - N)}{K} \qquad (1.6)$$

which expresses, in the simplest possible way, the growth of a population increasing in a biological manner to an upper asymptote K, continuously and without catastrophes. The population increases at a rate determined by N, the number of individuals present, their maximum rate of increase r, and the proportion of the potential asymptotic population K which is still unrealized. The term $(K - N)/K$ giving this proportion would today be called negative feedback, but there is nothing in the equation to suggest how the feedback works.

The rate of increase rises slowly to a maximum as N reaches $K/2$, and then falls asymptotically to zero as N approaches K. Integration gives

$$N = \frac{K}{1 + e^{-rt}} \qquad (1.7)$$

time (t') being now measured backward and forward, from the time of the maximum value of dN/dt, when $N = K/2$.

Graphically the integral curve of a growing population is sigmoid, at first rising slowly but increasingly fast from any arbitrarily small starting population, inflecting when $N = K/2$ and then ever more slowly approaching the asymptote K (figure 1).

Experience has shown that most people prefer to think about the differential equation but to draw the integral graph. With regard to the latter it should be noted that if we artificially set $N > K$, the feedback term becomes negative, the population falling asymptotically to K, as indicated in the upper branch in figure 1. Richard Levins[4] pointed out that if in the usual form of the equation,

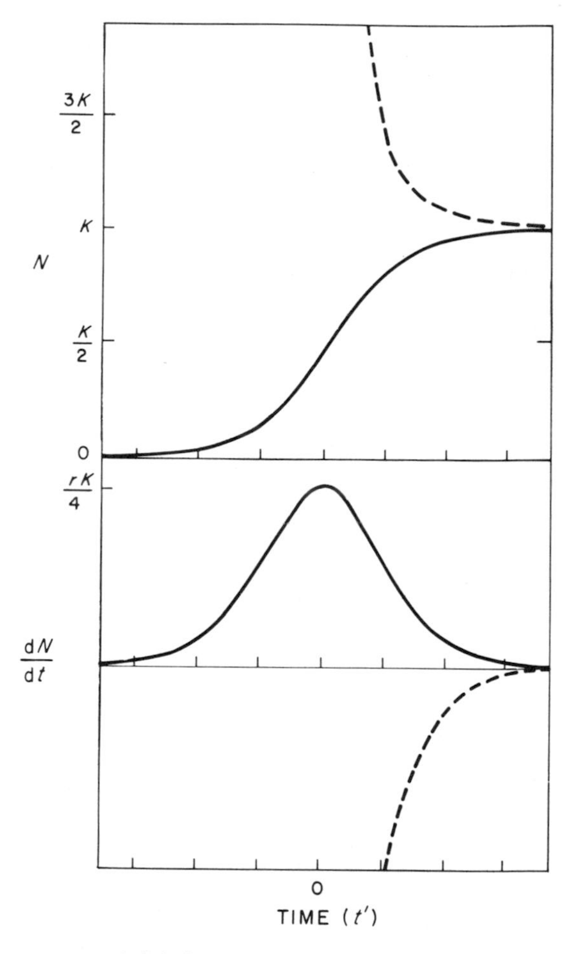

FIGURE 1. Solid line, upper panel: population increasing according to Verhulst's logistic; lower panel: rate of increase of such a population with a maximum corresponding to the inflection point in the upper curve, at $t' = 0$. Broken line, upper panel: hypothetical population declining from infinity at $t' = 0$ according to the logistic; lower panel, rate of such decline (in part after Lotka and after Pearl and Reed).

4. Richard Levins in an informal ecology seminar entitled "Some things that don't work," May 27, 1971, Osborn Memorial Laboratories, Yale University.

r, which is ordinarily regarded as an inherent birthrate (**b**) minus an inherent death rate (**d**), be negative, setting $N > K$ leads to an impossible result, for dN/dt is positive and increases indefinitely with N. An additional biological postulate is clearly required to imply that $r \geq 0$, so avoiding negative populations as well as impossible values of the rate of increase in a dying overcrowded world.

The very explicit presentation of Lotka's approach to the logistic, far more elementary and more detailed than his own, has been given to exhibit quite clearly what is behind most of the theoretical reasoning that we shall use. The argument is not axiomatic, there are plenty of places where it can go wrong. Williams[5] has pointed out that a much more rigorous derivation is possible. This involves several postulates as to the uniformity of the environment in space and time and the uniform distribution of organisms in it; these postulates, which express what one tries to achieve in a properly constructed experiment, are described by Williams as *instrumental*. It also involves two postulates of a biological, or as Williams calls it an *anacalyptic*, kind: that the properties of all organisms are the same at any time, and that the properties of a single organism are invariant with time. Like the convention of continuity, these postulates are strictly untrue; but in many circumstances the divergence from truth, within a genetically uniform population, is unimportant. If one adds the further anacalyptic postulate that all organisms with respect to their impact on the environment or on each other are identical through time, density cannot alter the reproductive rate and we can derive the exponential growth equation. If instead one postulates that growth rate declines linearly with increasing density, one gets the logistic. In the treatment based on Lotka, stopping at the second term to give an upper bound is in fact equivalent to the postulate of an inverse linear dependence of growth rate on density.

A historical digression

While it was convenient to present the logistic equation in Lotka's way, so making clear what it is about, it is also interesting to see how the ideas on which it is based came into being, particularly as their history is rather curious.

Scientific demography may be said to have begun in 1662, with the publication of John Graunt's *Natural and Political Observations mentioned in a following index and made upon the Bills of Mortality*.[6] Graunt was mainly a collector and classifier of facts, which he obtained initially by tabulating the Weekly Bills of Mortality that were published, from the early sixteenth century onward, recording the deaths in the City of London, in the first instance as an early warning of the incidence of plague. Christenings were also reported, at least from the later years of the sixteenth century, and in the next century indications of causes of death other than plague were given. The reports were submitted to each Parish Clerk by two searchers, honest and discrete matrons sworn truly to search the body of every person dying in the parish; no great level of competence in pathology, even by seventeenth-century standards, can be expected (figures 2, 3). In addition to the

5. F. M. Williams, Mathematics of microbial populations, with emphasis on open systems. *Trans. Conn. Acad. Arts Sci.*, 44:397–426, 1972.

6. *Natural and Political Observations mentioned in a following index, and made upon the Bills of Mortality by John Graunt Citizen of London* . . . London. Printed by Tho: Roycroft, for John Martin, James Allestry, and Tho: Dicas at the sign of the Bell in St. Paul's Churchyard. MDCLXII.

John Graunt (1620–74) was by trade a haberdasher; evidently a public-spirited man, and though a captain and later a major in the trained band, he was well known for his peaceable disposition. He was elected to the Royal Society as the result of a "Recommendation which the *King* himself was pleased to make, of the judicious Author of *The Observations on the Bills of Mortality*; In whose Election, it was so far from being a Prejudice, that he was a Shop keeper of *London*; that his Majesty gave this particular Charge to his Society, that if they found any more such Tradesmen, they should be sure to admit them all, without any more ado" (T. Sprat, *The History of the Royal Society*. London, 3rd ed., 1722, p. 67).

With regard to Graunt's calculation, the fact that we now would put the first human couple, however we defined them, at a point much further back in time, merely strengthens the argument.

London 9 From the 11 of February to the 18. 1661

Parish	Bur.	Pl.	Parish	Bur.	Pl.	Parish	Bur.
St. Albans Woodstreet	1		St. Gabriel Fenchurch	1		St. Martins Iremongerla.	
St. Alhallowes Barking	2		St. George Botolphlane			St. Martins Ludgate	
St. Alhallowes Breadstreet	1		St. Gregories by St. Pauls	4		St. Martins Orgars	1
St. Alhallowes Great	2		St. Hollens			St. Martins Outwitch	
St. Alhallowes Honylane			St. James Dukes place	1		St. Martins Vintrey	1
St. Alhallowes Lesse			St. James Garlickhithe			St. Matthew Fridaystreet	
St. Alhall. Lumbardstreet			St. John Baptist	2		St. Maudlins Milkstreet	
St. Alhallowes Staning			St. John Evangelist			St. Maudlins Oldfishstr.	
St. Alhallowes the Wall			St. John Zachary			St. Michael Bissishaw	
St. Alphage	4		St. Katharine Coleman			St. Michael Cornhill	2
St. Andrew Hubbard	2		St. Katharine Creechurch	2		St. Michael Crookedlane	1
St. Andrew Undershaft	1		St. Lawrence Jewry			St. Michael Queenhithe	1
St. Andrew Wardrobe	4		St. Lawrence Pountney			St. Michael Quern	1
St. Anne Aldersgate			St. Leonard Eastcheap	1		St. Michael Royall	
St. Anne Black-Fryers	2		St. Leonard Fosterlane			St. Michael Woodstreet	
St. Antholins Parish			St. Magnus Parish	2		St. Mildred Breadstreet	1
St. Austins Parish			St. Margaret Lothbury	2		St. Mildred Poultrey	
St. Bartholom. Exchange			St. Margaret Moses	1		St. Nicholas Acons	2
St. Bennet Fynck	1		St. Margaret Newfishst.			St. Nicholas Coleabby	1
St. Bennet Grace-Church			St. Margaret Pattons			St. Nicholas Olaves	
St. Bennet Paulswharfe			St. Mary Abchurch	1		St. Olaves Hartstreet	1
St. Bennet Sherehog			St. Mary Aldermanbury			St. Olaves Jewry	2
St. Botolph Billingsgate			St. Mary Aldermary	1		St. Olaves Silverstreet	
Christs Church	2		St. Mary le Bow			St. Pancras Soperlane	1
St. Christophers			St. Mary Bothaw			St. Peters Cheap	
St. Clement Eastcheap	1		St. Mary Colechurch			St. Peters Cornhil	
St. Dionis Backchurch			St. Mary Hill	2		St. Peters Paulswharf	1
St. Dunstans East	2		St. Mary Mounthaw			St. Peters Poor	
St. Edmun. Lumbardstreet			St. Mary Summerset			St. Stephens Colemanstr.	2
St. Ethelbourgh			St. Mary Staynings	1		St. Stephens Walbrook	2
St. Faiths			St. Mary Woolchurch	1		St. Swithins	2
St. Fosters	1		St. Mary Woolnoth	1		St. Thomas Apostle	4
						Trinity Parish	

Buried within the 97 Parishes within the Walls, of all Diseases — 75 *Whereof, of the Plague* — 0

Parish	Bur.	Pl.	Parish	Bur.	Pl.	Parish	Bur.
St. Andrews Holborne	11		St. Botolph Aldgate	13		St. Saviours Southwark	23
St. Bartholomew Great	1		St. Botolph Bishopsgate	9		St. Sepulchers Parish	[illegible]
St. Bartholomew Lesse			St. Dunstans West	5		St. Thomas Southwark	[illegible]
St. Brides Parish	12		St. George Southwark	7		Trinity Minories	[illegible]
Bridewel Precinct			St. Giles Cripplegate	22		At the Pesthouse	
St. Botolph Aldersgate	3		St. Olaves Southwark	11			

Buried in the 16. Parishes without the Walls, and at the Pesthouse — 130 *Whereof, of the Plague* — 0

Parish	Bur.	Pl.	Parish	Bur.	Pl.	Parish	Bur.
St. Giles in the Fields	22		Lambeth Parish	4		St. Mary Islington	[illegible]
Hackney Parish	2		St. Leonard Shoreditch	10		St. Mary Whitechappel	[illegible]
St. James Clerkenwell	6		St. Magdalen Bermond.	7		Redriffe Parish	
St. Katharine Tower	6		St. Mary Newington			Stepney Parish	24

Buried in the 12 out Parishes in Middlesex and Surrey — 113 *Whereof, of the Plague* — 0

Parish	Bur.	Pl.	Parish	Bur.	Pl.	Parish	Bur.
St. Clement Danes	18		St. Martins in the Fields	26		St. Margaret Westminster	16
St. Paul Covent Garden	4		St. Mary Savoy	1		Whereof at the Pesthouse	0

Buried in the 5 Parishes in the City and Liberties of Westminster — 65 *Whereof, of the Plague* —

FIGURE 2. Raw data for the demography of the seventeenth century: bill of mortality for the week 11–18 February, 1661 (o.s., i.e., 1662 n.s.), giving burials in all the parishes of London (Guildhall Museum Library, London, by kind permission).

The Di[illegible] and Caſualties this Week.

Abortive	4	Lethargy	1
Aged	39	Overlaid	1
Ague	3	Pluriſie	2
Bruſed	1	Quinſie	3
Burnt (an Infant) by accident at St. Alhallowes Great	1	Rickets	7
Cancer	1	Riſing of the Lights	4
Childbed	4	Rupture	1
Chriſomes	15	Scowring	2
Conſumption	78	Scurvy	2
Convulſion	25	Spleen	1
Diſtracted	1	Spotted feaver	2
Dropſie	23	Stilborn	10
Feaver	37	Stopping of the Stomach	6
Flox and Small-pox	21	Suddenly	2
Flux	1	Surfet	4
French-pox	1	Teeth	29
Gowt	2	Threw himſelf out of a window (being Diſtracted) at St. Clement Eaſtcheap	1
Griping in the guts	20	Thruſh	1
Jaundies	4	Tiſſick	2
Impoſtum	2	Ulcer	1
Infants	14	Winde	2
Kingſevill	2		

Born and Chriſtned { Males 109, Females 102, In all 211 } Buried { Males 194, Females 189, In all 383 } Plague 0

Increaſed in the Burials this week — 34

Pariſhes Clear of the Plague — 130 Pariſhes Infected — 0

The Aſſiſe of Bread ſet forth by Order of the Lord Maior and Court of Aldermen, A Penny Wheaten loaf [illegible] contain 5 Ounces and three half penny White loaves the like weight.

FIGURE 3. Verso of bill illustrated in figure 2, tabulating deaths by causes and giving a summary of baptisms and burials. No plague was recorded during this week (Guildhall Museum Library, London, by kind permission).

weekly bills, more convenient annual summaries were also prepared and published (figure 4).

Graunt also studied the records of birth and death, presumably from the parish register, in a country parish, identified by Hull as Romsey, near Southampton. Some of his results, notably those on the sex ratio at birth, will be noted later. Much of Graunt's work consisted of a study of the incidence of various causes of death. He made in addition a theoretical estimate of the rate of growth of the City of London, from the proportion of people of reproductive age and their supposed fertility. Though such a study was preliminary and unsatisfactory, it provided a starting point for further research. Graunt concluded that the population of London was doubling every 64 years, which compared favorably with an empirical estimate of doubling in 56 years. No adequate allowance, however, was made for immigration, which probably accounted for a great part of the empirical rate.

Graunt allowed himself to speculate that if the descendants of Adam and Eve, created according to Scaliger's chronology in 3948 B.C., had doubled every 64 years, they would have filled the world with "far more People, then are now in it" (p. 59). He did not give the number and might well have been terrified by it. In the period between 3948 B.C. and 1662 A.D., there would have been 87.7 doublings, giving a population of $2^{87.7}$ or about 10^{26}, which is about a 100 million people on each square centimeter of habitable land. This example shows that Graunt was familiar with both the potential geometrical increase and some limitation imposed on that increase, but he paid no further attention to the general problem.

Sir Matthew Hale,[7] in a remarkable book posthumously published in 1677, seems to have been the first to use the expression "Geometrical Proportion" for the growth of a population from a single family. He was primarily interested in showing that the human population of the world was not an infinitely old one, living on an infinitely old earth, as some philosophers had maintained. This led him to an extensive study of pestilence, famine, wars, conflagrations, and floods as sporadic limitations to human populations. He rightly concluded that such factors were not enough to bring the population to "Equability," as he termed a stable state, though his empirical data were for the most part very inadequate. He used the Domesday Book to show that the population of Gloucester had increased in historic times, and believed the increase to be general. He gave no discussion of an upper bound. He did, however, believe that in animals, notably insects, various natural, if providential, calamities reduce the numbers to low levels intermittently, so maintaining usually appropriate populations, and producing a balance of nature. This idea was greatly developed by William Derham in his Boyle lectures published (probably in 1713) as *Physico-theology; or, a demonstration of the being and attributes of God from his works of creation.*[8]

Graunt's great friend, Sir William Petty, seems early to have been involved in the work on bills of mortality, continuing it after Graunt's death and extending it to a study of Dublin. Of a more speculative, though less accurate turn of mind than Graunt, Petty engaged in rather more extensive intellectual flights of imagination about the population of the world and its growth.

In *Another essay in Political arithmetick* of

7. Sir Matthew Hale, *The Primitive origination of mankind considered and examined according to the light of nature.* London. Printed by William Godbid, for William Shrowsbery at the sign of the Bible in Duke Lane, CIↃLXXVII.

Sir Matthew Hale (1609–76) was an eminent lawyer who became Chief Justice of the King's Bench in 1671. Though brought up as a strict puritan, he was a tentative evolutionist so far as the fauna of the New World was concerned. I have discussed this aspect of his work in: The influence of the New World on the study of natural history. Philadelphia Academy of Natural Sciences, Bicentennial Symposium (in press).

8. Rev. William Derham [1713(?)], *Physico-theology; or, a demonstration of the being and attributes of God, from his works of creation. Being the substance of sixteen sermons, preached in St. Mary-le-Bow Church, London; at the Honourable Mr. Boyle's lectures, in the years 1711 and 1712.*

A generall Bill for this present yeere, ending the 16. of *December* 1641. according to the report made to the Kings most excellent Matie.

By the Company of Parish Clearks of *London*, &c.

	Bur.	Pl.		Bur.	Pl.		Bur.	Pl.		Bur.	Pl.
Albans Woodstreete	52	13	Christophers	21	3	Margaret Lothbury	34	6	Michael Bassishaw	64	14
Alhallowes Barking	129	21	Clements Eastcheape	20		Margaret Moses	15	1	Michael Cornhill	[illegible]	4
Alhallowes Breadstreet	28	1	Dionis Back-church	19		Margaret Newfishstr.	33	3	Michael Crookedlane	[illegible]	1
Alhallowes Great	105	21	Dunstans East	119	16	Margaret Pattons	13		Michael Queenhithe	54	7
Alhallowes Honilane	6		Edmunds Lumbardst.	23	5	Mary Abchurch	[illegible]7	3	Michael Querne	21	[illegible]
Alhallowes Lesse	38	2	Ethelborough	43	9	Mary Aldermanbury	44	2	Michael Royall	42	[illegible]
Alhall. Lumbardstreet	25	3	Faiths	40	2	Mary Aldermary	40	10	Michael Woodstreet	[illegible]	
Alhallowes Staining	74	23	Fosters	34	6	Mary le Bow	42		Mildred Breadstreet	47	3
Alhallowes the Wall	70	13	Gabriel Fen-church	28	1	Mary Bothaw	16		Mildred Poultrey	19	
Alphage	69	12	George Botolphlane	11	1	Mary Colchurch	6		Nicholas Acons	13	
Andrew Hubbard	27	2	Gregories by Pauls	72	5	Mary Hill	[illegible]	2	Nicholas Coleabby	39	5
Andrew Vndershaft	45		Hellens	38	3	Mary Mounthaw	19	2	Nicholas Olaues	22	1
Andrew Wardrobe	123	15	Iames Dukes place	58	11	Mary Summerset	77	10	Olaues Hartstreet	77	2
Anne Aldersgate	92	34	Iames Garlickhithe	54	5	Mary Staynings	[illegible]	13	Olaues Iewry	25	1
Anne Blacke-Friers	130	13	Iohn Baptist	30	4	Mary Woolchurch	26	5	Olaues Siluerstreete	[illegible]	9
Antholins Parish	23	1	Iohn Euangelist	7		Mary Woolnoth	[illegible]	3	Pancras Soperlane	8	
Austins Parish	34	16	Iohn Zacharie	32	9	Martins Iremonger	24		Peters Cheape	20	3
Barthol. Exchange	34	2	Katherine Coleman	73	13	Martins Ludgate	79	9	Peters Cornhill	22	1
Bennet Fynch	26		Katherine Creechurch	142	39	Martins Orgars	29	2	Peters Pauls Wharfe	22	3
Bennet Grace-church	18		Lawrence Iewry	34	1	Martins Outwitch	22	3	Peters Poore	29	0
Bennet Pauls Wharfe	82	14	Lawrence Pountney	34	3	Martins Vintrey	72	8	Steuens Colmanstreete	[illegible]	[illegible]
Bennet Sherehog	11	1	Leonard Eastcheap	7		Matthew Fridaystreet	10		Steuens Walbrooke	19	1
Botolph Billingsgate	41	2	Leonard Fosterlane	156	74	Maudlins Milkstreet	10		Swithin	[illegible]	3
Christs Church	192	36	Magnus Parish	29	1	Maudlins Oldfishstreete	57	11	Thomas Apostle	[illegible]	6
									Trinitie Parish	35	2

Buried in the 97. Parishes within the walls, —— 4268 *Whereof, of the Plague* —— 643

	Bur.	Pl.		Bur.	Pl.		Bur.	Pl.		Bur.	Pl.
Andrew Holborne	967	183	Bridewell Precinct	27	7	Dunstans West	348	36	Sauiours Southwarke	544	74
Bartholmew Great	146	17	Botolph Aldersgate	244	46	Georges Southwarke	441	22	Sepulchres Parish	1276	327
Bartholmew Lesse	35	7	Botolph Algate	802	78	Giles Cripplegate	1817	363	Thomas Southwarke	104	21
Brides Parish	599	180	Botolph Bishopsgate	619	67	Olaues Southwarke	1022	169	Trinity Minories	26	2
									At the Pest-house	109	96

Buried in the 16. Parishes without the walls —— 9126 *whereof, of the Plague* —— 1697

	Bur.	Pl.		Bur.	Pl.		Bur.	Pl.
Clement Danes	504	99	Katherines Tower	295	41	Mary Whitechappel	810	158
Giles in the Fields	961	118	Leonards Shorditch	483	45	Magdalens Bermond	310	25
Iames at Clarkenwel	377	72	Martins in the Fields	1058	152	Sauoy Parish	103	17

Buried in the nine out-Parishes, in Middlesex and Surrey, —— 4901 *Whereof, of the Plague* —— 727

The total of the burials this yeere —— 18295
Whereof, of the Plague —— 3067
The totall of all the Christnings —— 10370

The Diseases and Casualties this yeere.

Abortive & Stilborne	511	Executed	8	Mother	4
Aged and bed-rid	793	Falling sicknesse	6	Ouer-laid and starved at nurse	29
Ague and Feaver	1434	Fistula and Gangrene	25	Palsie	29
Appoplexie	11	Flocks and small Pox	2483	Plague	3067
Bleach and Swine pox	6	French Pox	25	Planner	2
Bleeding	1	Frighted	1	Plurisie	36
Bloody flux, scowring & flux	512	Gout	7	Purples and spotted Feaver	204
Burnt and scalded	7	Griefe	18	Quinsie and sore throate	35
Cancer and Wolfe	17	Hanged & made away themselvs	9	Rickets	80
Canker Sore mouth & Thrush	84	Iaundies	56	Rising of the lights	180
Childbed	228	Iawfaloe	6	Rupture	13
Chrisomes and Infants	1897	Impostume	117	Scurvey,	33
Cold, and Cough	93	Kild by severall accidents,	62	Sores, Issues Vlcers, broken and Bruised limbs	28
Collicke and winde	52	Kings Evill	49	Stone & Strangury	47
Consumption and Tisficke	2738	Leprosie and Shingles	3	Stopping of the Stomacke	31
Convulsion and Crampe	780	Lethargy	2	Suddenly	53
Cut of the Stone	3	Livergrowne & Spleene	96	Surfet	415
Dead in streets, fields, &c.	19	Lunatique	7	Teeth and Wormes	1217
Dropsie, & Timpany	499	Meagrome	26	Vomiting	9
Drowned	44	Measles	48		

Christened	Males 5292	Females 5078	In all 10370		
Buried	Males 9539	Females 8756	In all 18295	Of the Plague	3067

Increased in the Burials in the 122 Parishes and at the Pesthouse this yeere —— 5524
Increased of the Plague in the 122 Parishes and at the Pest-house this yeere —— 1617

	Christned	Buried	Plague
S. Margret Westminst.	582	1014	132
Lambeth	132	270	46
S. Mary Newington	144	337	69
Redriffe	[illegible]	[illegible]	[illegible]
S. Mary Islington	60	133	17
Stepney	959	1380	110
Hackney	55	79	1

The totall of the Burials in the 7 last Parishes this yeere —— [illegible]
Whereof of the Plague —— [illegible]
The totall of the Christnings —— [illegible]

FIGURE 4. A general bill of mortality, for the year ending 16 December 1641, giving all the data for London in a year in which plague was the most prevalent cause of death.

1683,[9] Petty noted (p. 45) that few or no countries had a population density of more than 2 persons for 5 acres of land. He estimated that the inhabitable parts of the earth had an area of under 50 thousand million acres, or 50 billion acres in modern American parlance. In metric terms this would be under 2.10^8 km^2, which figure is almost exactly twice the modern estimate of that part of the terrestrial land surface that is neither ice-covered nor desert. Petty's estimate of the total carrying capacity would thus be under 20 thousand million persons. He obtained a maximum rate of human increase by assuming that in any population of 600, there would be 180 "*Teeming* Females between 15 and 44" (p. 13), who ideally could bear a child every 2 years, thus giving 90 children per year. For sickness, spontaneous abortion, and natural barrenness he subtracts 15, and then another 15 to balance the expected death rate, thus obtaining a rate of increase of 60 or 10 percent per year. This he expresses as a doubling time of 10 years. Such instantaneous doubling times are used by Petty rather inaccurately over inappropriately long periods of time, blithely neglecting any consideration of the demographic equivalent of compound interest; the correct time would be just over 7 years.

In the City of London the actual rate of growth was estimated from the trend in the death rate, taken to be proportional to the population when comparison is made of sufficiently long intervals, in practice about 20 years. This approach indicated that between 1604 and 1682 a doubling of the population took place in about 40 years. This growth was largely attributable to immigration from the country. For the whole of England, study of the difference between recorded birthrates and recorded death rates led Petty to pitch on, if we may use his favorite expression for making an informed guess, doubling in 360 years, though he admits that some of the data could imply a much slower doubling, in 1,200 years. He points out that if the trends for London and for England as a whole continued unchanged, London would have absorbed practically the whole population of England, of 10,917,389 persons, in 1842, leaving but 198,509 employable in agriculture to feed the city. Actually he expected London to reach a maximum size of 5 million in 1800.

Petty clearly realized that under different circumstances the rates of increase would be different. Starting with the Deluge and the reestablishment of the human race from the 8 inhabitants of Noah's Ark, whom he supposed left that vessel in 2700 B.C., Petty concluded that it would take between 100 and 150 doublings to give the population of the world, which he put at 320,000,000, an alleged contemporary estimate the source of which seems not to have been identified. Actually if one starts with 2^3, the figure should be about 25 doublings. Since his figures of 100 to 150 fell between the estimated maximum rate of doubling in 10 years and that for England of doubling in between 360 and 1,200 years, Petty concluded that the rate of reproduction fell off from the maximum that prevailed in immediate postdiluvian times to the low value characteristic of his own time. He published a tentative table to show this (figure 5), which he wisely left "to be Corrected by Historians, who know the bigness of *Ancient Cities*, *Armies*, and *Colonies* in the respective Ages of the World" (p. 24). He pointed out that to achieve a saturation population of 20 thousand million would take 6 doublings of his currently estimated 320,000,000, "And then, according to the Prediction of the Scriptures, there must be Wars and great Slaughter, &" (p. 17). It will be noted that in spite of sloppy arithmetic, a commitment to the historical accuracy of the Book of Genesis, and the unavailability of the infinitesimal calculus, Petty did get the idea of a declining rate of increase, though he gives

9. Sir William Petty, *Another essay in Political arithmetick concerning the growth of the city of London: with the measures, periods, causes, and consequences thereof.* London, Mark Pardoe, 1683. A second edition appeared in 1686 as *Multiplication of mankind with another essay in political arithmetick concerning the growth of the city of London.* London, Mark Pardoe.

Sir William Petty (1623–87), who made his reputation as a surveyor in Ireland, was a great friend of Graunt; there has been some argument as to what part Petty may have had in Graunt's work.

A Table ſhewing how the People might have doubled in the ſeveral Ages of the World.

Periods of doubling	*Anno* after the Flood.	
In 10 Years	1	8 Perſons.
	10	16
	20	32
	30	64
	40	128
	50	256
	60	512
	70	1024
	80	2048
	90	4096
	100	8000 and more.
In 20 Years	120 Years after the Flood.	16 Thouſand.
	140	32
30	170	64
	200	128
40	240	256
50	290	512
60	350	1 Million and more.
70	420	2 Millions.
100	520	4 Millions.
190	710	8 Millions.
290	1000	16 In Moſes Time.
400	1400	32 About Davids Time.
550	1950	64
750	2700	128 About the Birth of Chriſt.
1000	3700	256
In 300/1200	4000	320

FIGURE 5. Sir William Petty's table of the estimated increase of the human population of the world after the biblical deluge, which he took as occurring in 2700 B.C., to show his idea of a decline in the rate of doubling.

no analysis of its possible causes.

During the eighteenth century a great many writers[10] discussed human population, almost exclusively from an economic or political point of view. They were greatly hampered by the lack of any adequate censuses; some considered that alarming decreases in population were taking place. There was, however, a quite usual belief that population adjusted to the carrying capacity of the land at any given state of agricultural development and that prudence played a decisive part in the adjustment, points of view in line with the characteristic faith in reason during the period of the Enlightenment.

At the very end of the century the problem of population was brought into far sharper focus by the Rev. Thomas Robert Malthus[11] in *An essay on the principle of population*,[12] the first edition of which was published anonymously in 1798. In his memoir of Malthus in the *Dictionary of National Biography*, Sir Leslie Stephen wrote, "Although more or less anticipated, like most discoverers, Malthus gave a position to the new doctrine by his systematic exposition, which it has never lost." To biologists his book is of additional interest on account of the influence that it had on Darwin and Wallace. It is therefore appropriate to

10. This literature may be most easily explored by using as guides:

J. Bonar, *Theories of Population from Raleigh to Arthur Young*. London, George Allen and Unwin, 1931, 253 pp.

J. J. Spengler, *French Predecessors of Malthus: A Study in Eighteenth-Century Wage and Population Theory*. Durham, N. C., Duke University Press, 1942, ix, 398 pp.

J. Stassart, *Malthus et la population*. Collection scientifique de la Faculté de Droit de l'Université de Liège, no. 6, 1957, 342 pp. Stassart gives a good account of the discussions as to whether the populations of the countries of Europe had decreased or increased in modern times.

Robert Wallace was probably the most important forerunner of Malthus. My friend Mary P. Winsor tells me she hopes to publish in the near future new material that will emphasize his role in the development of population theory.

11. The Reverend Thomas Robert Malthus was born near Guilford on 17 February 1766. His father, Daniel Malthus, knew Rousseau and is said to have been the latter writer's English executor. Malthus was educated by his father and by tutors, coming up to Jesus College, Cambridge, in 1784. He graduated in 1788 and after some further private study was elected a fellow of his college in 1793. He seems not to have resided much in Cambridge, and from 1796 had a curacy at Albury, Surrey, where he lived with his father. He married Harriet Eckersall of Claverton House near Bath in 1804, and on that account, according to the statutes of the time, had to relinquish his fellowship. Late in 1805 he was appointed Professor of History and Political Economy at East-India College, Haileybury, Hertfordshire, an institution established by the East India Company for the education of their civil servants. Malthus was elected F.R.S. in 1819, and was made a Royal Associate of the Royal Society of Literature by George IV in 1824; the latter position carried a stipend of 100 guineas, but William IV declined to continue the Royal Associateships on his accession. Just before his death Malthus became one of the original fellows of the Statistical Society. He died at Claverton House on 29 December 1834 and was buried in Bath Abbey. In view of the opprobium often heaped upon him, it may be of interest to reproduce the epitaph on his monument in the north-west porch of that church.

examine what Malthus actually said.

He was aware, like his predecessors from Graunt onward, of the obvious tendency for a population to increase exponentially. Largely on the basis of early information about the United States, Malthus concluded (pp. 20–21) that in a favorable and nearly unrestricted environment the doubling of a human population might occur every 25 years. In more anciently settled countries where all good land had long been under cultivation, he believed that at best the rate at which food supplies could be increased, as more and more people sought them with increasing energy, might be taken to be a constant. Writing of Britain he says, "Let us take this for our rule, though certainly far beyond the truth; and allow that by great exertion the whole produce of the Island might be increased every twenty-five years, by a quantity of subsistence equal to what it at present produces." It is evident that the position taken aphoristically by Malthus, that while population tends to increase geometrically with time, food supply does so arithmetically, is based on a hypothesis which he believed to be overoptimistic and adopted solely for the sake of argument. Presumably he meant by his curious state-

---

SACRED TO THE MEMORY | OF THE REV. THOMAS ROBERT MALTHUS | LONG KNOWN TO THE LETTERED WORLD | BY HIS ADMIRABLE WRITINGS ON THE SOCIAL BRANCHES OF | POLITICAL ECONOMY | PARTICULARLY BY HIS ESSAY ON POPULATION | ONE OF THE BEST MEN AND TRUEST PHILOSOPHERS | OF ANY AGE OR COUNTRY | RAISED BY NATIVE DIGNITY OF MIND | ABOVE THE MISREPRESENTATIONS OF THE IGNORANT | AND THE NEGLECT OF THE GREAT | HE LIVED A SERENE AND HAPPY LIFE | DEVOTED TO THE PURSUIT AND COMMUNICATION | OF TRUTH | SUPPORTED BY A CALM BUT FIRM CONVICTION OF THE | USEFULNESS OF HIS LABORS | CONTENT WITH THE APPROBATION OF THE WISE AND GOOD | HIS WRITINGS WILL BE A LASTING MONUMENT | OF THE EXTENT AND CORRECTNESS OF HIS UNDERSTANDING | THE SPOTLESS INTEGRITY OF HIS PRINCIPLES | THE EQUITY AND CANDOUR OF HIS NATURE | HIS SWEETNESS OF TEMPER URBANITY OF MANNERS | AND TENDERNESS OF HEART | HIS BENEVOLENCE AND HIS PIETY | ARE THE STILL DEARER RECOLLECTIONS OF HIS FAMILY | AND FRIENDS. [rule] | BORN FEB: 14. 1766. DIED 29. DEC: 1834.

Malthus had three children; one of his two daughters died during his lifetime. Though he thus left a zero population growth family, absurd stories as to the number of his descendants were evidently circulated. The Yale University Library copy of the sixth edition of the *Essay* bears below his name "Père de quartorze [*sic*] enfants, Father of fourteen!" This probably implies a confusion with Francis Place (see note 16) who reached his opinions on population in the hard way.

12. The first edition of Malthus's *Essay* was published anonymously:

[Malthus, T. R.], *An essay on the principle of population, as it affects the future improvement of society, with remarks on the speculations of Mr. Godwin, M. Condorcet, and other writers.* London, J. Johnson, 1798, v ix, 396 pp. (A reprint, edited by J. Bonar, appeared under the title, *First essay on population 1798.* London, Macmillan, 1927, for the Royal Economic Society.)

Later editions of Malthus's essay were entitled:

An essay on the principle of population; or, a view of its past and present effects on human happiness; with an inquiry into our prospects respecting the future removal or mitigation of the evils which it occasions. With this title there were published, during the author's lifetime,

1803: A new edition, very much enlarged, viii, 610 pp. (quarto).

1806: Third edition, 2 volumes, xvi, 505 (60 unpaginated index to both volumes); vii, 559 pp. The second volume ended in an appendix answering objections; it was advertised as also available in quarto for the convenience of those who had acquired the 1803 edition.

1807: Fourth edition, this differs from the third of the previous year in having the index transferred to the second volume, the original first chapter of which now appears in the first volume.

1817: Additions to the fourth and former editions of *An essay on the principle of population.* London, John Murray, iv, 327 pp.

1817: Fifth edition, in 3 volumes incorporating the additions published separately in the same year (I, xvi, 496 pp.; II, iv, 507 pp.; III, iv, 500 pp.).

1826: Sixth edition, in 2 volumes (I, xviii, 535 pp.; II, iv, 528 pp.).

Johnson published the first, second, third, and fourth editions; John Murray the fifth and sixth, with the volume of additions that went to form the fifth edition. I have seen no evidence, other than the advertisement in the third edition, of the existence of the separate quarto version of the appendix and cannot be quite sure that it materialized.

The second or new edition of 1803 is greatly rewritten and much longer than the first, being in effect largely a new book. All the other editions, except the fourth, moreover, contain significant new material.

The first American edition, based on the third English, was published in George Town, J. Mulligan, at J. March's Bookstore, R. Chew Wieghtman, Printer, 1809.

ment (appendix, written 1806, printed in the third and subsequent editions; p. 453 of vol. II of sixth edition which I, like Darwin, have used), "the first of these propositions [i.e., geometrical tendency of population growth] I considered as proved the moment the American increase was related, and the second [i.e., the arithmetical tendency of increase of food] as soon as it was enunciated" that the first proposition was based on observation while the second could be taken as a postulate. Malthus believed (appendix, sixth edition, vol. II, p. 451) that the "power of the earth to produce subsistence is certainly not unlimited, but it is strictly speaking indefinite; that is, its limits are not defined, and the time

---

There was a German translation by F. H. Hegewisch, *Versuch über die Bedingung und die Folgen der Volksvermehrung* (Altona, J. F. Hammerich, 1807, 2 vols.), also based on the third English edition.

Two years later a French translation of the same English edition by P. Prévost appeared as *Essai sur le principe de population ou exposé des effets passés et presens de l'action de celle cause sur le bonheur du genre humain, suivi de quelques recherches relatives à l'espérance de guérir ou d'adoucir les maux qu'elle entraîne.* Paris and Genève, P. Paschoud, 3 vols.

Malthus obtained his data on the population growth in North America from two sources. One was Benjamin Franklin's *Observations concerning the increase of mankind. Peopling of Countries, &c.*, originally published as *Observations on the late and presenj Conduct of the French . . . To which is added, wrote by another hand; Observations concerning the increase of mankind, peopling of countries, &c.* Boston, printed and sold by S. Kneeland in Queen Street, 1755, 4 l., iv, 1–47, 1–15. The dedication on the first four, unnumbered, leaves is signed by William Clarke, who wrote the main part of the pamphlet; no indication is given of the authorship of the last fifteen pages, "wrote by another hand." The work was reprinted in London, John Clarke, in the same year.

Franklin's observations were also published in *The interest of Great Britain considered with regard to her colonies and the acquisitions of Canada and Guadaloupe, To which are added, Observations concerning the increase of mankind, peopling of countries, &c.* London, T. Becket, 1760. This was reprinted in Boston, B. Mecom, in the same year. This American edition is described on the title page as by the "very ingenious useful and worthy Author . . . [B____n F____n, LL.D.]." Some doubt, however, has been expressed as to the authorship of *The interest of Great Britain considered*; the editors of *The Papers of Benjamin Franklin* (New Haven and London, Yale University Press, vol. 9, pp. 53–58) conclude that Franklin was indeed the author, though his friend Richard Jackson, to whom the work is sometimes ascribed, may have given him useful material. There is no doubt about the authorship of the *Observations concerning the increase of mankind.*

These pamphlets are rare, though originally they must have had a wide circulation on both sides of the Atlantic. *Observations concerning the increase of mankind* has been reprinted in *The Magazine of History with Notes and Queries*, extra number 63, Tarrytown, N.Y., William Abbot, 1918 (extra numbers 61–64 are bound as vol. 16 of such reprints, which should not be confused with vol. 16 of the magazine; the title page used is not a facsimile of the original one). A more modern edition with useful notes is in *The Papers of Benjamin Franklin*, ed. Leonard W. Labaree and Whitfield J. Bell. New Haven and London, Yale University Press, vol. 4, pp. 225–34, 1961. A French translation was added to [John Dickinson], *Lettre d'un fermier de Pensylvanie, aux habitans de l'Amérique Septentrionale* (Traduites de l'anglois. Amsterdam [Paris] 1769), in which the observations are stated to have been "adressées à la Société royale de Londres." Franklin's point of view on these matters must have been quite well known in England.

Franklin indicates that at the time of writing the population consisted of over "One Million English Souls." These were derived from not more than 80,000 immigrants, which implies just under four doublings. If the average immigrant had arrived in 1651, the doubling time would be just over 25 years. Malthus's remarks on animals and plants increasing to the limits of their resources were taken from the same page (p. 9) of this little work.

The second source was the Rev. Ezra Stiles, *A discourse on the Christian Union. The substance of which was delivered before the reverend congregation of the Congregational Clergy in the Colony of Rhode Island; assembled at Bristol, April 23, 1760* (Boston, Eder and Gill, 1761, 139 pp.). Ezra Stiles discussed in detail the role of growth of population in Rhode Island in connection with the polity of the Congregational Churches. He concludes that the doubling time in parts of the state may be rather less than 25 years. He is fully aware, with biblical illustrations reminiscent of Graunt and Petty, that what he called *patrial* rates of doubling are not destined to continue indefinitely. It is evident from the correspondence in William Godwin's *On Population. An enquiry concerning the power of increase in the numbers of mankind, being an answer to Mr. Malthus's essay on that subject* (London, Longman, Hurst, Rees, Orme, and Brown, 1820. p. 122) that Malthus obtained the information from details given by Richard Price in his *Observations on Reversionary Payments*, and from parts of Ezra Stiles's work which he had seen. The last-named author is spelled Styles by Malthus and Price, though he is given as Stiles on the title page of *A discourse on the Christian Union*; his uncle Jared used Styles. Godwin evidently had a poor opinion of Stiles as a "puritanical preacher in Connecticut."

will probably never arrive when we shall be able to say, that no further labor or ingenuity of man could make further addition to it." This statement is very like those that until recently were the stock-in-trade of economists and businessmen. There is no hint of an upper limit here; in the form in which Malthus phrased his basic conception, it is the rate of increase rather than the population that tends to a constant limiting value.

The very categoric statement on the increase in food supplies actually grossly underestimated the agricultural possibilities of the British Isles during the century and a half after he wrote it. This underestimate is doubtless responsible for some of the criticism, so lacking in foresight, that Malthus's ideas have received from unbiologically minded economists and sociologists.

Malthus clearly believed that the observed low rates of population increase in Europe, much lower than those in the "United States of America, where the means of subsistence have been more ample, the manners of the people, more pure, and consequently the checks on early marriage fewer" (p. 20), were due primarily to the operation of misery and vice. By misery he evidently meant so low a standard of living that deprivation became a significant demographic factor. He is less explicit about vice, but imagines a man considering whether he can afford to marry and have children or whether "he may not see his offspring in rags and misery, and clamouring for the bread that he cannot give them. . . . These considerations are calculated to prevent . . . a very great number . . . from pursuing the dictate of nature in an early attachment to one woman. And this restraint almost necessarily, though not absolutely so, produces vice" (p. 28). Presumably he is saying that prudence leading to a postponement of marriage forced many young men into masturbation, homosexuality, or sexual relations with one of a limited number of prostitutes, leaving many young women virginal and unreproductive when physiologically they could have been bearing children. The first version of the *Essay* seems to hide beneath its demographic argument the erotic dilemma of a young man brought up to virtue. Malthus concludes that a population that is growing fast, with early marriages, is a happy population, but inevitably becomes unhappy as it increases, and its growth rate tends to decrease.

Inevitably a happy state gives place to an unhappy one, and the only checks on the process are, at best, unpleasant and, at worst, vicious. There is an incredulous rejection (p. 154) of Condorcet's view that prejudice and superstition would cease to invest morality with a corrupting and degrading austerity. "Improper arts," which presumably may include any method of contraception or abortion, are mentioned in the second edition in connection with concealing the consequences of irregular connections. In the appendix to the fifth edition, of 1817, Malthus vigorously and correctly denies that he "ever adverted to the checks suggested by Condorcet without the most marked disapprobation. Indeed I should always particularly reprobate any artificial and unnatural modes of checking population, both on account of their immorality and their tendency to remove a necessary stimulus to industry. If it were possible for each married couple to limit by a wish the number of their children, there is certainly reason to fear that the indolence of the human race would be greatly increased; and that neither the population of individual countries, nor of the whole earth, would ever reach its natural and proper extent." His own advocacy of restraint, added to misery and vice, in the second edition of 1803, as a third check and not merely as a prelude to vice, he sees in 1817 as not only sanctioned by religion but also tending to decrease indolence, marriage being the reward "of habits of industry, economy, and prudence." Basically, since he knew that misery and vice were the lot of so many of his contemporaries, his panacea of restraint is rather unconvincing. Obviously his cultural environment and religious belief made it impossible for him to adopt the only available release from a deep personal and sociological dilemma.[13] Malthus had no doubt been able himself to justify the qualifying "not absolutely" before he was rewarded by marriage in the year after the publication of the second edition.

Malthus, like most of his eighteenth-century predecessors, viewed population as an economist, rather than as a biologist, as Graunt had done intuitively, or for that matter as a mathematician, as Verhulst was to do 40 years after the first edition of the *Essay* had been published. Nevertheless Malthus's analysis called attention to the continuous operation of seemingly inevitable processes reducing the innate potential growth rate. Though he was primarily thinking about populations of *Homo sapiens*, he did in fact indicate, at first somewhat inexpertly, that his approach also applied to other organisms, and that among "animals and plants its effects are waste of seeds, sickness, and premature death" (p. 15). At the beginning of all later editions he notes that Dr. Franklin had called attention to the fact that animal and plant populations expand until limited by lack of food or space.

The immediate effect of Malthus was the generation of interest in family limitation, not so much as a means of postponing an ultimate demographic catastrophe, but as a way of avoiding immediate misery, if not vice. Birth-control methods, though in the English mind largely associated with prostitution and primarily with prevention of venereal disease, had clearly been practiced in parts of France, notably in Normandy, for a long time and on a scale sufficient to have some demographic effects. Moheau[14] in a book on the population of France published

13. The question of what Malthus thought about birth control is discussed by N. E. Himes in an appendix "Note on Malthus' attitude toward birth control" to his edition of Francis Place, *Illustrations and Proofs of the Principle of Population* (see note 16). The answer to the question is clearly, if slightly ungrammatically, "as little as possible." The present writer, brought up in a liberal Anglican academic family a century after Malthus wrote, can in retrospect and without too much effort identify with him, but for most people today it is difficult to realize the gigantic blind spot that existed even among people who by nature and education could have done most to improve understanding of this matter.

The views of Condorcet to which Malthus took exception (both on p. 154 of the first edition of the *Essay* and later in the appendix to the fifth edition) were expressed by their originator, Marie-Jean-Antoine-Nicolas de Caritat, Marquis de Condorcet, in *Esquisse d'un tableau des progrés de l'esprit humain* (Paris, Agasse l'an III de la République, 1794, viii, 389 pp.; see p. 358), after he had indicated the distant possibility of a saturation population on the earth, in the following words:

> Mais en supposant que ce terme dût arriver, il n'en résulteroit rien d'effrayant, ni pour le bonheur de l'espèce humaine, ni pour sa perfectibilité indéfinie; si on suppose qu'avant ce temps les progrès de la raison ayent marché de pair avec ceux des sciences et des arts, que les ridicules préjugés de la superstition, ayent cessé de répandre sur la morale, une austérité qui la corrompt et la dégrade au lieu de l'épurer et de l'élever; les hommes sauront alors que, s'ils ont des obligations à l'égard des êtres qui ne sont pas encore, elles ne consistent pas à leur donner l'existence, mais le bonheur; elles ont pour objet le bien-être général de l'espèce humaine ou de la société dans laquelle ils vivent; de la famille à laquelle ils sont attachés; et non la puérile idée de charger la terre d'êtres inutiles et malheureux[.]. Il pourroit donc y avoir une limite à la masse possible des subsistances, et par conséquent à la plus grande population possible, sans qu'il en résultât cette destruction prématurée, si contraire à la nature et à la prospérité sociale d'une partie des êtres qui ont reçu la vie.

Stassart, after a page-and-a-half discussion of this admirable, if indefinite, passage under the heading "Condorcet fait-il allusion aux procédés anticonceptuelles?" concludes that he probably envisioned, with approval, such procedures, but admits "le texte n'est pas clair."

Malthus, in the first edition of his *Essay* (p. 154), professes not to understand what Condorcet was implying, a reasonable position in view of the French author's obscurity. He goes on, however, to suppose that Condorcet was recommending "promiscuous concubinage, which would prevent breeding, or something else as unnatural." That Malthus should have rejected such vague views with so much horror is characteristic of the reaction that any mention of the subject, however oblique, was apt to generate in England. As Stassart points out, Malthus, after the first edition of the *Essay*, visited France and discussed population problems in that country. It is hardly credible that he was not well aquainted with the practice of contraception by 1817, much as he disapproved.

14. Moheau, *Recherches et considerations sur la population de la France*. Paris, Moutard, 1778, xv, 280, 157 pp. (2 books in one volume). Reprinted in 1912, R. Gonnard, ed., *Collection des économistes et réformateurs sociaux de la France*, No. 10, Paris, P. Geuthner. There seems to be great ignorance about the author of this quite important contribution to demography. Moheau has been identified as the secretary of J.-B.A.R. Auget, Baron de Montyon, an economist and founder of several prizes, who has been supposed to have collaborated on the books. Neither Moheau's Christian names, dates of birth and death, nor any other biographical details appear to have survived.

in 1776 wrote, "les femmes riches ... ne sont pas les seules qui regardent la propagation de l'espece comme une duperie du vieux temps: déjà ces funestes secrets inconnus à tout animal autre que l'homme, ces secrets ont pénétré dans les campagnes; on trompe la nature jusque dans les village" (p. 258). The matter was later discussed by Verhulst,[15] though with great discretion as to the methods employed. We have noticed that Malthus was horrified at being associated with Condorcet in this matter, even though the relevant passage of the French writer is quite vague and entirely useless as a practical guide.

The first important proponent of birth control in any English-speaking country was Francis Place,[16] the most significant radical theorist in Britain in the early part of the nineteenth century. His *Illustrations and proofs of the principle of population* was directly inspired by Malthus, as its title indicates, but went much further, openly advocating contraception, though without indicating how it was to be achieved. Several small publications encouraging such a practice and aimed specifically at manual workers of both sexes appeared in the 1820s.[17] Part of Place's argument was a very simple one of supply-and-demand economics. If the supply of labor were reduced by restricted reproduction, wages would obviously rise. It is very interesting to contrast this early nineteenth-century attitude with the later official Marxist criticism of Malthus, only now beginning to abate.[18] The English publications that followed Place's little book are now of considerable rarity, but an American writer, Robert Dale Owen, U.S. Ambassador to the Court of the King of the Two Sicilies and U.S. Senator, the son of Robert Owen of New Harmony who had undoubtedly been influenced by Place, achieved in his *Moral phy-*

15. Verhulst, in his second memoir, of 1847 (see note 23), gives the average number of births per marriage in 30 cities of France, from estimates made by Chevalier Des Pommelles in 1789. The numbers varied from 3.52 to 5.04. The lowest value was from Rouen. Ten cities gave figures of less than 4.0, including the 2 others in Normandy. Other investigators had also noted that the fertility in Normandy was apt to be lower than elsewhere in France. Verhulst quotes a M. Villermé who had made enquiries in that region and concluded that the lower fertility was the result of a practice "condamné par tous les casuistes, notamment par le père Sanchez dans son traité *De sancto matrimonii sacramento*; lib. ix, p. 224, §12." Tomas Sanchez was a Spanish Jesuit, living from 1550 to 1610. His gigantic work, the full title of which is *De Sancto Matrimonii Sacramento Disputationum Tomi Tres*, consisting of about a million and a half words, is ordinarily, though not always, bound in 3 volumes, as its title suggests. The work is divided into 10 books, which in most editions are separately paginated; each book contains a set of numbered disputations of which the paragraphs are also numbered. In none of the editions that I have seen (Madrid, 1605; Venice, 1693, 1712; Viterbo, 1754) does p. 224 bear a paragraph 12 of any disputation in Book ix, nor can I with any confidence recognize any of the paragraphs numbered 12 in any disputation in this book, which deals with marital duty, as the one mentioned. Sanchez is of course perfectly definite (*Lib.* ix, *Disput.* 20, §13) that any procedure that prevents the semen from reaching its natural destination (*ad quem natura ipsum destinavit*) is a deadly sin against nature (*peccatum lethale contra naturam*). I have not noticed, in a fairly cursory study of the 165,000 words of the text of book ix, any mention of mechanical devices, though there is a reference to giving of an oral contraceptive (*dans potionem ... quo conceptio prolis impediatur. Lib.* ix, *Disput.* 20, *Summarium*), unfortunately of an unspecified nature, and of attempts to expel the semen; these procedures are condemned. It is reasonable (see note 19) to assume Verhulst meant *coitus interruptus*, specifically forbidden in *Lib.* ix, *Disput.* 17, §13. It is probable that §12 in Verhulst is a misprint for §13 of *Lib.* ix, *Disput.* 17. J. Kenyon (The staging of the sex war. *Times Lit. Suppl.*, 76:102, 1977) suggests that the practice was by no means unknown in seventeenth- and eighteenth-century England.

16. Francis Place, *Illustrations and proofs of the principle of population including an examination of the proposed remedies of Mr. Malthus, and a reply to the objections of Mr. Godwin and others.* London, Longman, Hurst, Rees, Orme, and Brown, 1822, xv, 280 pp. (reprinted with additional material collected by N. E. Himes, London, Allen and Unwin, 1930).

17. F. H. Amphlett Micklewright, The rise and decline of English Neo-Malthusianism (*Popul. Stud.*, 15:32–51, 1962) gives an account of the early history of the movement. The *Diabolical Handbill* of 1823 circulated by J. E. Taylor, the radical editor of the *Manchester Guardian*, the *Black Dwarf* edited by T. J. Woolner, a friend of Jeremy Bentham, from 1817 to 1824, and the *Republican* edited by Richard Carlile from 1819 to 1826; all, after 1822, at times were involved in promoting contraceptive practices. Carlile's article in the *Republican* entitled "What is Love?" was expanded as *Every Woman's Book; or, What is Love? containing the most important instructions for the prudent regulation of the principle of love, and the number of a family.* Of this I have seen a late edition (London, R.

siology[19] a very wide public. In Britain the book was reprinted, and under attack by the Society for Suppression of Vice, as late as 1878. Owen wrote, "We are . . . compelled to admit to Malthus, that, *sooner or later*, some restraint or other to population *must* be employed; and compelled to admit to his aristocratic disciples, that if no other better restraint than vice and misery can be found, then *vice and misery must be*." He unhesitatingly recommended *coitus interruptus* and wrote (p. 62 fn.), a "Frenchman belonging to the cultivated classes, would as soon be called a coward, as to be accused of causing the pregnancy of a woman, who did not desire it; and that too, whether the matrimonial

---

Carlile Jr., 1834, 31 pp.). Carlile recommends a damp sponge attached to a tape that can be inserted as a block into the vagina. Carlile indicates that this method was widely used in France and Italy and among the English aristocracy.

I have not had an opportunity to examine the earlier periodical publications; what is significant about them is that they appeared shortly after Place's book, under the auspices of well-known radical editors or publishers. They mark the beginning of active propaganda for birth control in English-speaking countries. At first the movement was radical, secularist, and opposed to the orthodox establishment.

18. An interesting attempt at reconciliation may be found in H. E. Daly, A Marxian-Malthusian view of poverty and development. *Popul. Stud.*, 25:25–37, 1971.

19. Robert Dale Owen, *Moral physiology; or, a brief and plain treatise on the population question.* According to R. W. Leopold's *Robert Dale Owen: A Biography* (Cambridge, Mass., Harvard University Press, 1940, x, 470 pp.), the first edition was published on December 1, 1830, but he was unable to find a copy. *The British Museum Catalogue of Printed Books* records the third edition as published in New York in 1831; Thomas Skidmore reprinted the fourth edition with critical notes as *Moral Physiology exposed and refuted* (New York, Skidmore and Jacobus, 1831). There was a fifth edition published on August 13, 1831. However, Leopold says that 1,500 copies were sold in five months. These were presumably of the first three or four editions, which must therefore have been quite small. The eighth American edition, the only one that I have seen, appeared in 1835; there seems to have been an eighth English edition as early as 1832, mentioned in the *B. M. Catalogue of Printed Books.* A new English edition appeared in 1870. Leopold says that the first three editions recommended *coitus interruptus*, a sponge or a condom; the second and third methods were relegated to a footnote in the fourth and omitted in the fifth edition.

Moral physiology became the stimulus to Dr. Charles Knowlton of Boston to write and publish a fairly elaborate account of sexual physiology as known at the time, set in a materialist frame and entitled *Fruits of philosophy; or, the private companion of young married people* (Boston, 1833).

This work, of which I have only seen the second English edition, advocated a douche containing in solution any innocuous salt that acts chemically on the semen. Zinc sulphate, alum, or pearlash (potassium carbonate) are recommended, lead acetate specifically is not.

Knowlton's book had a similar history to that of *Moral physiology*, being published in England soon after its American appearance. It was imitated by a small pamphlet, *The Connubial Guide; or, married peoples best friend* . . . , by Intercidona (London, John Wilson, 78 Long Acre, n.d., 12 pp.). This recommended an alkaline douche, which the pseudonymous author was prepared to supply, with instructions for its use, through the publisher, for the then exhorbitant price of half a sovereign.

Both *Moral physiology* and *Fruits of philosophy* produced *causes célèbres.* In 1877 Charles Bradlaugh, M.P., a well-known secularist, and Mrs. Annie Besant, the estranged wife of an Anglican clergyman, were prosecuted in the Queen's Bench Division, before the Lord Chief Justice, for the distribution of *Fruits of philosophy.* The prosecution was instigated by the Society for the Suppression of Vice. Both defendants were convicted, but the conviction was quashed on the technical grounds of a faulty indictment and no new trial was held.

Done out of its earlier victims, the Society for the Suppression of Vice proceeded to bring Edward Truelove, the English publisher of R. D. Owen's *Moral physiology*, into court. Truelove later issued: *In the High Court of Justice. Queen's Bench Division, February 1, 1878. The Queen v. Edward Truelove for publishing the Hon. Robert Dale Owen's Moral Physiology" and a pamphlet entitled "Individual, Family, and National Poverty"* [*sic*] (Specially Reported. London, Edward Truelove, 1878, viii, 125 pp.). Truelove's account of the trial, which perhaps looks a little more official than it actually was, tells a fascinating story. The case resulted in a hung jury; at a new trial, Truelove was convicted and served a four-month's prison sentence. Shortly thereafter, the secretary of the Society for the Suppression of Vice was professionally discredited and disbarred from the practice of law. No further prosecutions of this kind took place. (See Micklewright, note 17.)

The pamphlet *Individual, Society, and National Poverty* is unknown to me. The existence of a considerable number of ephemeral and fugitive nineteenth-century publications on family limitation is evident. Their nature and relation to each other is not easily understood as even the larger libraries are very deficient in such works.

law had given him legal right over her person or not." It is interesting to see how the republican Owen compares his own views and those of the educated people of France with the attitude of Malthus's aristocratic followers, whoever they were. Malthus would perhaps have felt a little differently about the purity of American manners had he read Owen.

The whole contraceptive movement of the nineteenth century has generally been designated neo-Malthusian to indicate the fundamental difference between the attitude of its socialist and secularist proponents and that of Malthus himself. The most important organization promoting these ideas in Britain was, however, called the Malthusian League,[20] the word Malthusian being used in a sense very different from that employed on pp. 3, 4. Although all this activity, which was undertaken at some personal risk, has been of enormous importance, the study of the more theoretical and particularly the biological aspects of population languished.

In 1830 Michael Thomas Sadler[21] put forward what he presumed to call the Law of Population, that the "prolificness of human beings, otherwise similarly circumstanced, varies inversely as their number" (p. 352). He did not of course mean that it was linearly inversely proportional, which would make all populations of whatever size increase at a constant rate, but he did provide evidence that as the population increases, so dN/dt declines. He did not, as did no one until Verhulst, understand the advantage of a model with a maximum rate of increase ($r N$) and a feedback function of N reducing that rate. His work is, however, of importance as the first clear indication of biological density dependence, "however occult as to its primary causes" (p. 569). Though intended to be anti-Malthusian, the difference between Sadler's point of view and that of Malthus appears less great now than it would have done in 1830.

Quetelet, the first important statistician to devote much effort to the biological problems of mankind, and indeed the inventor of the *average man*, naturally turned his attention to population growth. Observing, as Malthus and his predecessors had done, that the rate of growth decreases as the population increases, he suggested that the resistance to growth varies as the square of the rate of increase. Fourier appears to have come up with the same idea simultaneously in 1835 or perhaps earlier. This unsupported and, in fact, inapplicable conclusion was presumably adopted solely as an analogy to the resistance encountered by a large body such as a projectile moving rapidly through a fluid medium. More kinds of resistance in the world of physics being now known than were familiar to Quetelet, the analogy now loses its charm. At the same time Quetelet[22] seems to have interested his younger friend and colleague, Pierre-Francois Verhulst (figure 6), in the

20. In the 1930s a Malthusian Ball was held in London. The calendar published in the London weekly *Time and Tide* (vol. 14, p. 334, March 18, 1933) indicates that in that year the ball was to be held on March 22 in the Dorchester Hotel at 9:30.

21. Michael Thomas Sadler, M.P., *The law of population: a treatise, in six books; in disproof of the superfecundity of human beings, and developing the real principle of their increase*. London, J. Murray, 1830, vol. I, xvi, 639 pp.; vol. II, 690 pp.

Lamont C. Cole, in his classical paper, The population consequences of life history phenomena. (*Quart. Rev. Biol.*, 29:103–37, 1954), seems to have been the only recent author to realize the importance of Sadler's work.

22. Adolphe-Lambert-Jacques Quetelet (1796–1874) was a Belgian scientist of great distinction and an enormous range of interests, encompassing mathematics, statistics, astronomy, meteorology, and the more quantitative aspects of physical anthropology and the social sciences.

His most important demographic and sociological work was *Sur l'homme et le développement de ses facultés; ou, essai de physique sociale*. Paris, Bachelier, 2 vols., 1835. (There also seems to have been a Belgian edition, but neither has been available to me.) A revised edition was published as *Physique sociale; ou, essai sur le développement des facultés de l'homme*. Bruxelles, Muquardt; Paris, Baillière; St. Petersburg, Issakoff; 2 vols., 1869, I, viii, 503; II, 485 pp.

The 1835 edition was translated into English as *A treatise on man and the development of his faculties*. Edinburgh, 1842, 126 pp., 7 bl. This is reprinted in facsimile (New York, Burt Franklin, 1968).

There is an extensive treatment of Quetelet by F. H. Hankins, Adolphe Quetelet as a statistician. *Studies in History, Economics, and Public Law*, 31, no. 4, 134 pp. New York, Columbia University, 1908.

P. F. VERHULST.

FIGURE 6. Pierre-François Verhulst (1804–49). From the portrait published in Quetelet's obituary notice.

whole subject. Verhulst tackled the problem in an essentially modern manner, constructing the simplest mathematical model of a continuously growing population with an upper limit. In his second paper, in which the term *logistic* was introduced, he assumed that the limiting effect of the environment did not come into play immediately and that in an unsettled country the initial expansion would be exponential. This led him to a rather complicated treatment of actual cases for which the data were hardly good enough to make the distinction. He was fully aware that life at the upper limit would be a condition of misery and felt that only a controlled economy could cope with the social problems created by an equilibrium population.

Verhulst's work on the logistic seems to have made no impression on any of his contemporaries. Quetelet in his obituary of Verhulst published in 1850 indicates the general form of his conclusions, but makes no reference to the logistic as such. He does say that some of Verhulst's ideas about population were, to say the least, bold. The fact that they had no physical analogies clearly worried Quetelet. Verhulst was an interesting man,[23] who in his youth had traveled in Italy for his health and had had the temerity to propose a constitution for the Papal States.

Fourier is said by Quetelet to have arrived at the view that the resistance to population growth varies as the square of the rate of increase in 1835. *The Recherches statistiques sur la ville de Paris et le départment de la Seine*, 6 vols., initially published in 1821–29, are said by Hankins to contain introductory essays by Fourier and to have been reprinted in the 1830s. Quetelet refers to this work as the place of publication of Fourier's thoughts on the subject. Fourier may have priority for this curious idea. I have been unable to verify this.

23. Verhulst's three papers on population are:

Notice sur la loi que la population suit dans son accroissement. *Correspondances Mathématiques et Physiques*, 10:113–21, 1838.

Recherches mathematiques sur la loi d'accroissement de la population. *Mem. Acad. Roy. Belg.*, 18 (no. 1): 1–38, 1845.

Deuxième mémoire sur la loi d'accroissement de la population. *Mem. Acad. Roy. Belg.*, 20 (no. 3): 1–32, 1847.

The first of these papers gives the derivation of the law as the simplest analytic expression fulfilling our biological postulates, though Verhulst is explicit only about the upper bound. As Nils Tongning points out to me, the logistic is a special case of the well-known Riccati equation. In the second paper of 1845, the term logistic is used, but the equation for practical purposes is modified, so that in its early part it is exponential and becomes logistic only after a definite lapse of time.

Verhulst was born in Brussels, October 28, 1804. After a very successful, if somewhat irregular, education in his home city and at the University of Ghent, he translated Sir John Herschel's *Treatise on Light* and taught under Quetelet at the Brussels Museum. He became interested in the theory of probability and certain problems connected with repayment of public debt by lot and with state lotteries, in the late 1820s. This was doubtless the beginning of his interest in social and demographic problems. After his sojourn in Italy, which occurred at the time of the Belgian revolution of 1830, he became unsuccessfully involved in politics and also published one original historical study. By 1834 he had returned to science and began giving instruction in mathematics at the *École militaire*, where he later became professor. Quetelet regarded his *Traité élémentaire des fonctions elliptiques* of 1841 as his most important work. Though his contemporaries ignored the logistic, his interest in population studies led him into sociological problems, as for instance in his second memoir, and the Belgian government appointed him to commissions on the relief of poverty and on state insurance policy. He died, presumably of tuberculosis, on February 15, 1849.

The primary source of information on Verhulst's life is Quetelet's obituary: Notice sur Pierre-François Verhulst. *Ann. Acad. Roy. Sci. Lettr. Beaux Arts Belg.*, 1850, pp. 97–124. This has been translated into English by J. R. Miner, Pierre-François Verhulst, the discoverer of the logistic curve, (*Hum. Biol.*, 5:673–89, 1933), together with an English version of the relevant account in *Mémoires de la Reine Hortense* (publiés par le Prince Napoléon, Paris, Plon, 1927, vol. III, pp. 210–12).

Queen Hortense, originally Hortense de Beauharnais (1783–1837), was the daughter of Josephine, later Napoleon I's Empress, by her first husband, the Vicomte de Beauharnais. Hortense was married to Louis, Napoleon's brother, whom the Emperor made King of Holland. This unhappy marriage produced three sons, one of whom survived to become Napoleon III. Hortense also had, more happily, a son by her lover, the Comte de Flahaut. She was the author of several songs. She enjoyed the company of artists and intellectuals and must have been a woman of considerable talent and great charm. During her stay in Rome she clearly mothered the young Belgian *savant*. The incident of the constitution for the Papal States leading to Verhulst's being forced out of Rome seems to have taken place early in 1831. There may be further information, as Quetelet seems to imply, in the newspapers of the period.

He received some help and advice in this venture from his friend Queen Hortense, but was soon expelled from Rome as a dangerous character. Quetelet's effective dismissal of the logistic from the obituary leaves the latter merely celebrating the achievements of a quite ordinary mathematician of considerable personal charm. In the whole of the rest of the recital of his interesting life and largely derivative works, there is nothing that has any strictly scientific interest today.

Quetelet continued to believe in his hydrodynamic analogy, still putting forward the idea about the square of the velocity of increase in the new edition of his *Physique sociale* in 1869, more than 30 years after Verhulst's first paper. This lack of support for his colleague and friend seems very odd and may well have contributed to general neglect by the scientific public. Miner, who republished Quetelet's obituary in an English translation, notes only one later specific reference before 1920 to the equation and its integral curve, in a curious little paper by Du Pasquier[24] in 1918.

Actually the logistic had been used and had become fairly widely known, though not under that name, as an equation expressing autocatalysis and as the result of Brailsford Robertson's[25] studies of the autocatakinetic growth of individual organisms. It is a curious irony, as Lloyd[26] points out in an excellent historical study of the equation, that Robertson used Quetelet's growth data to fit his curve in his first contribution of 1909, without realizing Quetelet's ambivalent role in promoting Verhulst's original, but by then forgotten, discovery.

A formally equivalent equation was, moreover, suggested for the growth of a population of bacteria in a limited medium by M'Kendrick and Kesava Pai[27] in 1911, and was used for yeast by Carlson in 1913.

In 1920 Pearl and Reed[28] again rediscovered the equation, using it in a paper on the population growth of the United States. A year later Pearl[29] acknowledged Verhulst's priority; Pearl and Reed were certainly well acquainted with Robertson's conclusions about the growth of individual organisms. Pearl, Reed, and Kish published a later curve in 1940, with an extrapolation far into the century; this and the actual curve that the population of the United States has followed are shown in figure 7. It is evident that the logistic does not provide a predictive model for human demography. So far during the twentieth century, Malthus's belief that "the time would not come when no further labor or ingenuity of man could make further additions" to the value of K, has held true, though it is quite clear to us that this happy state cannot continue indefinitely. When the possible value of K is continuously increasing, Verhulst's equation loses its value. By 1920, however, work was beginning on the experimental demography of laboratory animals. Here Verhulst's formulation came into its own.

Experimental study of the logistic

The construction of the model has aimed at simplicity, but its experimental falsification or verification involves some technical complexities to maintain the simple conditions demanded by theory. Ideally, if we are to have a persistent equilibrium population, a continual influx of energy in the form of food, which just maintains the equilibrium population, is needed. As an individual be-

24. L. G. Du Pasquier, Esquisse d'une nouvelle théorie de la population. *Viertejahrschr. Naturforsch. Ges. Zürich*, 63:236–49, 1918. This author gives an equation for a population at first increasing and then decreasing.

25. Brailsford Robertson, On the normal rate of growth of an individual, and its biochemical significance. *Arch. Entwicklungsmech.*, 25:581–614, 1908. The idea that growth of an individual is an autocatalytic process was taken up by various other investigators (see Lloyd, note 26; Pearl, note 33).

26. P. J. Lloyd, American, German, and British antecedents to Pearl and Reed's logistic curve. *Popul. Stud.*, 21:99–108, 1967.

27. A. G. M'Kendrick and M. Kesava Pai, The rate of multiplication of microorganisms: A mathematical study. *Proc. Roy. Soc. Edinb.*, 31:649–55, 1911.

T. Carlson, see note 32.

28. R. Pearl and L. J. Reed, On the rate of growth of the population of the United States since 1790 and its mathematical representation. *Proc. Nat. Acad. Sci. U.S.A.*, 6:275–88, 1920.

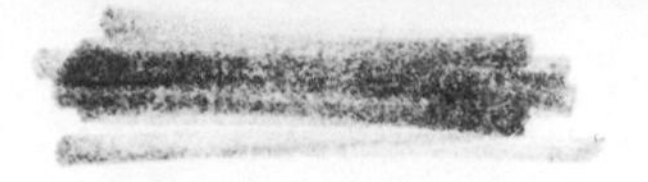

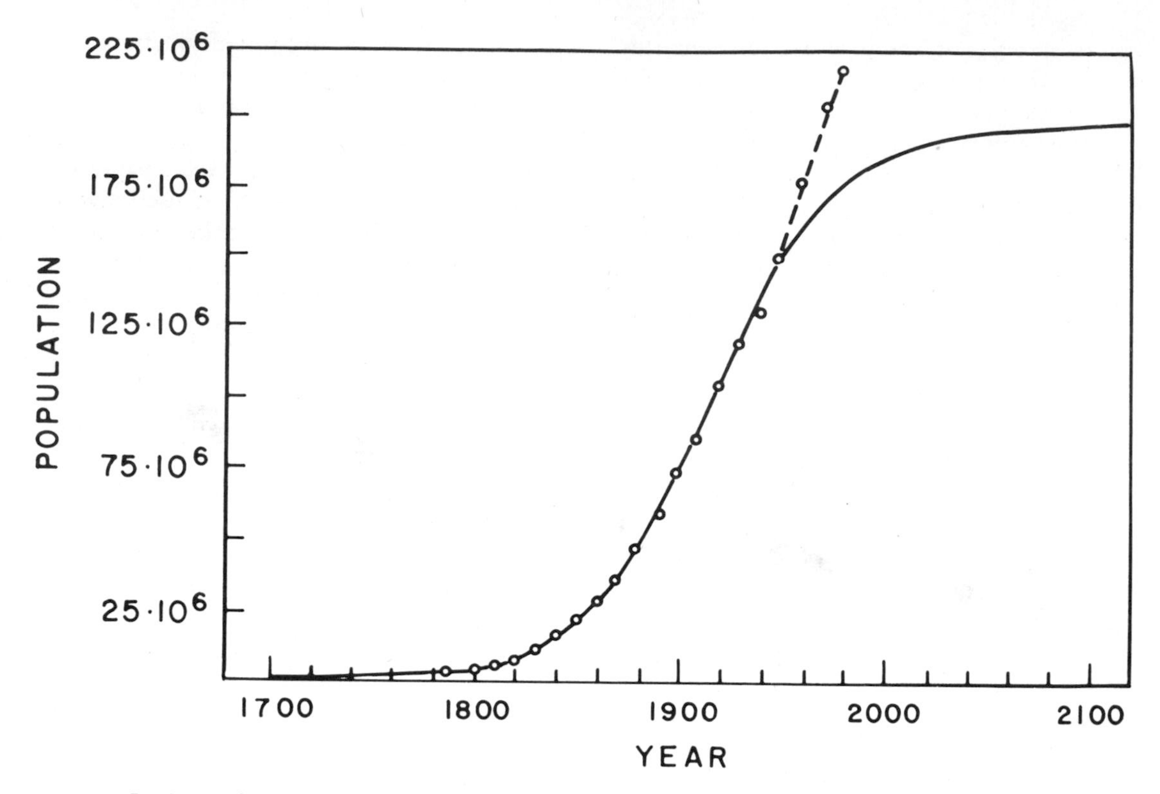

FIGURE 7. Logistic (solid line) fitted to the human population of the United States from censuses up to and including that of 1940, by Pearl, Reed, and Kish, with actual points from later censuses, following the broken line.

comes senescent and dies, a new one can be born and can survive to replace the one that has died. If we contrast this with what happens if we start a population in a closed system with an initially limited amount of food, we find that in both cases a sigmoid growth curve will develop, but in the closed temporally limited system the population rises to a maximum, which is lower than that in the open unlimited system, and then declines. Food, moreover, is not the only important variable. Inhibition by the increasing concentration of waste products of metabolism may limit growth, as for example ethanol in a yeast culture. In the open, temporally unlimited system, such material is removed as food is added; in the closed system, it accumulates and becomes toxic. A great many other, more complicated types of limitation may be expected to occur in populations of the more advanced metazoa.

Two further warnings about experimental verification or falsification must be given. First, although our mathematical procedure looks reasonable, we may legitimately question the justification of our hope to obtain a single equation expressing growth in a limited universe. Should we not consider the possibility of continuous, even perhaps unrestricted exponential, growth until the population has reached a limiting value and then a catastrophic decline, due to starvation or overcrowding, and not connected analytically with the initial rise of the population? Such a *relaxation*, to use the euphemistic word of the applied mathematician, certainly can occur as the result of special enviromental changes such as the introduction of a pathogenic virus or bacterium into the culture. Relaxations must always be kept in mind as

29. R. Pearl, The biology of death: V. Natural death, public health and the population problem. *Sci. Month.*, 13:193–213, 1921 (Verhulst is mentioned in footnote 2, p. 206).

R. Pearl, L. J. Reed, and J. F. Kish, The logistic curve and the census count of 1940. *Science*, 92:486–88, 1940.

possible alternatives to single, neat, analytical expressions.

The second warning is this. When we know a reasonable amount about the organism being studied, deriving a growth equation that is confirmed by very carefully arranged experiments may tell us practically nothing that we did not know before. What we have indeed done is to construct a rather inaccurate analogue computer for giving numerical solutions of our equation, using organisms for its moving parts. When we find that we have confirmed the logistic, what we have mainly confirmed is that the reduction in the rate of population growth is linearly dependent on the relative density of organisms. Actually, the beautiful S-shaped integral curve may be too insensitive a result to tell us how well we have established this conclusion.

Feller[30] has indeed pointed out that none of the confirmatory experiments enable one to distinguish some arbitrary curve based on an equation for an S-shaped curve involving three arbitrary constants (corresponding to K, r, and the constant of integration, however expressed) from the logistic.

This is of course a usual situation in science, for any set of data can be fitted by an infinite number of curves. We choose the logistic to study because it is general and because it is realistic in that it makes simpleminded biological sense. It is an excellent base from which to set out on further, more elaborate theoretical and, we hope, more accurate investigations. It is therefore abundantly worthwhile to see if it can give good approximations to population growth in the world of physical rather than purely conceptual organisms.

The first observations on populations that tested the validity of an equation that can be put in the form of the logistic were those of M'Kendrick and Kesava Pai,[31] who worked in the Pasteur Institute of Southern India in Madras, on *Escherichia coli*. One of their technically more satisfactory experiments gave the results set out in figure 8.

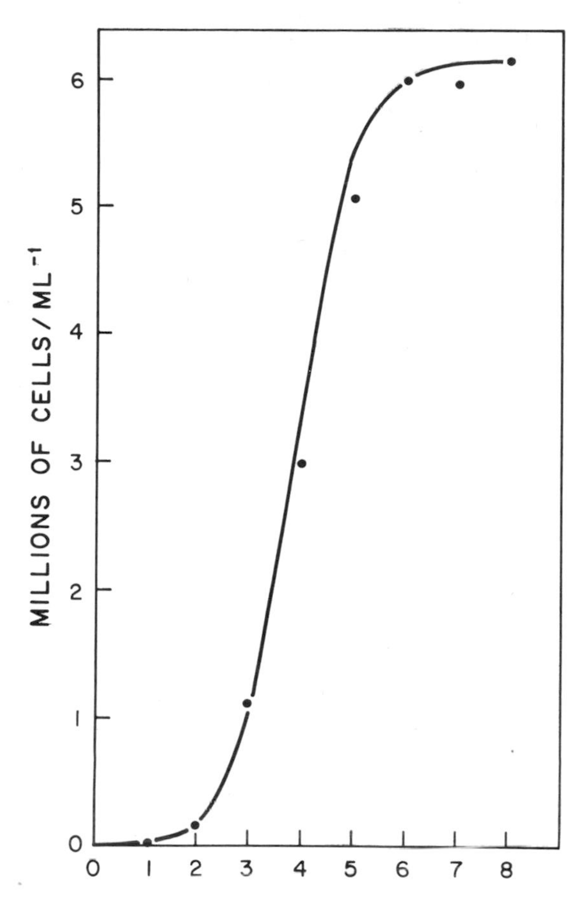

FIGURE 8. Growth of *Escherichia coli* at 37°C in peptone broth, from the data of M'Kendrick and Kesava Pai, with their calculated curve, which they plotted logarithmically, for which r is taken as implying division every 22.3 minutes.

A little later Carlson,[32] in Stockholm, made an excellent study on the growth of the yeast *Saccharomyces cerevisiae*. He examined the mathematics of growth, considering the possibility that the limitation might be proportional to $(K - N)^n$, where n is greater than unity, so giving an asymmetrical curve for the rate of growth as a function of population size. In his discussion of the inflection point, however, he reverts to M'Kendrick and Kesava Pai's equation. His data, unlike those of M'Kendrick and Kesava Pai, who used a logarithmic scale, were plotted arithmetically and so gave the first sigmoid curve

30. W. Feller, On the logistic law of growth and its empirical verifications in biology. *Acta Biotheoretica*, 5:51–66, 1940.

31. See note 27.

32. T. Carlson, Über Geschwindigkeit und Grösse der Hefevermehrung in Würze. *Biochem. Z.*, 57:313–34, 1913.

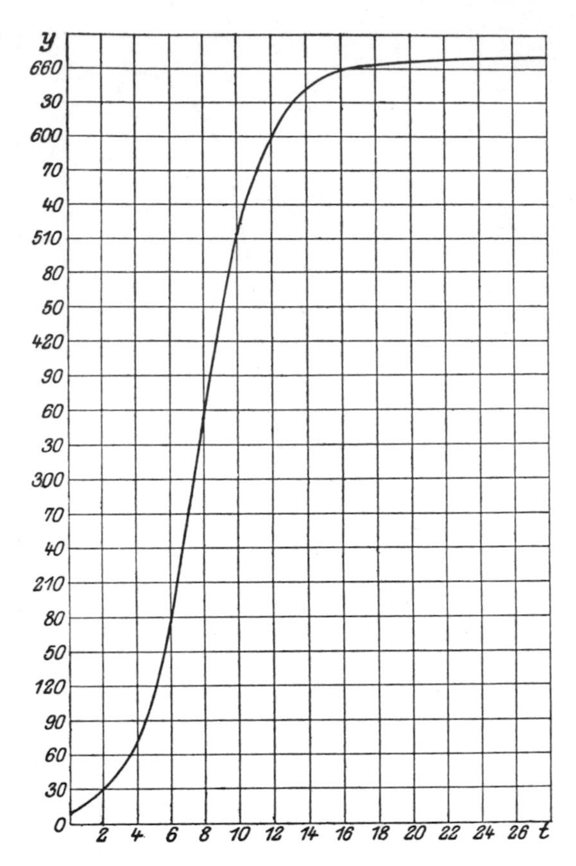

FIGURE 9. Growth of *Saccharomyces cerevisiae* as plotted by Carlson. This curve, plotted arithmetically, seems to be the first one for an actual population drawn to show the sigmoid form.

of population growth (see figure 9).

After Pearl and Reed's study of the human population of the United States, and the rediscovery of Verhulst's papers, the logistic began to be used extensively. Pearl himself, undoubtedly influenced by Brailsford Robertson's work on the growth of individual organisms, and later[33] clearly familiar with Carlson's study, started his own experiments with *Drosophila melanogaster*, which was already famous as the organism that had enabled Morgan, Bridges, and Sturtevant to establish the chromosome theory of inheritance, which rapidly became the theory of the

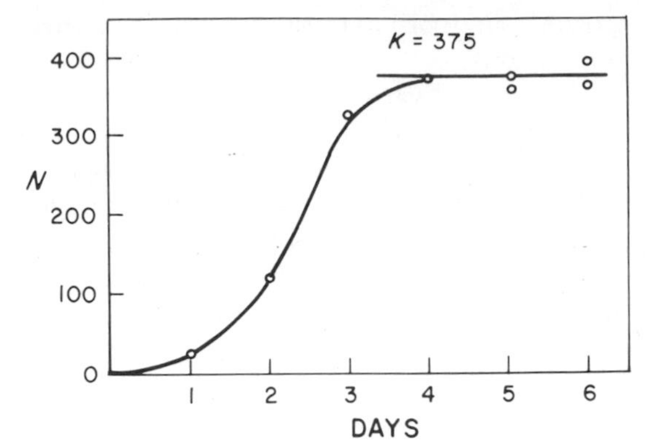

FIGURE 10. Growth of a population of *Paramecium caudatum*, fitted to a logistic curve (Gause).

gene. A little later Terao and Tanaka[34] did interesting work with the water flea, *Moina macrocopa*, and Gause[35] with *Paramecium* (figure 10). In general, organisms with simple life histories and reproduction by division or by parthenogenetic pseudovivipary give smoother curves than do bisexual animals with complicated life histories. Certain features of this early work have proved to be of particular importance and so have been restudied by later investigators.

The relation of the equilibrium population K to the parameters of the physical environment in any given set of experiments takes many forms. Slobodkin[36] found that approximately equilibrium populations of *Daphnia obtusa* depended linearly on the rate of supply of food (figure 11). Chapman[37] felt the same about the flour beetle, *Tribolium confusum*, but it is possible that in those of his cultures in which the food supply was great, equilibrium had not been reached when the experiment was terminated. Gause[38] thought that Chapman's results indicated an exponential rather than a linear dependence of population on food supply, though this does not seem to be particularly meaningful biologically.

It was evident in Pearl's early observations on the growth of *D. melanogaster* in two dif-

33. R. Pearl, *The Biology of Population Growth*. New York, A. A. Knopf, 1925, xiv, 260 pp.

34. A. Terao and T. Tanaka, Influence of temperature upon the rate of reproduction in the water-flea *Moina macrocopa* Strauss. *Proc. Imper. Acad.* (Japan), 4:553–55, 1928.

35. G. F. Gause, *The Struggle for Existence*. Baltimore, Williams and Wilkins, 1934. This book is one of the foundations of modern ecology.

36. L. B. Slobodkin, Population dynamics in *Daphnia obtusa* Kurz. *Ecol. Monogr.*, 24:69–88, 1954.

37. R. N. Chapman, The quantitative analysis of

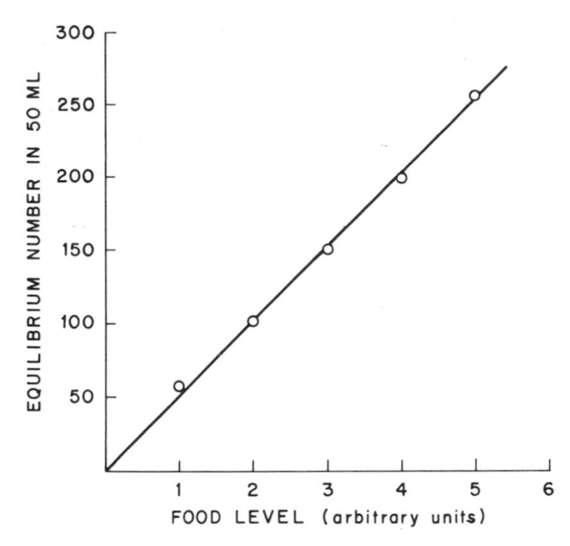

FIGURE 11. Relationship of K, the upper limit of population, to level of food, in arbitrary units, maintained in the course of experiments on *Daphnia obtusa* (Slobodkin).

ferent-sized bottles, shown in figure 12, curves (a) and (c), that the relationship of the size of the population to the environmental parameters of the culture vessel was not necessarily a simple proportionality. Pearl concluded that in his experiments, not population as such, but population density is proportional to the volume of space available.

Pearl[39] later concluded that most of the reduction in growth rate in growing *Drosophila* cultures is due to reduction in the number of eggs laid and that this is due to the increasing disturbance of the females as the population increases. Later studies suggest that the fundamental cause of this decline in natality is increasing inaccessibility of food in a crowded culture.

Much more recent work on *Drosophila serrata* by Ayala[40] showed that, in the experimental situation he used, if the rate of feeding

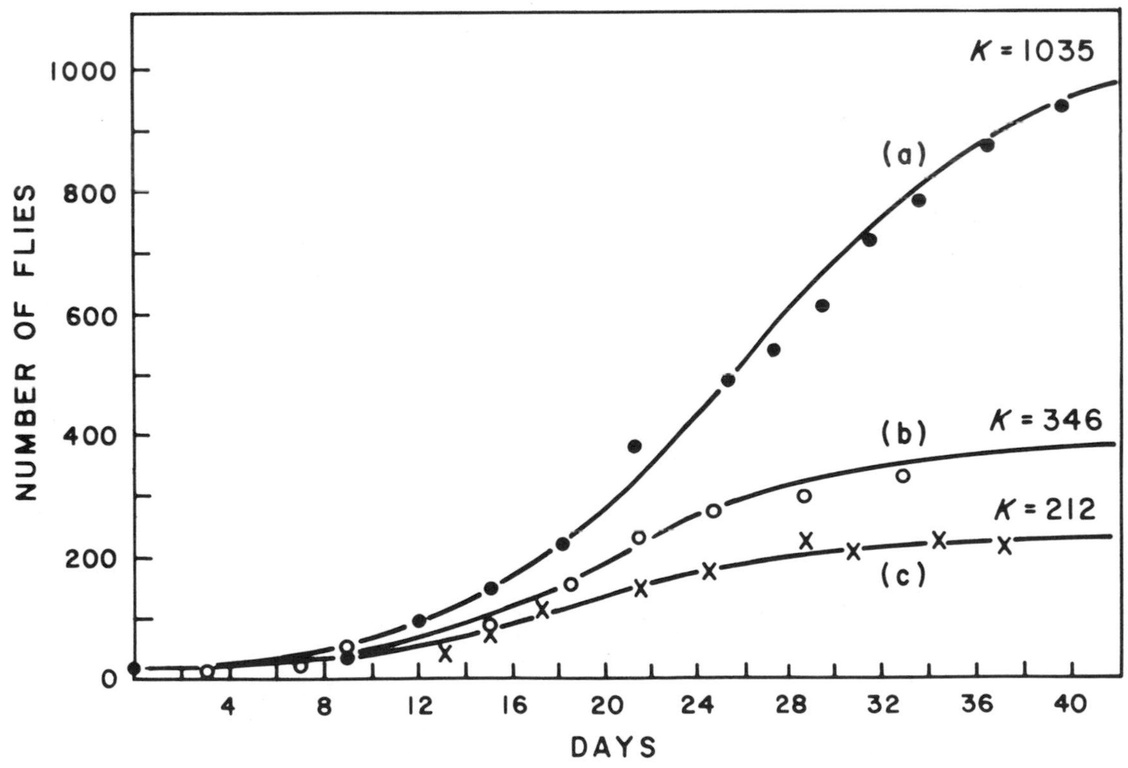

FIGURE 12. Growth of populations of *Drosophila melanogaster*, (a) wild type in a pint bottle; (b) stock homozygous or hemizygous for five recessives including vestigial wing, in same-sized bottle; (c) wild type in a half-pint bottle (Pearl).

environmental factors. *Ecology*, 9:111–22, 1928.

38. G. F. Gause, The influence of ecological factors on the size of populations. *Amer. Natural.*, 65:70–76, 1931.

39. R. Pearl, The influence of density of population upon egg production in Drosophila melanogaster. *J. Exp. Zool.*, 63:57–84, 1932.

40. F. J. Ayala, Genotype, environment, and population numbers. *Science*, 162:1453–59, 1968.

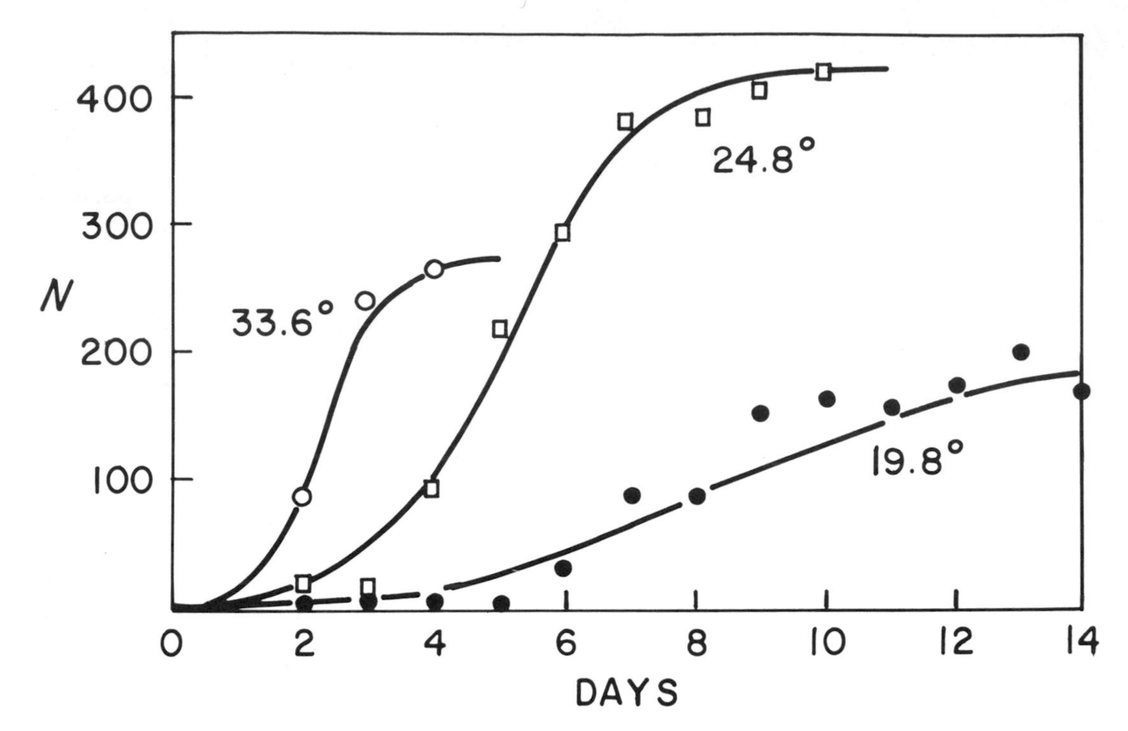

FIGURE 13. Growth of populations of *Moina macrocopa* at three temperatures.

was increased by 50 percent, an almost equivalent increase in the rate of production of young flies took place, but the adult population as a whole increased only by 17 percent. This implies that the expectation of life of adult flies fell as the recruitment into the population increased. This is probably due to overcrowding of adult flies, which averaged about 6.7 individuals per cubic centimeter in the air space of the culture vessel. In addition to these cases suggesting that increasing space or food increases K but not necessarily in an obvious and predictable manner, it will appear later (p. 57) that even in apparently nonsocial species, underpopulation as well as overpopulation can occur, a fact of considerable ecological interest.

Terao and Tanaka's work on the growth of populations of the water flea, *Moina macrocopa*, indicated something of the effect of temperature on population growth (figure 13). It is evident that the values of both r and K are temperature-dependent, r increasing over the whole range of temperatures studied, while K appears to have a maximum value at about 25°C. It is, however, possible that the slow-growing culture at 19.8°C had not reached equilibrium when the experiment was terminated. A priori it would seem likely that in a poikilotherm or "cold-blooded" animal, r would increase with increasing temperature, as is usual for physiological functions over the temperature range in which the organism is viable. K, however, depends fundamentally on the ratio of the total available energy flux to the amount needed per organism; as the latter need is likely to rise with increasing temperature, K would be expected to fall. This has, however, not been established in any general sense. Ayala recently found that in *Drosophila serrata* and *D. birchi* from New Guinea and Queensland, higher populations were developed at 25°C than at 19°C, while in *D. serrata* from the cooler climate of New South Wales the reverse is true. Clearly, the flies are adapted to their climatic environments rather than dependent on the logic of chemical kinetics.

Pearl[41] found that the genetic constitution of the stock of *Drosophila* could play a great part in determining the asymptotic maximum population. In figure 12, curve (b) shows the

41. See note 33.

growth of a population of flies homozygous or hemizygous for five recessives, including vestigial in which the wing is reduced and functionless, making the mode of life of such flies quite different from that of the wild type. The value of r in the mutant stock is 0.22 instead of 0.27, and the mean duration of life is much shorter. It is therefore not remarkable that the mutant stock realizes only about a third of the population ($K = 346$) that would be possible for the wild type ($K = 1{,}035$).

Genetic differences permitting r- *and* K- *selection*
Recent work, notably by Ayala,[42] has greatly extended our knowledge of the possible genetic bases of equilibrium populations. Ayala found, for instance, that under identical culture conditions, the equilibrium population densities of *Drosophila birchi* were always lower than those of its sibling species, *D. serrata*. Within either species different local populations, moreover, may differ in the saturation populations realized under identical conditions. There is also evidence from Ayala's work that the rate of production of new flies, dependent at least in part on r, though usually varying in the same sense as the size of the population, is not the only factor determining the latter. A difference of about 3 percent in productivity between the Queensland and New South Wales strains of *D. serrata* corresponds to a difference of almost 25 percent in population size; evidently in the Queensland population the longevity of the flies is greater. Since both r and K can vary within a species, it is evident that selection could operate on either variable independently. The distinction between r-selection and K-selection, developed originally by MacArthur and Wilson,[43] is fundamental to much of contemporary evolutionary ecology.

Ayala made the very interesting discovery that in his cultures of *D. birchi* from New Guinea, there was a slow upward drift in saturation number, of the order of just under one fly per day. Evidently K-selection was occurring (figure 14). Moreover, if the genetic

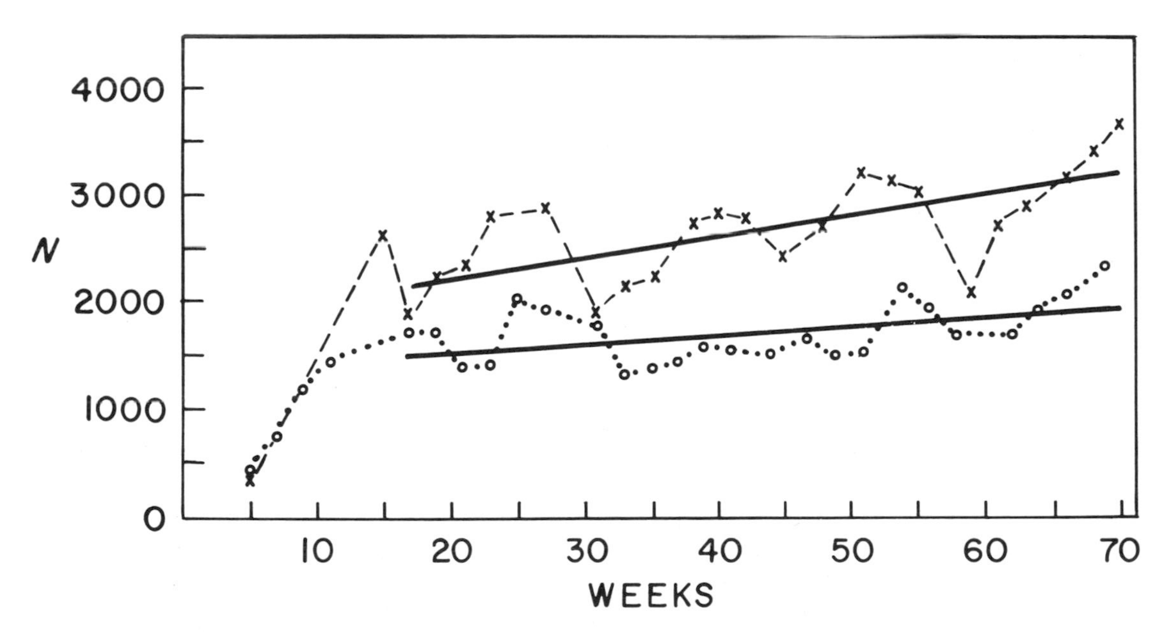

FIGURE 14. Populations of *Drosophila birchi*, reared at 19°C: one (o) from Popondetta, New Guinea; and one (x) a strain derived from mass hybridization of the Popondetta strain and of another from 200 km north of Sydney, Australia. The solid lines are the regression lines on time, starting 17 weeks after initiation of the populations, and indicate that although K-selection is occurring in both populations, it is more effective in the population of hybrid origin and therefore of greater genetic variance (Ayala).

42. See note 40.
43. R. H. MacArthur and E. O. Wilson, *The Theory of Island Biogeography*. Princeton University Press, 1967, xi, 203 pp.

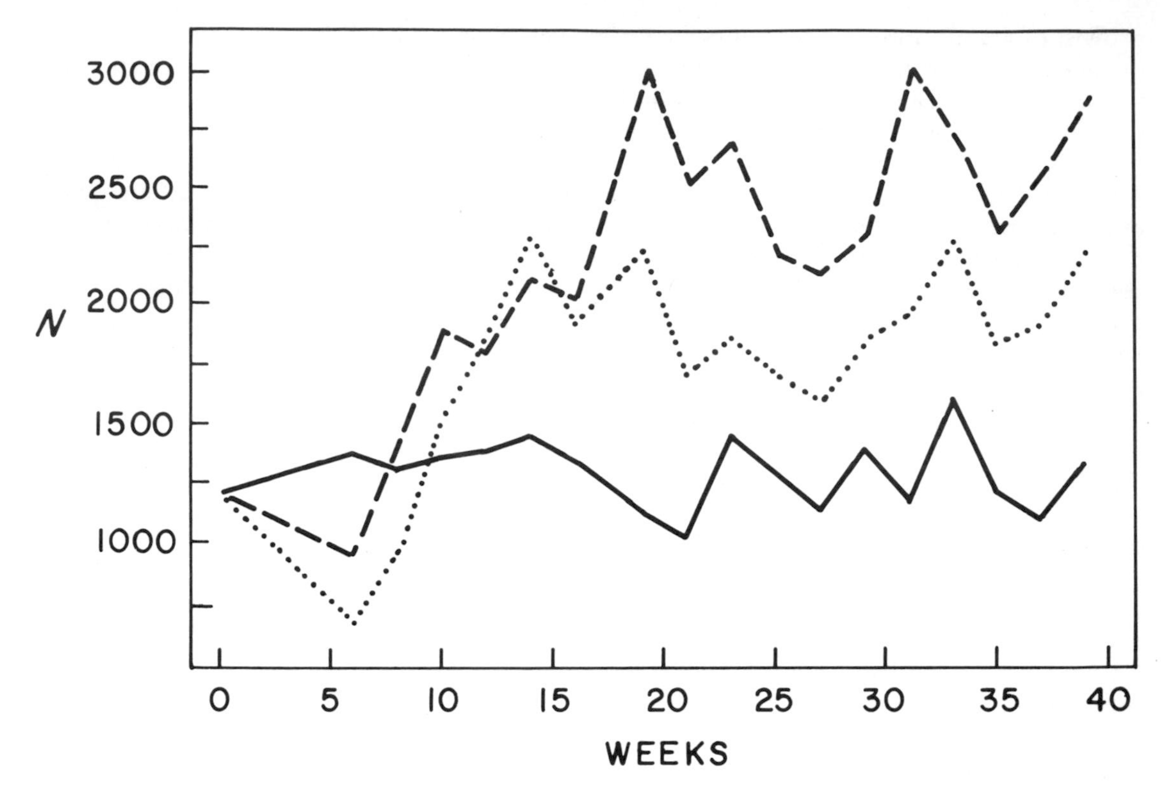

FIGURE 15. Growth of two irradiated populations (broken lines) of *Drosophila birchi* compared with an unirradiated control (solid line) at 25°C. The irradiated populations are at first low as radiation damage is eliminated, but later show striking *K*-selection, based presumably on their greater genetic variance (Ayala).

diversity was increased by hybridizing strains from two different localities, the saturation values were higher and tended to increase faster than in pure cultures of the parents.

Since jars containing suitable, but artificially prepared, food are not regularly parts of the habitat of wild *Drosophila*, it is clear that some adaptation to the laboratory environment may be needed when the domestication of the flies takes place in genetic experiments. The greater the genetic variance in fitness, the greater is the rate of increase in fitness. The hybrid populations thus adapt to the new situation more rapidly, though in nature, up against either parent in its own environment, they would presumably be less fit. Irradiated populations at first do much less well than the controls, but after 8 or 9 weeks they produce populations which may be two or three times as great as that of the unirradiated flies (figure 15). Again variance has been increased and *K*-selection made possible.

In general *r*-selection will be important in the establishment of new populations, particularly if the process involves intraspecific competition, while *K*-selection will be significant in the slow evolution of increasingly efficient utilization of resources.

The logistic in nature

While the Malthusian interaction between the exponential population growth and the limitation imposed by resources and crowding must operate in nature as well as in the laboratory, we should only expect to observe anything comparable to a full logistic curve in rather special circumstances. In what MacArthur[44] called *equilibrium species*, small displacements, due to particularly favorable or unfavorable environmental conditions, may occur above or below the mean number, but these tend to be followed by changes in

44. R. H. MacArthur, On the relative abundance of species. *Amer. Natural.*, 94:25–34, 1960.

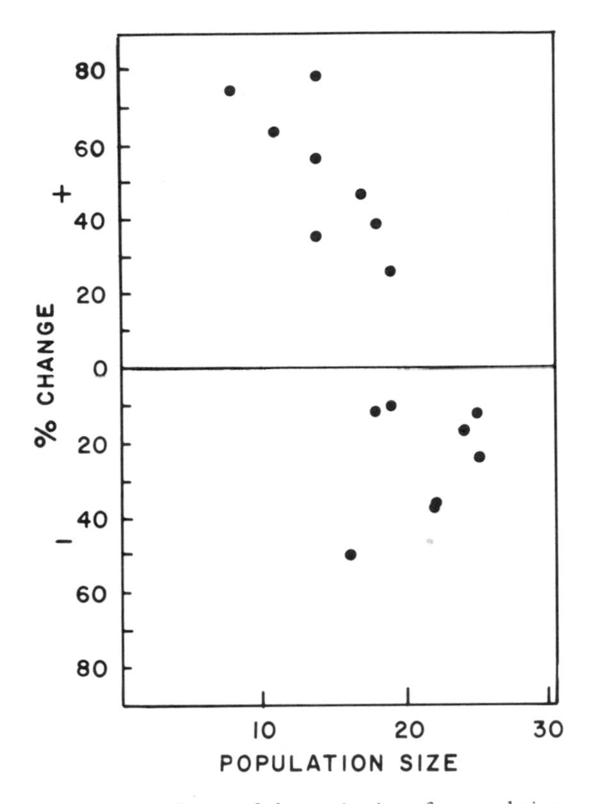

FIGURE 16. Relation of change in size of a population of ovenbirds (*Seiurus aurocapillus*) in a wood near Cleveland, Ohio, to the size of the population, showing a striking tendency for increases to occur when the population is low and decreases when it is high (MacArthur, from the data of Williams).

an opposite direction, returning the population toward the mean. Thus, increases will be commoner when the population is small, decreases when it is large (figure 16). Here *K*-selection is doubtless much more important than *r*-selection.

In contrast to this, *opportunistic species*, which are generally mobile and may for long periods have low densities, increase rapidly if circumstances become favorable, though the very great populations that may result are liable to crash, the species returning to its low density. The buildup of opportunistic species might be expected to imitate the early stages of a logistic population, but without a stable asymptote being approached. Innumerable cases of this sort are known in economic zoology, but they are probably rarely, if ever, studied under such constant environmental conditions that a very close approach to a logistic curve can be expected. We may expect *r*-selection to be of much greater importance in opportunistic species than in equilibrium species.

A very few cases are known where species have been introduced into, or have invaded, isolated areas and have been studied effectively. Unfortunately, on account of human failings, in the two best cases only the earlier stage of colonization is known.

The pheasant colony on Protection Island,[45] off the coast of the state of Washington, in its early history was clearly showing a sigmoid growth curve of the kind discussed below as a finite difference logistic (figure 17). Unfortunately, war and the presence of a hungry occupying force terminated the experiment. It is, however, quite clear that a reduction in rate of increase had been occurring in a way at least qualitatively resembling the logistic.

The other case is the invasion of Great Britain by the collared dove, *Streptopelia decaocto*.[46] This bird had been spreading westward in Europe and started to breed in Britain in 1954. For about ten years reasonably accurate censuses exist, but after this the commonness of the bird reduced its appeal to bird watchers and the data most tragically became inadequate. The available figures seem to suggest a pattern (figure 18) rather different from that of the logistic in its simple form, though they perhaps conform to Verhulst's idea that growth might initially be exponential for some time as the best sites were taken. During the movement of *S. decaocto* into Britain, a number of birds must have been living in essentially unrestrictive environments in which no intraspecific competition could occur simply because the

45. D. Lack, *The Natural Regulation of Animal Numbers*. Oxford, Clarendon Press, 1967, viii, 343 pp.; see p. 12. The data are obtained from A. S. Einarsen, Specific results from ring-necked pheasant studies in the Pacific Northwest. *Trans. N. Amer. Wildl. Conf.*, 7:130–45, 1942; and: Some factors affecting ring-necked pheasant population density. *Murrelet*, 26:39–44, 1945.

46. R. Hudson, The spread of the collared dove in Britain and Ireland. *Brit. Birds*, 58:105–39, 1965; and: Collared doves in Britain and Ireland during 1965–1970. *Brit. Birds*, 65:139–55, 1972.

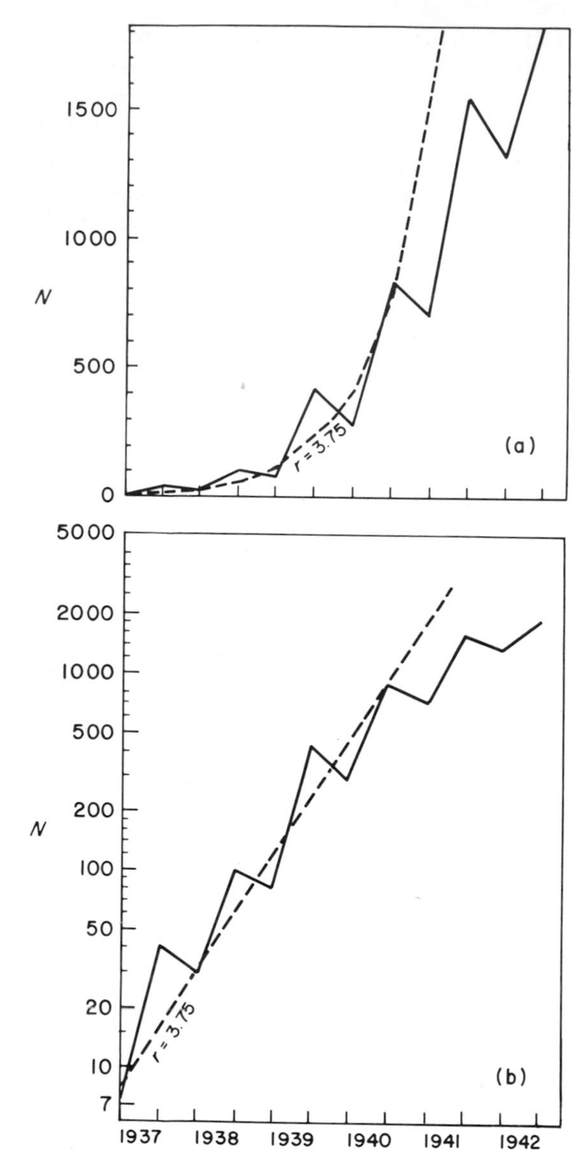

FIGURE 17. Population of pheasant (*Phasianus colchicus torquatus*) introduced onto Protection Island and followed during the early part of its development; (a) plotted arithmetically, (b) plotted logarithmically, against time (data of Einarsen). Broken lines indicate hypothetical exponential growth.

pairs in question were pioneers with no immediate neighbors. It is evident, however, that during the mid-1960s, when the dove was found in nearly every part of Britain, limitation was beginning to take place, and the estimate for 1970, rough as it is, quite literally is out of line with those for the period of logarithmic growth in the late 1950s and early 1960s.

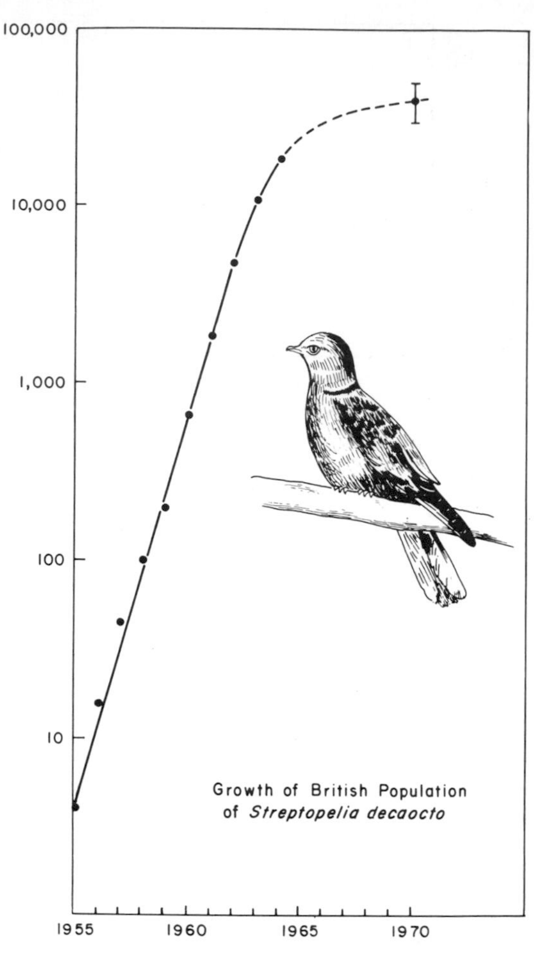

FIGURE 18. Extimates of population of the collared turtledove (*Streptopelia decaocto*) in Great Britain since 1955, logarithmic scale. Note exponential increase in first 8 years, with rapid decline in rate of increase after 1963.

Populations of very small, rapidly reproducing organisms, such as members of the plankton community, which overwinter as resting eggs, cysts, or the like, a limited number of which survive to start a new population in spring, may be expected to develop according to a logistic pattern, though predation would take its toll later in the process. The rise and fall in such populations may be likened to that exhibited by an opportunistic insect.[47]

47. J. Davidson, On the growth of insect populations with successive generations. *Austral. J. Exp. Biol. Med. Sci.*, 22: 95–103, 1944.

In the case of barnacles in which the space of settlement is strictly limited, very neat sigmoid curves for the number of larvae that have settled and metamorphosed may sometimes be obtained. An example from the work of Connell[48] on *Balanus balanoides* colonizing rock surfaces on the shore of Great Cumbrae Island in the Firth of Clyde, Scotland, is given in figure 19. A real saturation effect certainly limits further colonization by cypris larvae which are present, and still can settle on freshly exposed clean rock, well after colonization of occupied surfaces has ceased. It is not certain whether in the cases studied by Connell the initial rise in rate of colonization is due to surviving adults beginning to reproduce slowly, but at an increasing rate, so reaching a maximum some days after the first settlement occurs, or to the attraction that a settled cypris larva appears to have on a free-living one. Both effects may well occur. It is, however, obvious that the initial rise in the population has nothing to do with an exponential increase and its limitation as successive generations are produced, for only one generation is involved.

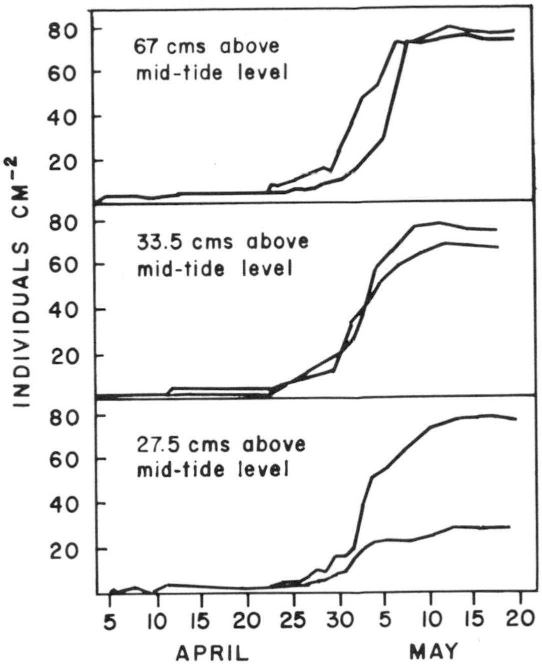

FIGURE 19. Sigmoid curve of settling of populations of *Balanus balanoides* on the shore of Great Cumbrae Island in the Firth of Clyde, April-May 1954. Populations on two stones are shown for three levels relative to mean tide level (Connell).

The most specialized cases in nature, perhaps more closely comparable to well-conducted laboratory experiments than are any other field examples, concern the development of colonies of some social insects.[49] Some of the best cases involve the growth of populations of honeybees in hives in which K is artificially fixed (figure 20), though swarming permits the regulation of the population by the colony. Among strictly wild species, the growth of a colony of the Brazilian ant, *Atta sexdens rubropilosa*, as measured by the number of craters (figure 21) which surround the nest entrances, is strikingly logistic. A more detailed analysis of social insect colonies in terms of life-history stages and castes reveal complicated growth patterns; the reader is referred to Wilson's[50] admirable summary in *The Insect Societies*.

Variations on the logistic

The important thing about any simple and easily understood formulation such as the logistic is that it can provide a stepping stone to a number of other, less simple, less general, but more accurate theories. This is in fact one of the significant uses of general theories in science. Several interesting cases of a more complicated and less general kind may now be considered.

48. J. H. Connell, Effects of competition, predation by *Thais lapillus*, and other factors on natural populations of the barnacle *Balanus balanoides*. *Ecol. Monogr.*, 31:61–104, 1961.

49. M Autuori, Contribuiçáo para o conhecimento da sáuva. *Arg. Inst. Biol.*, 12:197–228, 1941.

A. A. Bitancourt, Expressao matematico do crescimento de formigueiros "*Atta sexdens rubropilosa*" representado pelo aumento do numero de alheiros. *Arg. Inst. Biol.*, 12:229–36, 1941.

F. S. Bodenheimer, Population problems of social insects. *Biol. Rev.*, 12:393–430, 1937; and: Studies in animal populations: II. Seasonal population trends of the honeybee. *Quart. Rev. Biol.*, 12:406–25, 1937.

50. E. O. Wilson, *The Insect Societies*. Cambridge, Mass: Belknap Press of Harvard University Press, 1971, x, 548 pp.; see chap. 21.

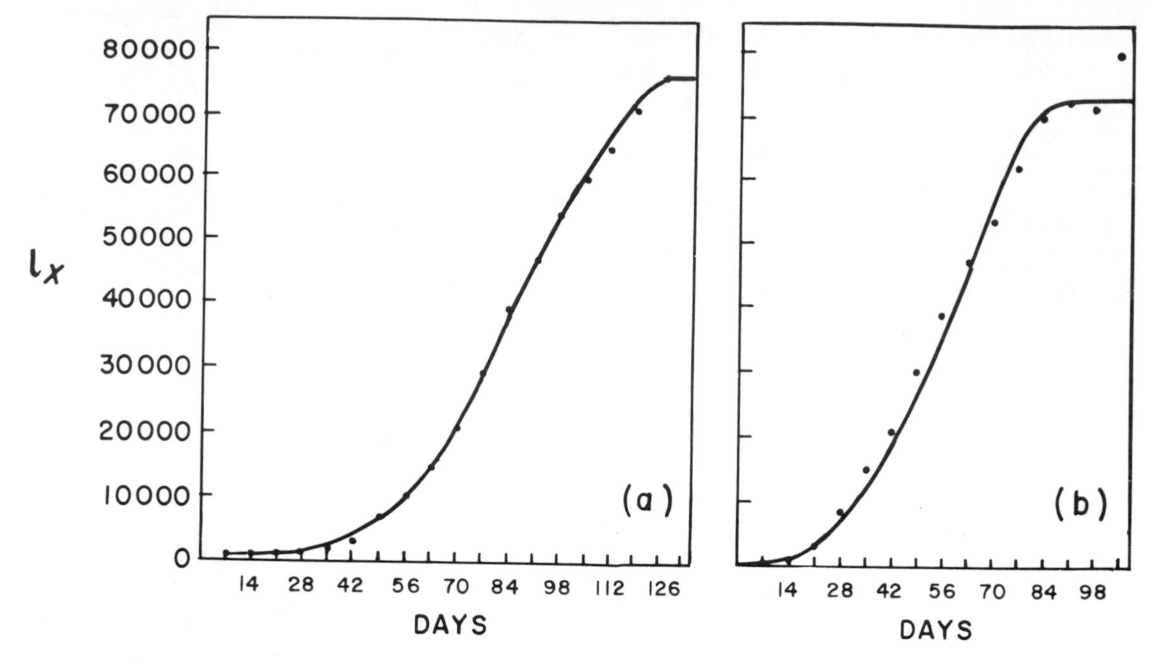

FIGURE 20. Logistic growth of colonies, (a) of Cyprian and (b) of Italian cultivars of the honeybee (*Apis mellifera*) near Baltimore, Maryland, showing almost identical values of K (76, 328, and 74,000) determined ultimately by hive size (Bodenheimer, from the data of Nolan).

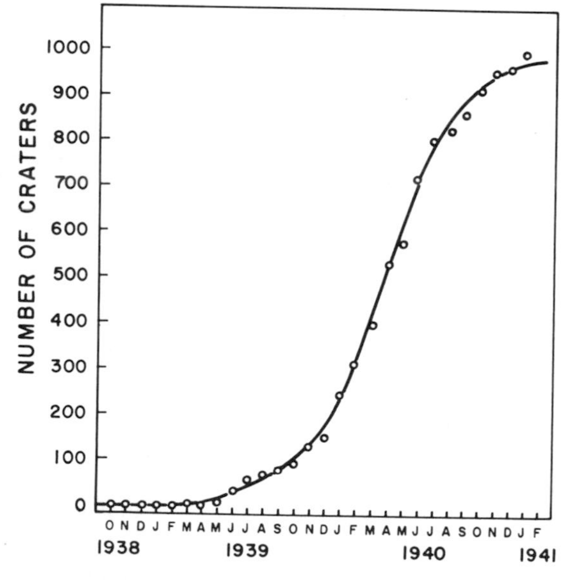

FIGURE 21. Logistic increase in number of craters marking entrances of a growing nest of *Atta sexdens rubropilosa* (Bitancourt, from the data of Autuori).

Smith's equation

As we have seen, the unmodified logistic implies the postulate that in setting an upper limit, the rate of increase per individual organism varies inversely with the density.

In a very ingenious series of experiments, F. E. Smith[51] tested this on a species of water flea or cladoceran, small freshwater crustacea that lend themselves to population studies by reproducing parthenogenetically. Smith cultivated *Daphnia magna* in constantly increasing volumes of medium, so that it was possible to obtain direct estimates of the rate of increase at particular densities. When such rates were plotted against density, the results appeared to fall around a concave curve, and not a straight line. Smith concluded that the valid logistic would be one describing the growth of the feeding rate (F) of a population, with a maximum at K_F, rather than the growth of number (N) or biomass (M), so that we can write

$$\frac{dF}{dt} = rF\left(\frac{K_F - F}{K_F}\right). \qquad (1.8)$$

51. F. E. Smith, Population dynamics in *Daphnia magna* and a new model for population growth. *Ecology*, 44:651–63, 1963.

A comparable relationship between specific growth rate and density was observed by O. W. Richards and A. J. Kavanagh, The course of population growth and the size of seeding. *Growth*, 1:217–27, 1937.

An animal uses more food when it is actively growing than when it is just maintaining itself in a semistarved condition at equilibrium. Uptake of food per animal therefore falls steadily as the growth rate declines with increasing density. Smith supposes that the relationship between feeding rate and biomass depends additively on the latter and on the growth rate from which

$$\frac{F}{K_F} = \frac{cM + \mathrm{d}M/\mathrm{d}t}{cK_M}.$$

This gives by substitution in the logistic for feeding rate a new equation

$$\frac{\mathrm{d}M}{\mathrm{d}t} = rM\left(\frac{K_M - M}{K_M + (r/c)\,M}\right), \qquad (1.9)$$

where c is the sum of rate of maintenance metabolism and of replacement of animals that die in the equilibrium population; it can be calculated from the loss of weight of unfed animals and from the observed death rate in equilibrium populations, and is expressed as mass per unit mass per unit time. The fraction r/c is therefore a dimensionless number. Since the equilibrium biomass, K_M, can be obtained directly from such equilibrium populations, and r can be determined from the reproductive rate of single individuals in large volumes of medium, the whole of the right-hand side of the equation can be evaluated for any value of M, and the rate of growth per unit of biomass calculated and compared with the observed values obtained from expanding cultures at uniform densities (figure 22a). The same type of calculation can be done for numbers (figure 22b) rather than biomass. It will be observed that though there is considerable scatter of the experimentally determined points, they tend to fall around the calculated curve rather than the straight line of the logistic.

Smith's differential equation is hard to handle mathematically but implies an asymmetrical rise of $\mathrm{d}M/\mathrm{d}t$ to a maximum at $M < K_M/2$, and then a fall, the integral curve rising progressively more slowly to the asymptote than does the logistic. A comparable form has been used in competition theory (see chap. 4).

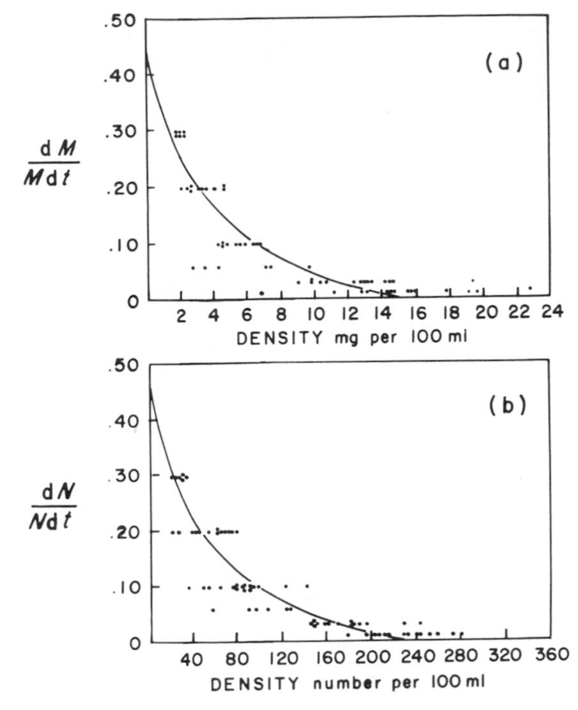

FIGURE 22. (a) Observed (dots) and calculated (line) rates of growth per unit biomass in populations of *Daphnia magna* growing at constant densities. (b) The same for numbers of individuals. Note that whatever the best fit may be, it is not a straight line corresponding to the logistic (Smith).

The logistic with a time lag

The classical logistic assumes that the rate at which a population is increasing at any time t depends on the relative number of individuals occupying potential spaces, at that time, in the system under consideration. In practice this is very improbable, for the process of reproduction is not instantaneous. In a *Daphnia*, for instance, a large clutch presumably is determined not by the concentration of unconsumed food available when the eggs hatch, but by the amount of food available when the eggs were forming, some time before they pass into the brood pouch.

Between this time of determination and the time of hatching many newly hatched animals may have been liberated from the brood pouches of other *Daphnia* in the culture, so swelling the population. In an extreme case, in fact, all the vacant spaces $(K - N)$ might have been filled well before reproduc-

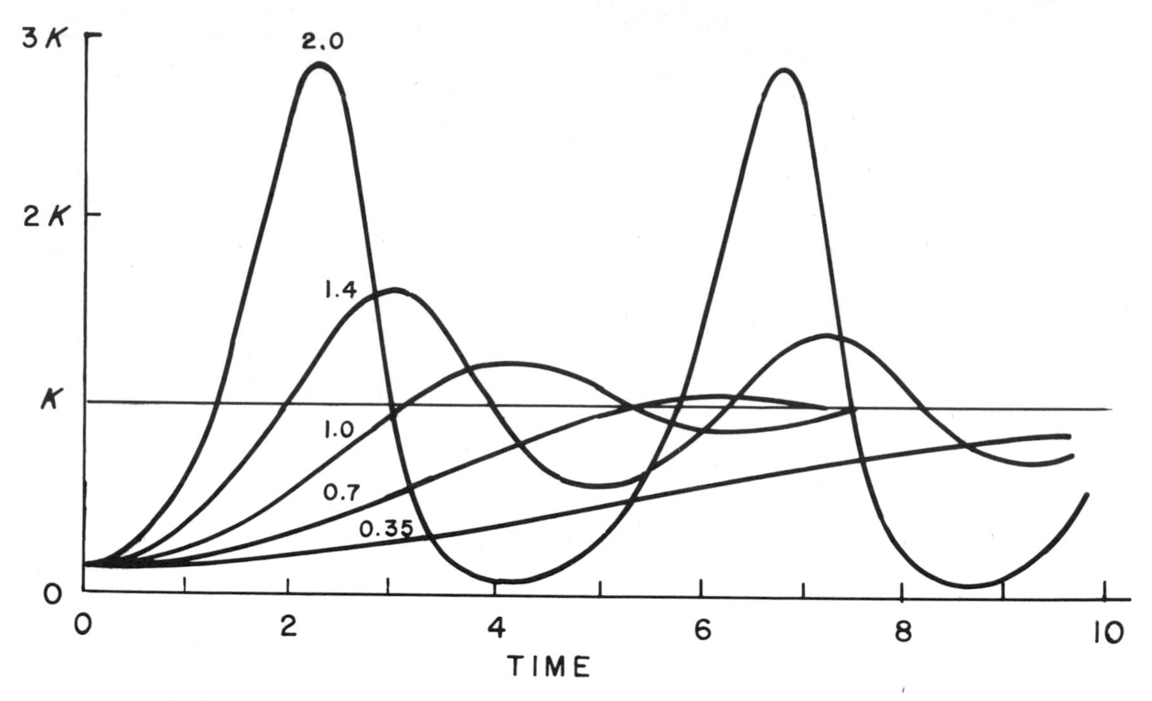

FIGURE 23. Ideal behavior of populations growing according to the logistic with a time lag τ, for different values of $r\tau$ (Cunningham).

tion stops. If we assume egg formation to occur τ units of time before hatching, we may write, in terms of numbers,

$$\frac{dN_{(t)}}{dt} = rN_{(t)}\left(\frac{K - N_{(t-\tau)}}{K}\right) \quad (1.10)$$

as a more realistic variation on the logistic.[52] This equation was first used in economics over 40 years ago, and was later independently introduced into population ecology. It is mathematically rather intractable, but its main properties are adequately understood.

It will be obvious that the time delay permits the development of a transitory population in excess of K, for if K were reached at time t_K the rate of reproduction would still be that appropriate to $(t_K - \tau)$. As the population overshoots the mark, it will cease the genesis of new eggs at $t_K + \tau$, but will still contain individuals in which the later phases of reproduction are occurring. The overpopulation will lead to a decline; and if τ is long enough, death will exceed birth for a sufficient time for the population to fall below K. Mathematical study indicates that if the product $r\tau < \pi/2$, the population will either approach K rather faster than in the case of the logistic, or with rather higher values of τ will oscillate, but the oscillation will be damped out. If, however, $r\tau > \pi/2$, an apparently stable oscillation or limit cycle is set up. At high values of $r\tau \simeq 2$ the peaks are high and short, the depressions low and long (figure 23). Under such circumstances

52. G. E. Hutchinson, Circular cause systems in ecology. *Ann. N. Y. Acad. Sci.*, 50:221–46, 1948.

W. J. Cunningham, A nonlinear differential-difference equation of growth. *Proc. Nat. Acad. Sci. U.S.A.*, 40:708–13, 1954.

The original formulation in economics was R. Frisch and H. Holme, The characteristic solutions of a mixed difference and differential equation occurring in economic dynamics. *Econometrica*, 3:225–39, 1935.

See also N. D. Hayes, Roots of the transcendental equation associated with a certain difference-differential equation. *J. Lond. Math. Soc.*, 25:226–32, 1950.

I am indebted to Robert M. May for these references. He has also discussed the matter in: R. M. May, Time-delay versus stability in population models with 2 and 3 trophic levels. *Ecology*, 56:315–25, 1973.

The adequately prepared reader may curl up and enjoy himself or herself with Robert M. May, *Stability and Complexity in Model Ecosystems*. Princeton University Press, 1973, ix, 235 pp.

the population is likely to become extinct if an unfavorable, random, external circumstance intervenes during a depression. Presumably, there is usually some selection tending to reduce the delay to the smallest practical biological value.

The simple time lag that we have introduced, though it is realistic from a physiological point of view when dealing with an animal such as *Daphnia*, is believed by May to be too restricted in conception. He suggests that instead of $N_{t-\tau}$, a weighed average of the population density up to the time t should be used. This is in some ways easier to treat mathematically than the original equation, but leads qualitatively to the same sort of results; with short time delays the population can approach a stable equilibrium, with long delays it oscillates, usually in a stable limit cycle.

Quite remarkable oscillations are well known in *Daphnia* cultures. It is, however, evident that their genesis may involve not merely time lags but also particular age structures.

Finite-difference logistic

If we think of a population growing by discrete generations, τ units of time apart, but with a continuous death rate and with the same sort of limitation by the existing population density as obtains with the logistic, we obtain

$$N_{t+\tau} - N_t = \tau r N_t \frac{[K - N_t]}{K} \qquad (1.11)$$

for the growth rate from one generation to the next. Such an equation will have some properties reminiscent of equation (1.10). If r is too great, a generation ultimately arises that overshoots K by a sufficient amount to produce instability, and a limit cycle results.

The subject has been extensively investigated by May,[53] who has examined several other equations of slightly greater complexity but having sounder biological properties. In general, if we measure time in generations, so that $\tau = 1$, the population can be stable if $0 < r < 2$. For higher values of r between 2 and 2.7 limit cycles appear, at first with period 2, to which is added, when $r > 2.5$, a cycle of period 4, and for higher values of r further cycles of period 8, 16, 32, etc. When $r > 2.7$, the system becomes chaotic and irregular; fluctuations or long periods of arbitrary length may appear. May points out that such behavior may be under the control of natural selection. He cites, as a possible case, variations in the population of the Colorado potato beetle, which has had little time to settle down in an evolutionary sense in its present agricultural environment.

Stochastic logistic

A number of investigators,[54] notably Feller, Pielou, Roughgarden, and Leigh, have considered the growth of populations in limited systems in which random variation is continually occurring. Leigh gives an equation which in our symbolism becomes

$$\frac{dN}{dt} = \gamma_1(t)\sqrt{N} + [r + \gamma_2(t)]\,N\left(\frac{K - N}{K}\right), \qquad (1.12)$$

where $\gamma_1(t)$ is a white noise or random set of disturbances of variance 1 and mean 0, which represents the internal accidents in a constant environment determining which individuals happen to reproduce and which do not, while $\gamma_2(t)$ is another set of disturbances of variance

53. See particularly R. M. May, Biological populations with nonoverlapping generations: Stable points, stable cycles, and chaos. *Science*, 186:645–47, 1974; and: Models for single populations. In: *Theoretical Ecology Principles and Applications*, ed. R. M. May. Philadelphia and Toronto, W. B. Saunders, 1976, viii, 317 pp.

54. W. Feller, Die Grundlagen der Volterraschen Theorie des Kampfes ums Dasein in Wahrscheinlichkeitstheoretischer Behandlung. *Acta Biotheoretica*, 5: 11–40, 1938.

E. C. Pielou, *An Introduction to Mathematical Ecology*. New York, McGraw-Hill, 1969, 532 pp.

J. Roughgarden, A simple model for population dynamics in stochastic environments. *Amer. Natural*, 109:713–36, 1975.

E. G. Leigh, Population fluctuations, community stability, and environmental variability. In: *Ecology and Evolution of Communities*, ed. M. Cody and J. M. Diamond. Cambridge, Mass.: Belknap Press of Harvard University Press, 1975, pp. 51–73.

σ^2 and mean 0, representing the external random variation in the environment as it influences reproduction. With such a system the population oscillates around K, the extent of its oscillations determining its ultimate fate. Leigh gives a fairly elaborate way of predicting extinction times for such a population while admitting that the data are inadequate to support any conclusions. Roughgarden points out that variations in the intrinsic rate of increase produce considerable differences in the general way that a stochastic population behaves. When r is low, the population primarily reacts, rather slowly, to long-term changes in carrying capacity; and when r is high, the short-term components are overemphasized. This is of course comparable to May's findings for the finite difference equation. The probability of extinction is clearly related to both the responsiveness of the population to a transitory change in carrying capacity, and to the stability or tendency to return to very near equilibrium. Usually, in a species that is a well-established member of a community, there will be continual but not very destructive variations around K, as in the experimental population of figure 11.

The kind of magnitude involved in this sort of situation can be appreciated by considering[55] a population of some organism reproducing by simple binary fission, so that at equilibrium after division each individual has a probability of 0.5 of contributing to future generations, or of leaving no descendants. The probability of random extinction is then given by $p_e = [t(1 + t)]^{N_0}$ where t is the time in generations, measured from the establishment of the population and N_0 is the initial number of organisms present, taken here as K. For large populations and long times we may approximate by $t = -N_0/\ln p_e$. For an initial population of 10^9 p_e would reach a value of 0.01 in 2.2×10^8 generations. Unless the population is very small, random extinction is usually not of great importance. In dealing with endangered species, the chance of extinction is of course of paramount interest, but in such cases we try to know so much about the causes of death or sterility as they affect single individuals, that the problem ceases to be stochastic. Introduction of a competitor, as MacArthur points out, may greatly increase the chance of random extinction.

The asymmetrical logistic

Pearl and Reed were worried by the perfect rotational symmetry of the integral curve and the bilateral symmetry of its derivative curve, about $N = K/2$. This implied that the forces producing the increase in rate up to the maximum are equal and opposite to those causing the subsequent retardation of the rate, which Pearl and Reed believed to be unlikely. Recently Gilpin and Ayala[56] have designed a generalized logistic

$$\frac{dN}{dt} = rN\left[1 - \left(\frac{N}{K}\right)^{\theta}\right] \tag{1.13}$$

which gives an inflection point at $N < K/2$ if $\theta < 1$ and at $N > K/2$ when $\theta > 1$. They suggest that invertebrate populations, when all stages are counted, are likely to have $\theta < 1$, and vertebrate populations, limited by territorial behavior, are likely to have $\theta > 1$. Though Gilpin and Ayala say that the interpretation of the parameters is straightforward, the biological meaning of θ is not so intuitively obvious as is that of K and r. Nevertheless, the equation may well prove useful. A comparable modification had, as we have seen, been considered by Carlson in 1913.

Special cases

It may be convenient to construct special equations for the rates of increase of particular species having complex life histories. An

55. J. G. Skellam, The mathematical approach to population dynamics. In: *The Numbers of Man and Animals*, ed. J. G. Cragg and N. W. Pirie. Edinburgh, Oliver & Boyd, 1955, pp. 31–45.

R. H. MacArthur, *Geographical Ecology*. New York, Harper and Row, 1972, xviii, 269 pp.; see pp. 121–26 for a more elaborate discussion.

56. M. E. Gilpin and F. J. Ayala, Global models of growth and competition. *Proc. Nat. Acad. Sci. U.S.A.*, 70:3590–93, 1973.

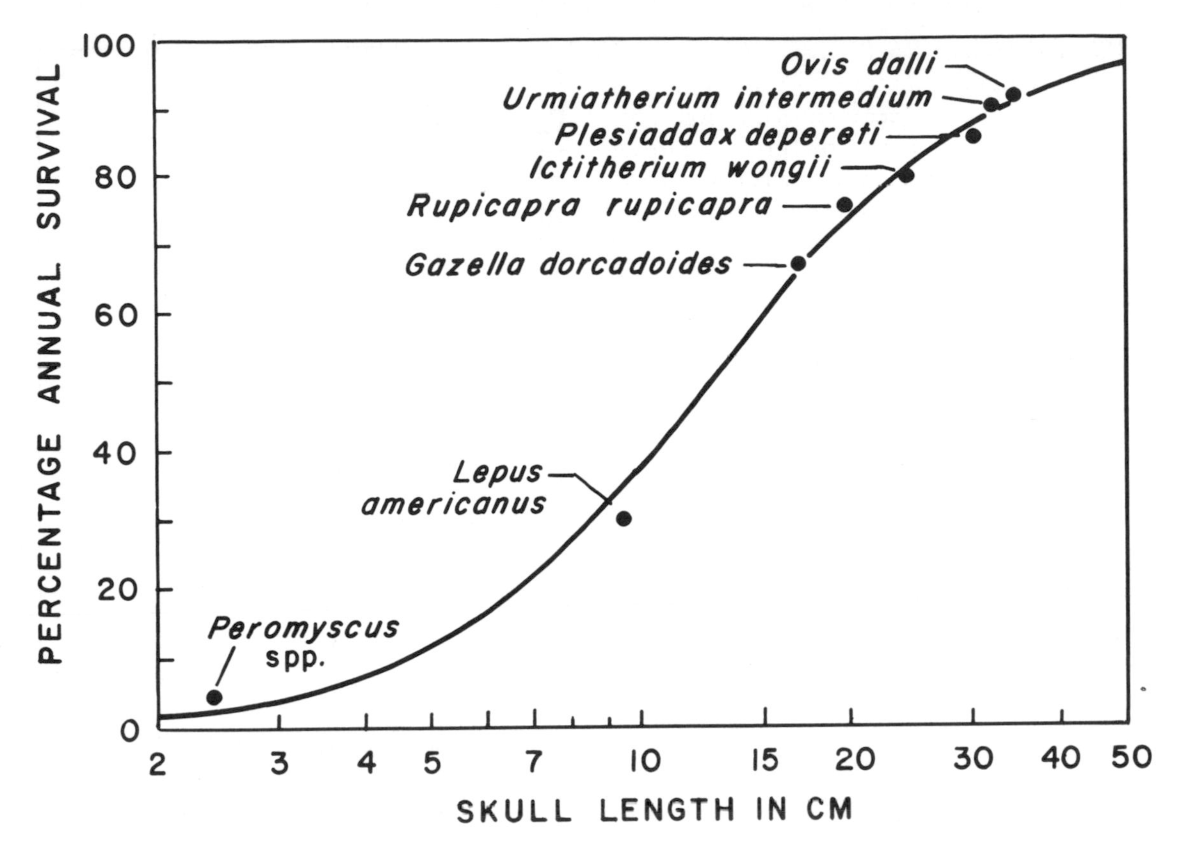

FIGURE 24. Annual percentage survival of mammals for which good data exist, plotted against skull length and fitted to a logistic curve (Kurtén). It is assumed that though the mean length of life increases with size, the rate of increase declines and that, with Socrates, all mammals are mortal.

example is afforded by Crombie's equation[57] for the increase in numbers (N) of adult graminivorous insects in which mortality is mainly due to cannibalism on eggs and pupae by all active stages of total number n,

$$\frac{dN}{dt} = N[rf(N)(1 - qn)^h(1 - pn)^k - m] \qquad (1.14)$$

where $f(N)$ is a function to be determined empirically relating the rate of egg production to density, $(1 - qn)^h$ is the proportion of eggs surviving being eaten for h days, where h is the length of the egg stage, and $(1 - pn)^k$ is the survival of pupae over k days of the pupal instar. If the function $f(N)$ is the same in two species and the small noncannibalistic death rate m is neglected, these equations can be used as the basis for constructing competition equations, just as can the ordinary logistic (see chap. 4).

Other uses of the logistic

Whenever we have a variable increasing with time toward an upper limit, it is tempting to think of a logistic type of increase. Cisne thus believes that diversity within a taxonomic group increases in this way. Kroeber thought that various aspects of human culture developed according to such a pattern till a particular activity or style of expression became saturated. Kurtén (figure 24) uses the curve to express the decrease in death rate with size that we shall find in the next chapter is characteristic of mammals.[58]

57. A. C. Crombie, Further experiments on insect competition. *Proc. Roy. Soc. Lond.*, 133B:76–109, 1946.

58. J. L. Cisne, Evolution of the world fauna of aquatic free-living arthropods. *Evolution*, 28:337–66, 1974.

A. L. Kroeber, *Configurations of Culture Growth.*

Williams's[59] saturation kinetics model
Williams points out the very obvious fact that increasing nutrient concentration does not indefinitely increase the growth rate of the population. Above a certain concentration the organism will be unable to handle any more nutrients, be they ions, plant cells, or other animals. He develops a theory based on the Michaelis-Menten equation used in enzyme kinetics, and also in studies of bacterial populations by Monod. This in its usual form may be written for a population as

$$\frac{dM}{dt} = M\mu_{max} \frac{C}{K_{\mu} + C} \qquad (1.15)$$

where μ_{max} corresponds to r in our earlier equations, C is the concentration of limiting nutrient, and K_{μ} is the value of C when $dM/dt = \frac{M\mu_{max}}{2}$. Williams considers two cases, the first a closed system in which there are no inputs or outputs; the second, a maximally open system in a chemostat in which there is an input of nutrients in the medium and an output of unused nutrients, metabolic products, and organisms. This open system thus differs from the ideal, natural, open system in which the input is usable energy and the output heat.

The mathematical treatment becomes quite formidable and no analytical solution, even for the simpler closed case, could be found. Numerical solutions can be obtained by computer simulation. The resulting growth follows a pattern which looks like a logistic but apparently at first rises more nearly exponentially and then flattens off to equilibrium rather suddenly. Williams indicates that in his experimental work this divergence from the logistic pattern is actually observed. A possible case from Riley's[60] work on the growth of a marine diatom under conditions comparable to those of Williams's closed system is given in figure 25. Very recent work along these lines, much of it unpublished, by P. and S. S. Kilham, by Tilman and by Rhee is providing most important insight into the development of phytoplankton populations.[61]

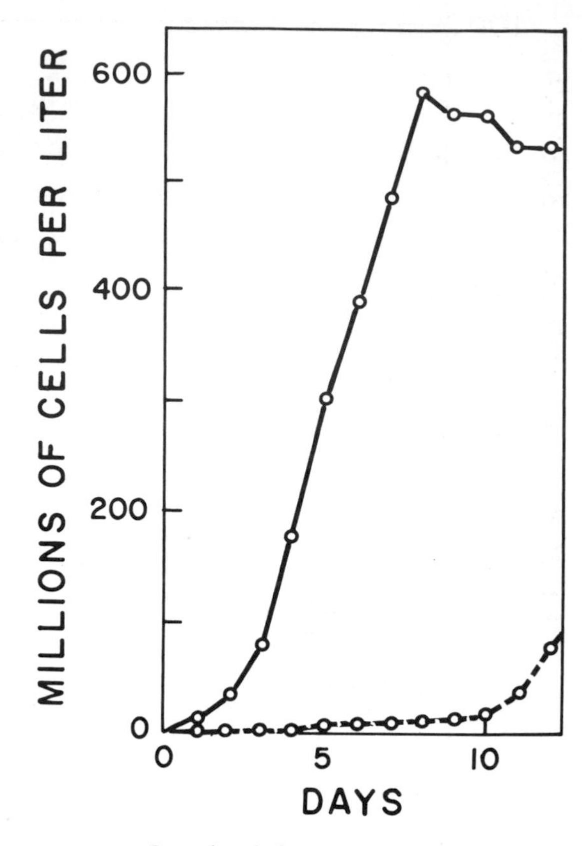

FIGURE 25. Growth of the marine diatom (*Nitschia closterium*) in culture, prior to the appearance of a significant number of senescent (broken line) cells (Riley).

Stanford, University of California Press, 1944.

See also G. E. Hutchinson, *The Itinerant Ivory Tower.* New Haven and London, Yale University Press, 1953; see pp. 70–78.

B. Kurtén, On the variation and population dynamics of fossil and recent mammal populations. *Acta Zool. Fenn.*, 76:1–122, 1953.

59. F. M. Williams, Mathematics of microbial populations, with emphasis on open systems. *Trans. Conn. Acad. Arts Sci.*, 44:395–426, 1972.

J. Monod, La technique de culture continue: théorie et applications. *Ann. Inst. Pasteur*, 79:390–410, 1950.

60. G. A. Riley, Physiological aspects of spring diatom flowerings. *Bull. Bingham Oceanogr. Coll.*, 8: 1–53, 1943.

61. D. Tilman and S. S. Kilham, Phosphate and silicate growth and uptake kinetics of the diatoms *Asterionella formosa* and *Cymbella meneghiniana* in batch and semi-continuous culture. *J. Phycol.*, 12:375–83, 1976.

SUMMARY

The logistic of P. F. Verhulst, of which the differential form is

$$\frac{dN}{dt} = rN\left(\frac{K - N}{K}\right),$$

is the simplest mathematical model of a population (N) growing from a very small number of individuals to an upper limiting population K, set by nutrient supply, space, accumulation of excretory products or other density-dependent factors. The idea that the rate of growth of a population declined with time as the population increased developed slowly during the seventeenth and eighteenth centuries, receiving some empirical confirmation from the more rapid rates of increase, greater than those obtaining in Europe, supposedly characteristic of the human population of early eighteenth-century North America. Malthus based his well-known conclusions on this supposed difference, but arrived at a result that was quantitatively unsatisfactory. Verhulst produced the first treatment that was both general and realistic in 1838, but it excited practically no interest. A comparable form was used in the study of the growth of individual organisms early in the present century and was independently applied to bacterial populations in 1911 by M'Kendrick and Kesava Pai. Later in 1920 Pearl and Reed rediscovered the logistic and in 1921 Pearl learned of, and recognized, Verhulst's priority. After this time most investigators of experimental populations used Verhulst's equation or some modification thereof.

In laboratory studies the equation gives a good fit to data obtained for the growth of small, rapidly growing, asexual or parthenogenetic organisms. Since they can be treated as reproducing almost continuously in large populations, and apparently have their reproductive activity retarded linearly by the term $(K - N)/K$, they approximate to the moving parts of an analogue computer set up to give solutions of the equation.

In experiments with *Drosophila*, though the results are often rather irregular, there is good evidence that the population can adapt in a single experiment to the conditions imposed in the laboratory, so that K slowly increases. There is also evidence that r can increase during long-continued domestication. These observations give support to the distinction of r-selection and K-selection, due to MacArthur and Wilson, as an important concept in population ecology.

In nature few populations have been observed quantitatively developing from a very small colonizing group. In the one case in which one would expect a good logistic, the population, of pheasants, was eaten by the army before K had been reached, but the general form of the early part of the curve looked encouraging. Where a species is invading an area over a broad front, initially one would not expect a pure logistic until the whole area is sparsely colonized, as part of the population would always be growing exponentially. The best documented case, of the spread of the collared dove, *Streptopelia decaocto*, into Britain, suggests a prolonged exponential increase before a decrease in the rate of population growth occurred.

A number of variants of the logistic have been put forward.

In the logistic the specific rate of increase of $\frac{1}{N} \cdot \frac{dN}{dt}$ is proportional to $(K - N)$. Cases have been found, notably in yeast and in *Daphnia magna*, in which this relationship does not hold, and Smith has put forward a modified logistic equation which allows the specific rate to be lower at intermediate values of N than initially or finally. This appears reasonable in that the overall consumption of food per organism will be greater when the population is growing than when it is at equilibrium.

A time lag in the operation of the feedback term $(K - N)/K$ may be expected, since young will be produced as a result of a certain rate of food intake at time t, τ units of time

---

G. Y. Rhee, Effects of N/P atomic ratios and nitrate limitation on algal growth, cell composition, and nitrate uptake. To appear in *Limnology and Oceanography*.

I am much indebted to these authors for advance copies of their papers, in the bibliographies of which other references will be found.

before the young are liberated, at which time a larger population will be competing for food. If τ is great enough, oscillations may be set up and in extreme cases persist as a limit cycle.

A variant of the logistic using finite differences rather than continuous functions is appropriate to organisms with discrete, usually annual, breeding seasons, even though death may occur at any time. The finite difference logistic may lead to oscillations, as in a sense it involves time lags.

A stochastic treatment is possible; in slow-breeding species, random long-term trends in K are important, while fast-breeding species tend to react selectively to transitory chance changes, as might be expected.

If it be felt that the symmetry of the logistic growth curve is too good to be true, Gilpin and Ayala have indicated (1.13) how to introduce an asymmetry into the logistic. The additional constant required, if it is validly conceived, may vary systematically from group to group, being below unity in many invertebrates but above unity in territorial vertebrates.

Williams has provided a theory which obviates one disadvantage of the logistic, namely, that it does not take care of the obvious biological fact that the growth rate cannot be increased above a certain inherent upper limit merely by increasing the food supply. The theory is developed in terms of uptake by, and transfer of nutrients by, specific sites on organisms and is primarily applicable to unicellular forms. The resulting mathematics is intractable, but numerical solutions can be obtained from a computer. Growth is comparable to that described by the logistic, but with a more rapid decline in rate as the asymptote is approached.

These examples make clear the value of Verhulst's equation. In population biology we are usually dealing with sets of organisms that are too large to warrant individual treatment, but too small and too lacking in uniformity to provide the accurate statistical mechanics of a volume of gas. Levins[62] has pointed out that theories ideally should be general, realistic, and accurate, but the nature of the subject makes simultaneous satisfaction of all three requirements nearly impossible. The logistic is general, and realistic in the biological meaning of its parameters. It can be made more accurate by various kinds of modification which, being designed for special situations, cause a loss of generality. The simplicity of the equation here is its great virtue; it is easily understood and so easily modified. Gilpin and Ayala wisely recommend modifications to be made in such a way that a small amount of modification produces a large improvement in realism and accuracy. More generally, even a potentially erroneous theory is an enormous advance over having no theory at all, for the incorrectness of the theory, when tested, is in a sense a measure of how far wrong are the postulates on which the theory is based. Once this has been determined, we can start modifying the theory; if we had no theory, there would be nothing to modify and we should get nowhere.

62. R. Levins, The strategy of model building in population biology. *Amer. Scientist*, 54:421–31, 1966. Reprinted, as the prologue, in *Readings in Population Biology*, ed. P. S. Dawson and C. E. King. Englewood Cliffs, N. J., Prentice Hall, 1971. This very valuable paper may be profitably contrasted with the somewhat more easygoing approach adopted in the present book.

Chapter Two

Interesting Ways of Thinking about Death

IN chapter 1 we looked at some models that seem to be useful in understanding growing populations. In one important respect, however, the treatment was undemographic, as we did not concern ourselves very much with the age structure of the populations. In order to understand them properly it is obvious that we should know what proportions of individuals of various ages occur and how these proportions change.

Ecologically it is convenient to begin by considering how long individuals are likely to survive, because the recruitment into a population will obviously depend on how many individuals are prereproductive, how many are reproducing adults, and how many are postreproductive. These proportions will depend on the inherent physiology of the organism and on its expectation of life. If we want to think intelligently about how population growth is controlled by natality, we find paradoxically that it is best to start by thinking about death.

The potential or maximal life-span of organisms
The longevity of organisms varies greatly, from $n.10^3$ seconds or some tens of minutes in bacteria reproducing at ordinary temperatures under good conditions, to $n.10^{11}$ seconds or several millenia in the big tree *Sequoiadendron giganteum* and the bristlecone pine, *Pinus longaeva*, both in western North America, and perhaps the dragon tree, *Dracaena draco*, of Teneriffe.[1]

In the plants that form the windswept krumholtz at the tree line on some mountains, in some mosses, species of *Lycopodium*, and grasses, and possibly in some reef-forming animals, growth may continue indefinitely in one direction while the organism dies at the opposite end. Such species clearly have a rather indefinite status in population ecology.

Within any restricted higher taxon such as a family or order, it is usual to find a striking correlation between maximum life-span and size (figure 26). This relationship is, however, not quite as simple as it might seem. Even within the vertebrates, there are two extreme and highly divergent possibilities, linked by all possible intermediate conditions. Teleost fishes, as every fisherman knows, may go on growing throughout their lives, though at a decreasing rate. In such cases a large individual is inevitably an old one, though how old depends greatly on both species and environment. In the majority of nidicolous birds,

1. A fairly representative sample of the data on the longevity of organisms is given in Life Spans: Animals, compiled by J. Raewales, C. F. E. Roper, O. Hartman, P. L. Altman, and D. S. Dittmer. *Biology Data Book*. Bethesda, Md., Fed. Amer. Soc. Exper. Biol., 2nd ed., 1972, vol. 1, pp. 229–335.

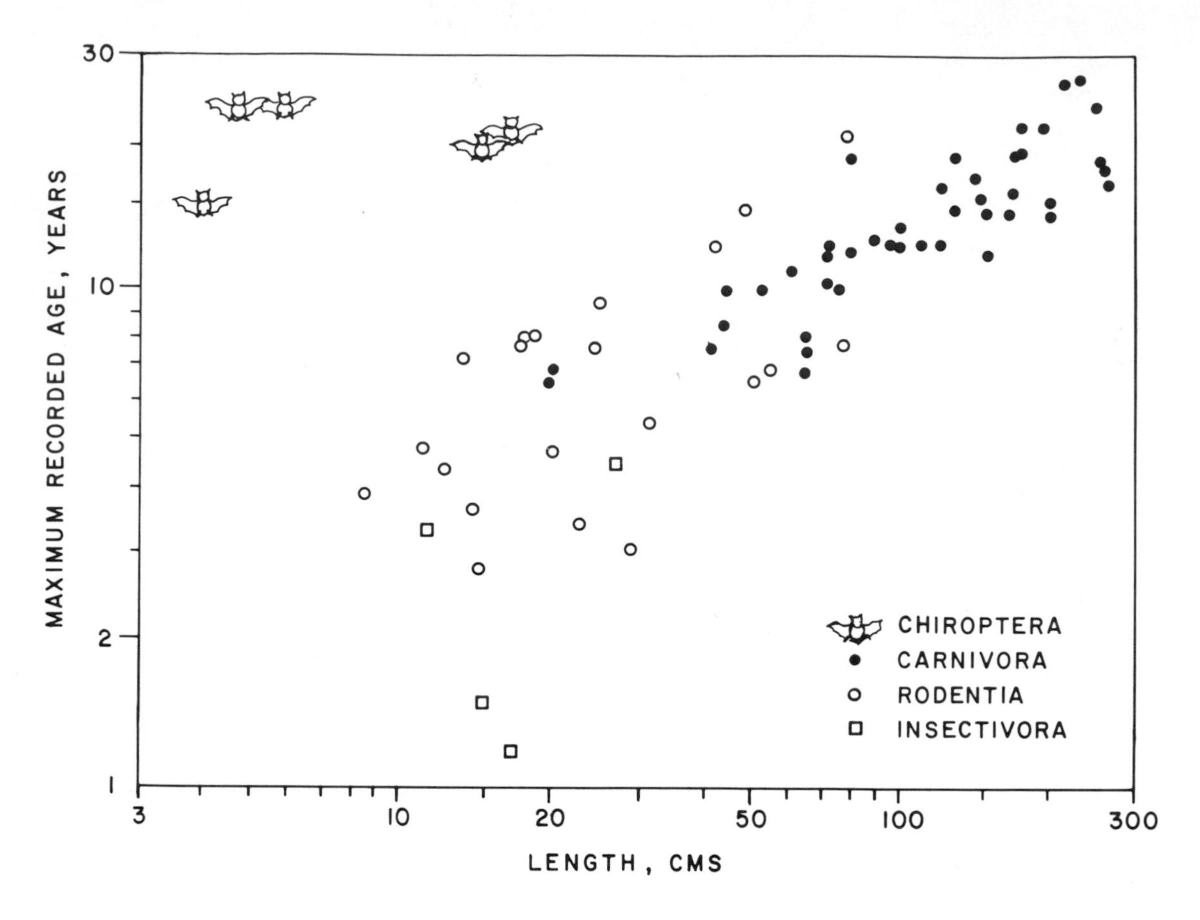

FIGURE 26. Maximum ages recorded in a number of bats, carnivores, rodents, and insectivores, showing a general trend of increasing maximal longevity with increasing size in the terrestrial mammals of the last three orders, but very great ages, relative to their sizes, in the Chiroptera (data from *Biology Data Book*, Kurtén, and various standard taxonomic works giving lengths of the species considered).

The maximum ages of the rodents, insectivores, and bats used in figure 26 are from this compilation, the lengths being obtained from a variety of standard taxonomic works. All the data for the carnivores are derived from Kurtén (see note 7).

The bristlecone pine has been the subject of some taxonomic discussion, the potentially very long-lived *Pinus longaeva* having been separated from *P. aristata.* The plant has become of major importance as a source of wood, dated by tree-ring counts, which can be used to correct carbon-14 dates for errors presumably due to variation in the cosmic-ray flux in the atmosphere. The results have produced a revolution in the archaeology of western Europe. The famous dragon tree, *Dracaena draco*, of Teneriffe, supposed to have been 6,000 years old, was blown down in 1868, according to J. C. Willis, rev. H. K. Airy Shaw, *A Dictionary of the Flowering Plants and Ferns*. Cambridge University Press, 1973, p. 388. There may be further literature on this case.

T. A. Stephenson, *The British Sea Anemones* (London, Ray Society, 1928, vol. I, p. 97), recorded specimens of *Cereus pedunculatus* (Pennant) collected prior to 1862 and healthy in 1927 after 65 years in captivity.

The longest invertebrate life-span reliably recorded seems to be the 116 years known in the freshwater pearl mussel; see J. Hendelberg, The freshwater pearl mussel, *Margaritifera margaritifera* (L.). *Inst. Freshw. Res. Drottningholm Rep.* (Fishery Board of Sweden), 41:149–71, 1960.

E. Frömming, *Biologie der mitteleuropäischen Süsswasserschnecken* (Berlin, Duncker and Humblot, 1956), mentions the alleged finding of ages over two centuries in Amazonian molluscs of unspecified identity by A. Beckh (*Aq. Terr. Z.*, 8, 1955). I have not been able to consult the original, but in view of the difficulty in aging tropical organisms by counting periodic structures, some skepticism may be pardoned.

maximum size is reached about the time that the fledgelings leave the nest, so that the growth period is very short, less than a 100th part of the maximum possible life-span. In such cases, there is no obvious simple-minded explanation of the relationship of life-span and size.

When highly diverse groups, even within a single class, are compared, there is often very little evidence of the effect of size. Bats appear to be potentially about ten times longer-lived than insectivores and rodents of comparable size.

Some sea anemones certainly have life-spans equal to, if not greater than, that of man, though they are three orders of magnitude smaller in weight and volume than we are. A simple physiology in such reasonably small animals may perhaps make them potentially immortal.

Life-span is clearly dependent not only on size, but also on temperature. In extreme cases, animals adapted to very cold habitats, where metabolism is slow and food supply limited, may be extraordinarily long-lived. The pearl mussel of Northern rivers may live for at least a century. The tiny shells of the clam, *Tindaria callistiformis*, from about 3,800 meters in the North Atlantic, show banding which, from the mean radiochemical ages of shells of various sizes, is legitimately regarded as annual.[2] The banding is presumably produced by seasonal events in the illuminated layers of the ocean, such events regulating the rate at which food particles are produced and then drop to the depths. Specimens of this clam 8.3–8.6 mm in diameter appear to be about 100 years old, and examination of smaller specimens suggests that sexual maturity is reached when the animal is about 50.

Most of the striking differences in longevity not obviously dependent on size, nor the inevitable consequence of living at very low temperatures in unproductive habitats, must be ascribed to physiological differences, largely unknown and in general outside the scope of this book. In population ecology the maximum life-span is of little explicit importance; what matters is how long an average individual does live, and not how long it might live if protected from the changes and chances of its mortal life. Yet it is usually implicitly believed that the expectation of life has something to do with the maximum possible longevity, so that accidental events are more likely to be lethal to an inherently short-lived than to an inherently long-lived species.

This is a matter of some significance when, as is usual in wild populations of birds, the mortality rate, referred to the surviving population at any age, remains constant throughout a large part of the life-span, and so is clearly determined by liability to accidental causes of death, yet varies considerably from species to species, which are thus evidently specifically accident-prone to a varying degree.

Life tables and survivorship curves

We will first consider a large number of individuals in a population, all born at the same time, or in practice very nearly so. Such a collection of equal-aged individuals is usually called a *cohort* in demography. Biologists often think of a cohort of 100 or 1,000, actuaries of one of 10,000, some human demographers of 1,000,000; Sir William Petty, as we have already noticed, pitched on 600.

Every day or month or year, depending on whether we are studying fast- or slow-growing organisms, we count the number of survivors in our cohort. We can then make a *life table*, in which the fraction $\mathbf{l}_x$ of our original cohort that has survived to time x is entered. This may be given as a decimal fraction of unity, or referred to any-sized cohort found convenient. In this book a cohort of 1,000 is usually employed. We can also record the data graphically, by plotting $\mathbf{l}_x$, either logarithmically or arithmetically, against x to get a *survivorship curve*. For exposition the graphical presentation is better than the tabular and will in general be used. The life table or survivorship curve obtained in this way is

2. K. K. Turekian, J. K. Cochran, D. R. Kharkar, R. M. Cerrato, J. R. Vaišnys, H. L. Sanders, J. F. Grassle, and J. A. Allen, Slow growth rate of a deep-sea clam determined by ^{228}Ra chronology. *Proc. Nat. Acad. Sci. U.S.A.*, 72:2829–32, 1975.

said to be *age-specific* since all surviving individuals at any time have like ages. It is also sometimes called *dynamic* because the population is followed as it changes with time, or *horizontal* as a contrast to vertical.

Investigations of the probability of a human being living to a certain age are obviously of great importance to insurance companies. Moreover, this particular kind of political arithmetic is of practical significance in government, as the life table gives fundamental information as to the proportion of men who are of appropriate age for military service, and how many children, and of what ages, need education, as well as providing the raw data for actuarial endeavors that go into social security and public medical aid.

The Romans appear to have made the first actuarial tables.[3] The best of such ancient tables, attributed to Ulpian, who lived in the third century A.D., gives the number of years for which provision (*quantitas alimentorum*) should be allowed after surviving to a given age. Taken at its face value, the table thus gives the number of years $\mathbf{e}_x$ of expected life remaining after reaching age x. As the result of Macdonell's[4] study of the ages recorded in the very large collection of epitaphs on Roman tombstones given in the *Corpus Inscriptionum Latinarum*, it is possible to compare Ulpian's table with the actual expectation of life $\mathbf{e}_x$ in Rome over a period including that at which table was derived. Macdonell calculated $\mathbf{e}_x$ not only for Rome itself but for the Iberian provinces of Hispania and Lusitania and for Roman North Africa. Survivorship curves based on these data are discussed later. It must be pointed out that the population of Rome represented is proletarian; it is unlikely that many of these Romans would have had occasion to use or be affected by Ulpian's table in litigation about annuities. The provincial populations were clearly healthier and may have been better

3. For the origins of actuarial procedure see C. F. Trenerry, *The Origin and Early History of Insurance*. London, P. S. King, 1926, xiv, 330 pp. The relevant text of Ulpian's table as given by Trenerry is (p. 10):

> a prima aetate usque ad annum vicesimum quantitas alimentorum triginta annorum computetur . . . [from the earliest age to 20 years, provisions for 30 years are to be computed].
>
> ab annis vero viginti usque ad vicesimum-quintum annorum viginti octo [but from 20 to 25 years, 28 years].
>
> ab annis vigintiquinque usque ad annos triginta, annorum vigintiquinque [from 25 to 30 years, 25 years].
>
> ab annis triginta usque ad annos trigintaquinque annorum vigintiduo [from 30 to 35 years, 22 years].
>
> ab annis trigintaquinque usquam ad annos quadriginta, annorum viginti [from 35 to 40 years, 20 years].
>
> ab annis quadraginta usque ad annos quinquaginta (tot) annorum computatio fit, quot aetati ejus ad annum sexagesimum deerit, remisso uno anno [from 40 to 50 years the computation is made by subtracting the age from 60 minus 1].
>
> ab annis vero quinquegesimo usquam ad annum quinquagesimum quintum, annorum novem [from 50 to 55 years, 9 years].
>
> ab annis quinquagentaquinque usque ad annum sexagesimum, annorum septem [from 55 to 60, 7 years].
>
> ab annis sexaginta cujuscunque aetatis sit, annorum quinque [above 60, 5 years whatever the age is].

Trenerry regarded this as an annuity table; but the values are given in years, so that, however it was used, it seems basically to be a table of life expectancy.

Trenerry is no doubt correct in believing that the most probable source of such empirical knowledge as lay behind Ulpian's table was derived from the experience of the officers of the collegia or societies that existed in Rome to provide funeral insurance for their members, probably including the cost of the tombstones bearing the epitaphs used by Macdonell (see note 4). These societies were in origin religious and bore the name of their titular deity or deities such as the Collegium Cultorum Dianae et Antinoi, though they later became institutions solely for the payment of *funeraticia*. Presumably the societies, named for a saint, which have played a similar role in Italian life in Christian times to the present day, are the lineal descendants of these Roman collegia. Until the middle 1920s, when Mussolini prohibited the appearance of anyone masked in public, the members of such an archiconfraternita took part in the funeral processions of members in the white hooded robe with eye-slits, more familiar in America as a cultural borrowing by the Ku Klux Klan.

The expectations of life from Macdonell, in figure 27, are mean values for the two sexes and are smoothed above the age of 20, to remove a marked tendency of the Romans to record a quite disproportionate number of deaths as taking place when the individual's age was a multiple of 5. This is presumably due to people keeping account of their ages in *lustra* rather than years, each *lustrum* being a 5-year period. At death these lustral ages must have been multiplied by 5 to give the age on the epitaph. Trenerry indicates that the officers of the collegia were selected for 5 years.

4. N. B. Macdonell, On the expectation of life in ancient Rome, and in the provinces of Hispania, Lusitania, and Africa. *Biometrika*, 9:366–80, 1913.

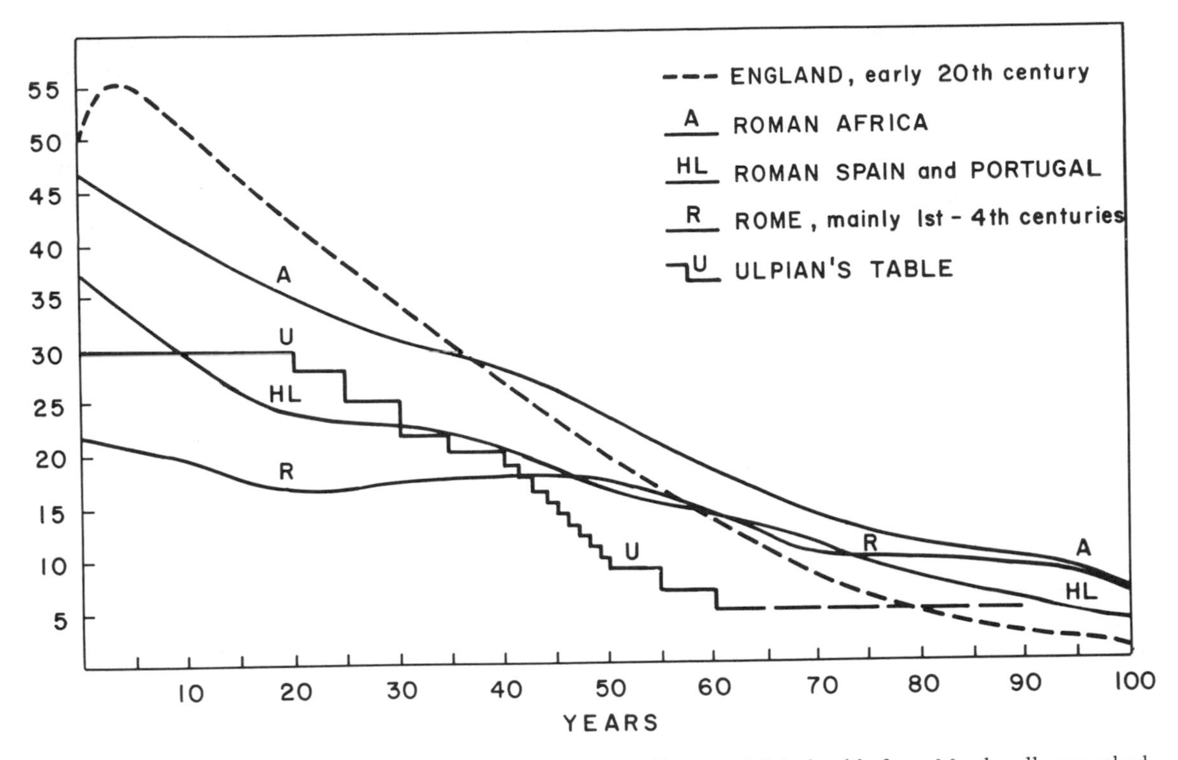

FIGURE 27. Expectation of life in the ancient Roman Empire. All except Ulpian's table from Macdonell, smoothed to remove effect of using 5 year *lustra* in recording ages.

off financially (figure 27). In the first 40 years of life, Ulpian's table gives a greater expectancy of life than the average Roman experienced, though the populations of the Iberian provinces behave very roughly as Ulpian supposed. The rapid decline in life expectancy between 40 and 50 is an obvious and quite unsubstantiated feature of the table, leading to excessively low values late in life. Here the maker of the table clearly had a false intuition. There is no evidence as to how the table was constructed; it is clearly just good enough for some kind of experience, perhaps of funeral insurance, to have informed the guesswork that must have been largely involved. In spite of its defects, Ulpian's table itself had a very long life, being used as the official annuity table in northern Italy till the end of the eighteenth century. By that time it was elsewhere superseded by something far more scientific.

The first age-specific life table in a modern form was constructed by the astronomer Edmund Halley, who published it in a paper in the *Philosophical Transactions of the Royal Society of London* in 1693.[5] Halley's life table is for the city of Breslau in Silesia, now Wrocław in Poland, for which city rather extensive sixteenth- and seventeenth-century records of births and deaths existed. They had been studied by the Rev. Dr. Caspar Neumann, who sent his results to Leibnitz, by whom they were apparently transmitted to Mr. Henry Justell, then Secretary of the Royal Society of London. Subsequently Justell seems to have corresponded directly with Neumann, and passed on what he learned to Halley. Halley, who was concerned largely with the practical significance of life tables, rightly felt that data from London

5. E. Halley, An estimate of the degrees of the mortality of mankind, drawn from curious tables of the births and funerals at the city of Breslaw; with an attempt to ascertain the price of annuities upon lives. *Phil. Trans. Roy. Soc. Lond.*, 17:596–610, 1694. (Reprinted with an introduction by L. J. Reed, *Two Papers on the Degrees of Mortality of Mankind.* Baltimore, Johns Hopkins University Press, 1942, vi, 21 pp.

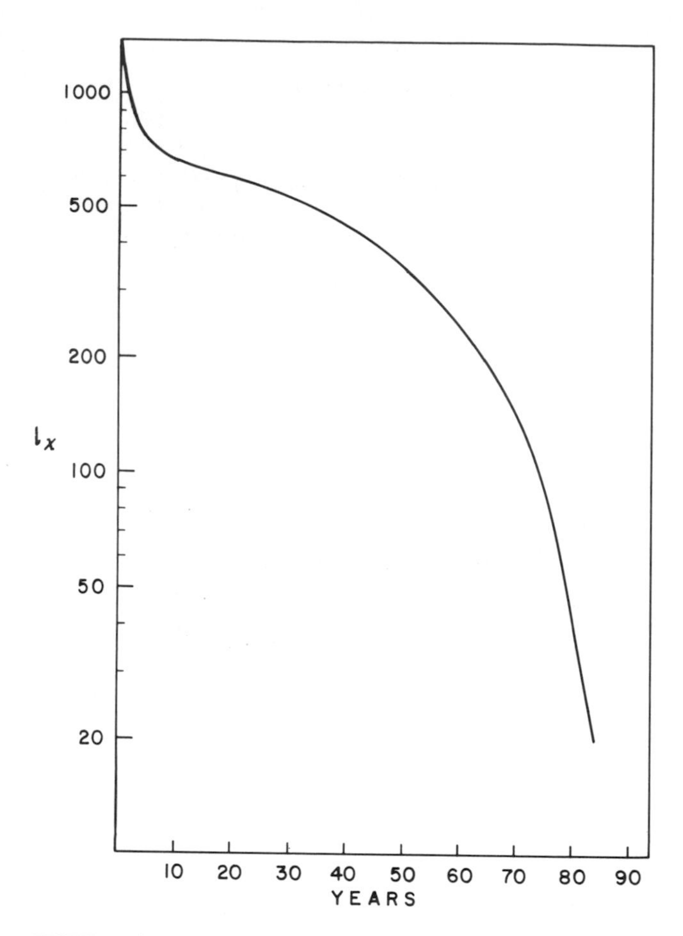

Age. Curt.	Per-sons.	Age. Curt.	Per-sons	Age. Curt.	Per-sons	Age. Curt.	Per-sons	Age Curt.	Per-sons	Age. Curt	Per-sons	Age.	Persons.
1	1000	8	680	15	628	22	586	29	539	36	481	7	5547
2	855	9	670	16	622	23	579	30	531	37	472	14	4584
3	798	10	661	17	616	24	573	31	523	38	463	21	4270
4	760	11	653	18	610	25	567	32	515	39	454	28	3964
5	732	12	646	19	604	26	560	33	507	40	445	35	3604
6	710	13	640	20	598	27	553	34	499	41	436	42	3178
7	692	14	634	21	592	28	546	35	490	42	427	49	2709
Age Curt	Per-sons.	Age. Curt.	Per-sons	Age. Curt.	Per-sons	Age Curt	Per-sons	Age. Curt.	Per-sons	Age. Curt.	Per-sons	56	2194
43	417	50	346	57	272	64	202	71	131	78	58	63	1694
44	407	51	335	58	262	65	192	72	120	79	49	70	1204
45	397	52	324	59	252	66	182	73	109	80	41	77	692
46	387	53	313	60	242	67	172	74	98	81	34	84	253
47	377	54	302	61	232	68	162	75	88	82	28	100	107
48	367	55	292	62	222	69	152	76	78	83	23		34000
49	357	56	282	63	212	70	142	77	68	84	20		Sum Total.

FIGURE 28. Age-specific survivorship curve for the city of Breslau or Wrocław from the table, here also reproduced, constructed by Edmund Halley from the data of Caspar Neumann. Halley gives the estimated mortality in the first year, which has been used in completing the graph.

were useless for his study as the immigration rate into the city was too great. Breslau had, he believed, an essentially stable population. Though his results are much smoothed and interpolated, involving some assumptions that are not explicit, the final table is realistic for a seventeenth-century city (figure 28).

If we knew the age of every individual in a sample of a population at a given time, we could construct a life table by a different method, simply entering the numbers (N_x) of individuals of age 0–1 units, 1–2 units, and so on, in the table, or by plotting N_x against x. This procedure implies that if we had a population made up of a large number of cohorts of equivalent demographic properties, each one started one unit of time before its successor, the sum of all individuals would at a given time produce a life table like that produced during time by following a single cohort.[6] Such a life table derived from a count of age groups at a given time is said to be *time-specific, static*, or, since it is not horizontal, *vertical*. It is only equivalent to an age-specific table or survivorship curve if the population does not change in age distribution.[7]

It is obvious that if a population is increasing in numbers due entirely to an increase in births, the time-specific estimate will contain a greater proportion of juvenile individuals than will any earlier age-specific cohort. If the population is increasing wholly by a reduction in mortality, the time-specific estimate will contain a greater proportion of old individuals. Usually both processes will occur together, though to an unequal degree. It is therefore impossible to indicate in any general way how a time-specific survivorship curve is likely to differ from an age-specific curve.

When we are dealing, as we usually are in biology, with a sample of individuals not all born at nearly the same time, and with records of dates of death spread out over many years, we may speak of a *composite* life table or survivorship curve, though it will have been obtained either by age-specific or time-

6. This is comparable to the celebrated ergodic theorem in statistical mechanics.

7. Good discussions, from the point of view of the biologist, of the different kinds of life tables are in:

E. S. Deevey, Life tables for natural populations of animals. *Quart. Rev. Biol.*, 22:283–314, 1947.

specific procedures, as indicated in the next section.

Examples of age-specific and time-specific analysis

These procedures may be demonstrated by considering a case of the kind that often confronts the palaeobiologist. It concerns the fossil bivalve crustacean, *Beyrichia jonesi*, presumably allied to the modern Ostracoda and usually placed in that subclass. The material comes from the Silurian (Mulde marl, upper Wenlock) of Gotland, and was initially published by Spjeldnaes, whose data were further considered by Kurtén.[8] The fossil remains fall into a number of discontinuous stages or instars of different sizes, initiated by molts as the animal grew. The first two or three stages, confined to the brood pouch of the female, are poorly preserved and cannot be considered, but from the beginning of the fourth up to the end of the eleventh sexually mature instar the fossils are well preserved and numerous. It is believed that the animal left the maternal pouch in the third instar, so we have a life table and survivorship curve for nearly all of its free life. As we know nothing about the life-span of *B. jonesi*, time must be measured by molts.

There are two possible modes of origin of the fossil association. Each fossil might represent a dead animal, so that the proportion of specimens of any instar represents the death rate in that instar. In such a case we can find what proportion of a sample of an original population dies in that instar. Subtracting this proportion from that present at the beginning of the instar tells us how many enter the next instar and so on until all the population is accounted for. This gives an age-specific table, though admittedly a very composite one (column [2] of table 1).

Alternatively we might suppose that the fossils represent exuviae and give a measure of the number that are alive in any instar. In such a case a time-specific table can be constructed simply by taking the proportion of individuals in each molt category.

Spjeldnaes thought that the fossils represented case exuviae, so that probably the time-specific table and its resulting survivorship curve (figure 29) is the right one to think about. The two interpretations do not produce greatly different results.

Another example of the difference between age-specific and time-specific treatments is given in figure 30. Here both modes of expression have meaning. The data are derived from Hollingsworth's[9] study of the demography of the British peerage, a group of individuals for whom there are reliable birth and death dates extending over at least four centuries. For the sake of clarity only the first quarters of the seventeenth, eighteenth, nineteenth, and twentieth centuries are considered in the figure, in which age-specific survivorship curves are given, for both sexes separately, for the composite cohort starting in each of these quarter centuries. There is an obvious increase in survival throughout the period, the greatest improvement taking place during the eighteenth and early nineteenth centuries. In addition to the age-specific curves, it is possible to draw rough time-specific curves, as has been done for the middle of the first quarter of the nineteenth century. At this time the infant death rate has fallen strikingly, but the old men, around 75 years of age, still retain some of their eighteenth-century mortality. The divergence of the two curves actually is greatest in the fifth decade of life. This is an artifact, due to the fertility of the

---

J. J. Hickey, Survival studies of banded birds. *Fish and Wildlife Service* (U.S. Dept. Interior), *Special Scientific Report*, 15, 117 pp., 1952.

B. Kurtén, On the variation and population dynamics of fossil and recent mammal populations. *Acta Zool. Fenn.*, 76:1–122, 1953.

D. Lack, *The Natural Regulation of Animal Numbers.* Oxford, Clarendon Press, 1954, viii, 343 pp., (reprinted 1967). See particularly chapters 8, 9, and 10.

Deevey's classic paper provided the first overall view of the subject. Though not the first worker to study the matter, Lack was in many ways the most important pioneer in the study of survivorship and mortality in birds.

8. N. Spjeldnaes, Ontogeny of *Beyrichia jonesi* Boll. *J. Paleontol.*, 25:745–55,. 1951. B. Kurtén, see note 7.

9. T. H. Hollingsworth, The demography of the British Peerage. *Popul. Stud.*, 18, no. 2, supplement; iv (preface by D. V. Glass), 108 pp., 1964.

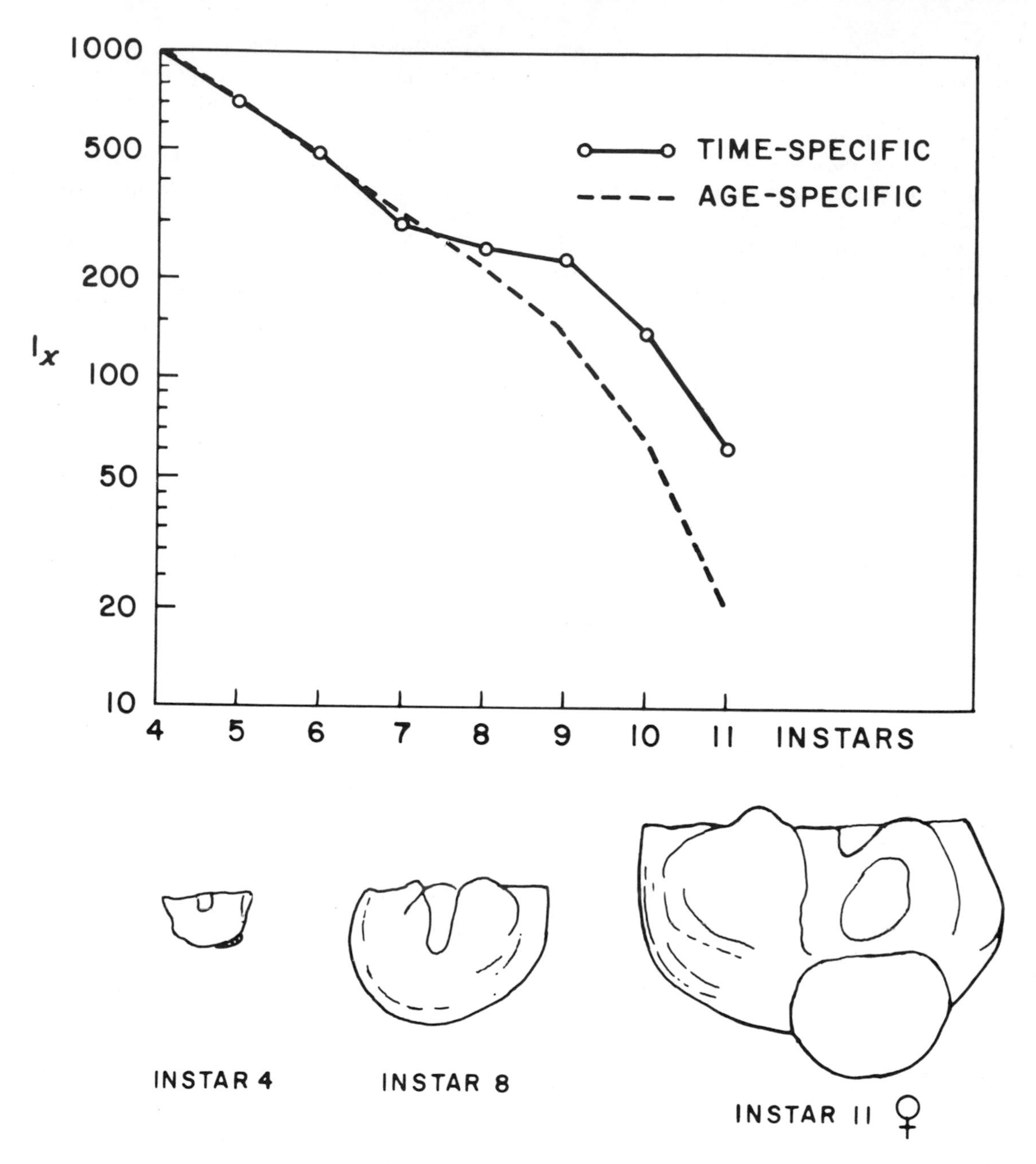

FIGURE 29. Survivorship curves for the Silurian fossil ostracod, *Beyrichia jonesi*, calculated either as age-specific or time-specific, the latter being more probable. Time measured in instars, and three individuals of different instars shown below the graph; the last is a female with a young individual in the brood pouch (data of Spjeldnaes, analyzed by Kurtén).

peerage having fallen, for undisclosed reasons, from about 1575 to 1724, so that the cohort born in 1725–49 was derived from a reduced population of peers, the decline in birth rate of the old families not being compensated by new creations. From about 1725 the fertility rose again.

Two other aspects of survivorship curves

Figure 30 also demonstrates two further points that are of considerable general importance in thinking about life tables and survivorship curves.

First, it is very easy to introduce distortions into the data in unexpected ways. Each of the

Instar (x)	1 Number of individual fossils (N_x)	2 $l_x = \frac{1000}{963}(963 - [N_x - N_{x+1}])$	3 $l_x = N_x \frac{1000}{309}$
4	309	1000	1000
5	210	679	679
6	146	461	472
7	92	307	268
8	75	214	243
9	70	136	227
10	42	63	136
11	19	20	61
	963		

TABLE 1. Age distribution by adult instars of the Silurian fossil, *Beyrichia jonesi*, in a sample [1], with resulting composite age-specific [2], and time-specific [3], life tables.

lowest pair of curves marked York 1538–1601 is the mean of four survivorship curves obtained by Cowgill[10] from a study of the sixteenth-century registers of six parishes in the city of York in England. The purpose of Cowgill's study was to reconstruct four cohorts born in spring (March, April, May) and autumn (August, September, and October) when the birthrates were maximal, and in summer (June, July) and winter (November, December, January, February) when they were minimal. Although there are probably social reasons for this seasonal variation in birthrate, which has tended to become less striking in modern times, it seemed to Cowgill interesting to inquire if the time of birth had any effect on mortality, either immediately or later in life. This is reasonable since respiratory infections are commonest in winter and various, partly insect-born, gastrointestinal infections in summer. A healthy child, perhaps growing into a healthy adult, might therefore be expected to have a more nearly equinoctial, rather than a solsticial, birthday. It might even be supposed that the seasonal differences in birthrate could have some sort of a genetic basis, due to natural selection. Cowgill matched birth and death dates, obtaining a cohort of people with known birthdays in the sixteenth century and known death days in that or the next century. This cohort, separated into four subcohorts by the season of birth, should give the required information. Rather surprisingly, there was no significant difference even in infant mortality.

For other studies Cowgill's cohort is unsatisfactory. It differs greatly from those of Hollingsworth for the aristocracy of the same period. The large difference in infant mortality is no doubt real, but the increasing divergence of the later parts of the curves must have another cause. Cowgill was able to identify the death dates in only 1,625 individuals out of 4,533 births recorded between 1538 and 1601. This means that about 66 percent of the population were not recorded as buried in the parishes for which the registers have been published. Some may have left York, others merely moved to parishes for which data are not available. Extensive carelessness in maintaining the registers could, as Cowgill points out, produce a like result. Emigration will affect very few newborn infants, but as life goes on the chances of having moved will increase. The cohort under consideration is therefore biased by a progressive underestimation of survival

10. U. M. Cowgill Life and death in the sixteenth century in the city of York. *Popul. Stud.*, 21:53–62, 1967. See also L. Henry, Some comments on U. M. Cowgill's article. *Popul. Stud.*, 22: 165–69, 1968; and U. M. Cowgill, A reply to M. Henry's comments on U. M. Cowgill's article. *Popul. Stud.*, 22: 170, 1968.

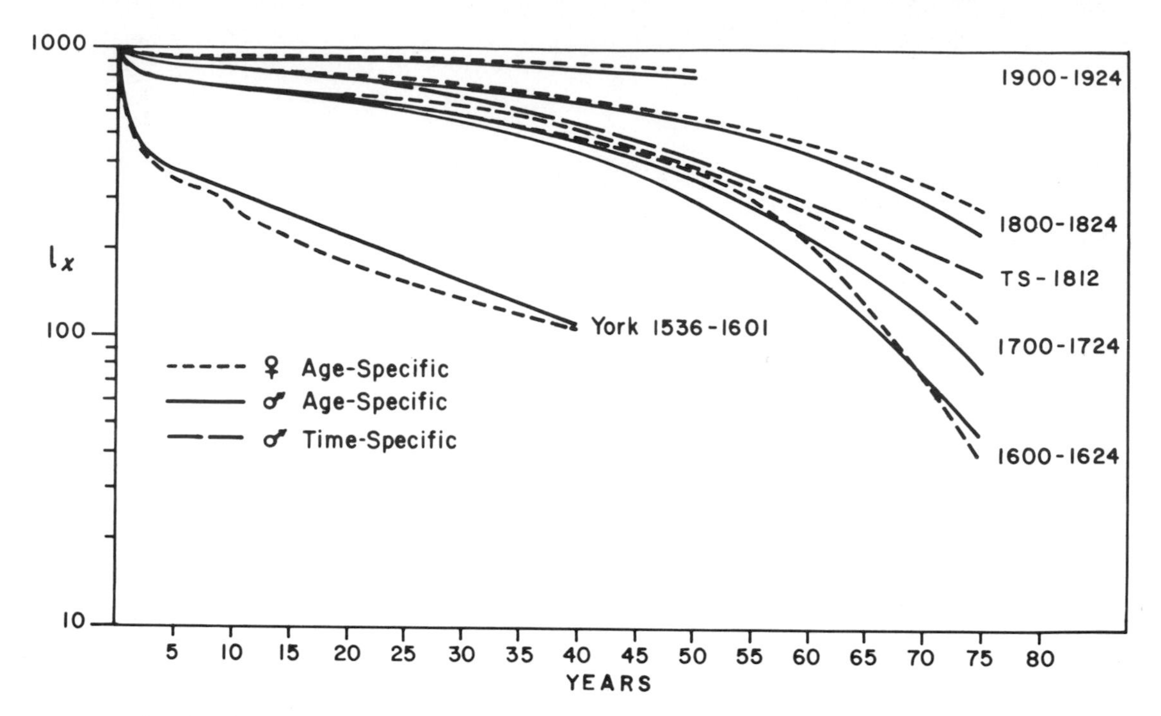

FIGURE 30. Survivorship curves for various British populations. Those designated only by date are age-specific curves for the cohorts of the aristocracy born in the first quarters of each century from 1600–24 onward. The curve TS-1812 is a time-specific curve based on the aristocratic males for the middle of the quarter century 1800–25. This is composed of men of varying ages, the oldest being of early eighteenth-century origin and the youngest expressing the improved public health of the nineteenth century. The curves for York represent a much more general population, but suffer from the effects of emigration from the parish of birth (data compiled by Hollingsworth and by Cowgill).

of the older age groups. Moreover, on marriage a couple is more likely to settle in the husband's rather than the wife's place of residence, which would emphasize the effect for women and depress their curve more than that of men. Cases are known in animal demography where a similar result of emigration distorts the demographic data.

Second, not every individual has the same potential longevity. In the upper curves, females tend to live longer than males, as is ordinarily the case in modern *Homo sapiens*; the reverse is true for some other species of mammals, as is apparent in figure 42.

In the survivorship curves that can be constructed from the distribution of mortalities by age classes given by Macdonell[11] (figure 31), the female survivorship in the city of Rome itself appears initially slightly better than the male, but as soon as reproductive ages (15–20) are reached the female curve descends rapidly below the male, though in postreproductive life ($\geqslant 50$) the curves run parallel. Death in childbirth was clearly of great demographic significance. From consideration of the 1,158 male epitaphs giving status or occupation, it appears that 675, or rather over half, are slaves (*servi*, *vernae*) or freemen (*liberati*). Only about 10 percent are professional and upper-class men (advocates, priests, augurs, members of the equestrian and senatorial orders); among these the advocates (*patroni*) were the most numerous class; 64 individuals had a mean age at death of 62 to 63 years, far greater than the average Roman.

The survivorship in the Iberian provinces was better than in Rome and that in Roman Africa better still. Here the female curve initially follows the male and drops very little below it when childbearing begins. The extraordinary number of centenarians in Roman Africa, rather over 10 percent of the population passed the age of 100, with a

11. See N. B. Macdonell, note 4.

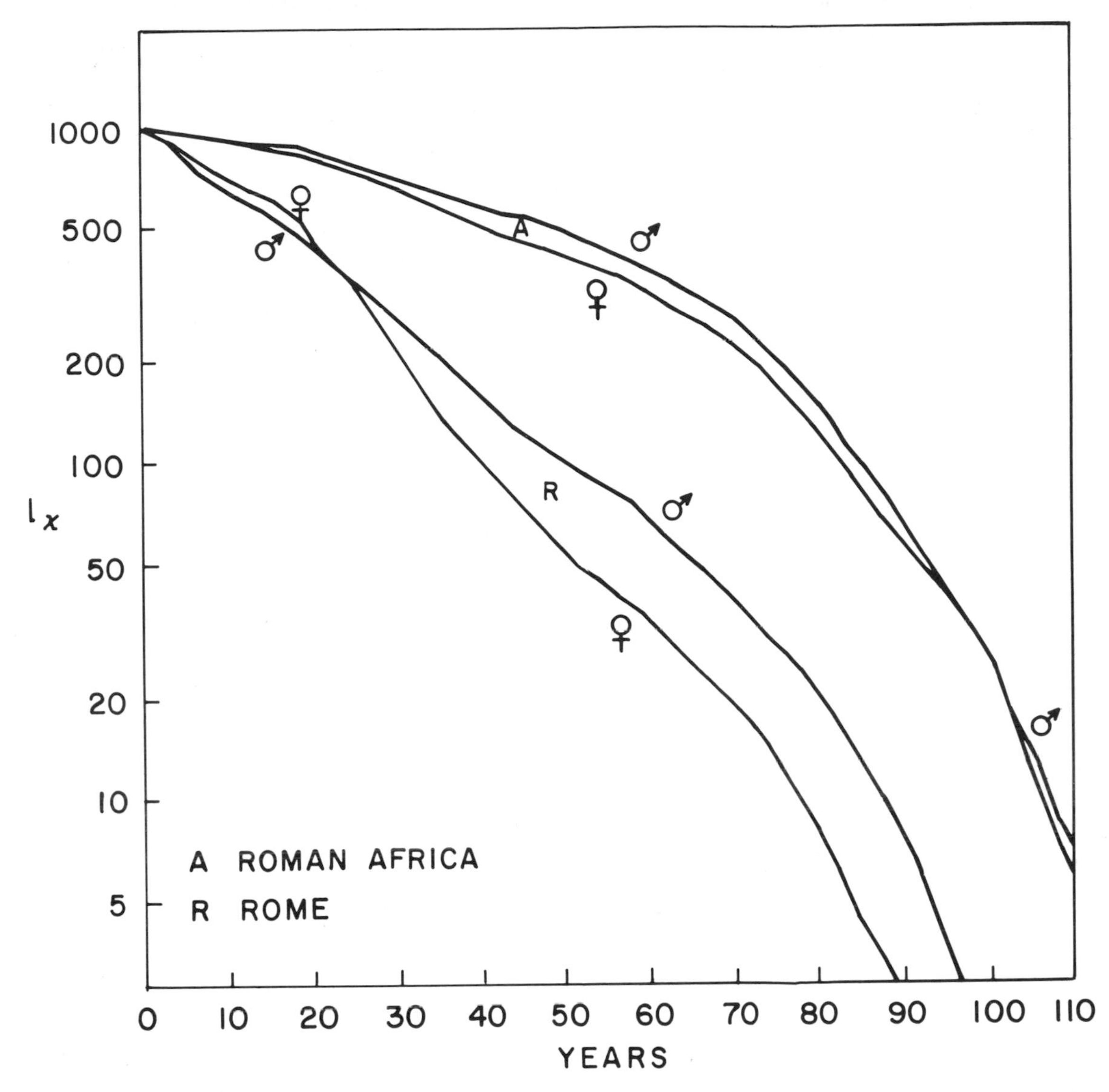

FIGURE 31. Survivorship curves for the Roman province of Africa (A) and for the city of Rome (R) from epitaphs recorded in the *Corpus Inscriptionum Romanarum*, analyzed by Macdonell, mostly from the first four centuries A.D.

maximum age of 132, would be remarkable if it were not suspicious. The data suggest that as an individual aged, there was an increasing probability of the age being over-estimated. It is, however, just conceivable that this African population was indeed one of great longevity, comparable to the few very long-lived local populations known today.

Comparison of the Roman and African curves is very instructive, showing how in a single species one population may provide what we shall call a *diagonal*, and another, a markedly *convex*, curve (see p. 53). When these are further compared with other species to be discussed later, one can only conclude that sometimes man is constrained to die randomly like a bird, but in other circumstances he may aspire to as ripe an old age as that of a wild sheep or an African buffalo.

Age-specific mortality

In addition to considering what proportion of a population survives to a given age, we often find it convenient to calculate the age-specific mortality $\mathbf{q}_x$, or the proportion dying

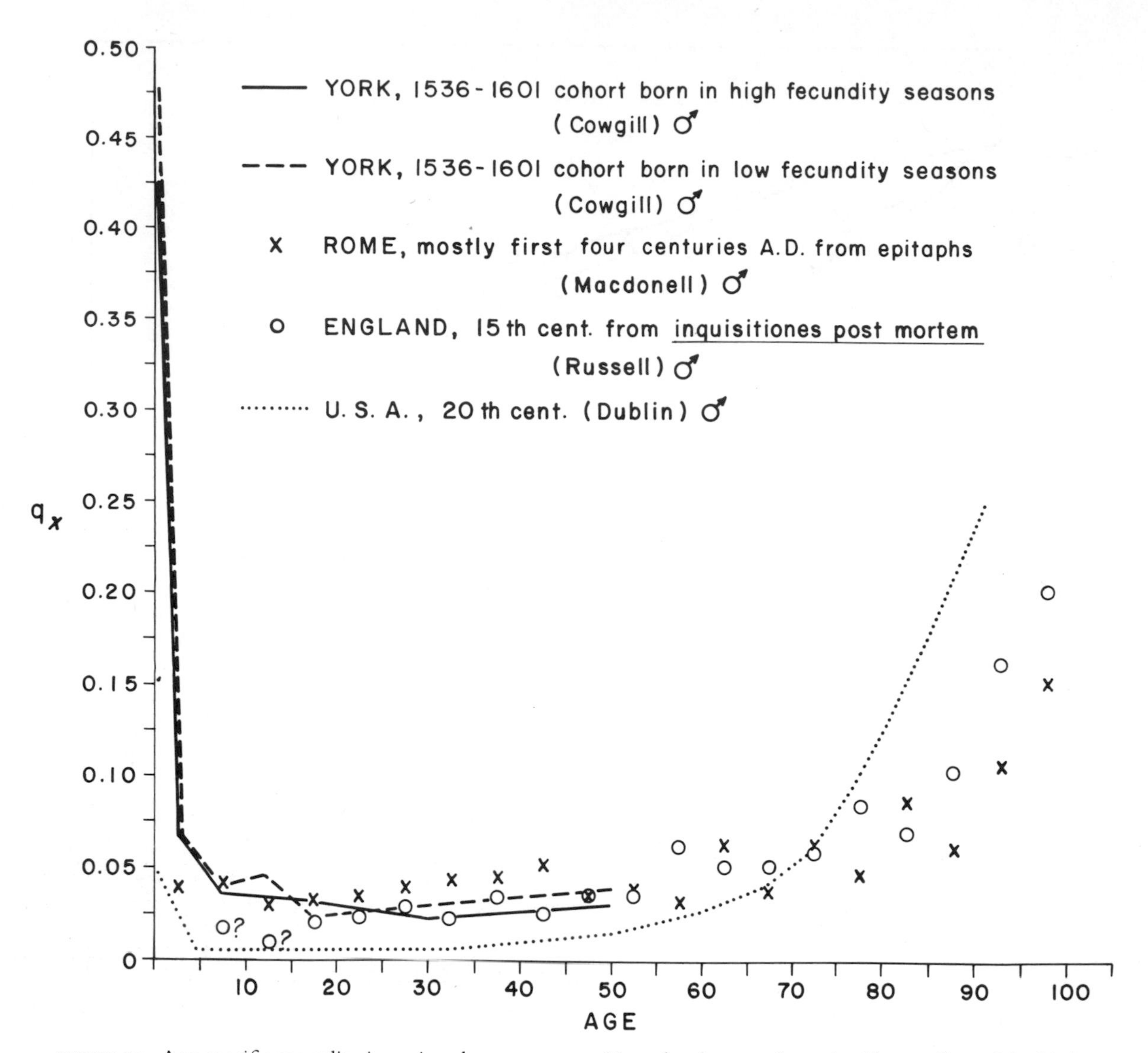

FIGURE 32. Age-specific mortality in various human groups. Note the absence of any significant effect of the birth seasons in the sixteenth-century York data, the close similarity of the Roman and English medieval and sixteenth-century data, and the remarkable tenacity of the surviving, more elderly Romans and fifteenth-century Englishman after 75, compared with a modern population (data of Cowgill, Russell, and Macdonell).

in the age class from x to $x + 1$, as a fraction of $\mathbf{l}_x$. In most cases this will be given as the proportion of those individuals surviving to age x years that will die in the succeeding year. A few examples are given in figure 32 of the variation of $\mathbf{q}_x$ with age in man. The curves are given for males only because there are no females recorded in the fifteenth-century data on the deaths of important landowners studied by Russell,[12] taken from a series of legal documents drawn up in connection with inheritance of land and termed *inquisitiones post mortem*. Russell gives a life table for upper-class fifteenth-century males based primarily on these sources. Its initial entry for children under 5 is, however, an emendation. There is an obvious deficiency in the data for this age group; Russell noted, however, that the later part of the curve followed closely that for modern rural India, so he adopted the Indian figure for children up to 5 years old for his life table. Since this procedure seems to me rather doubtful, I use

12. J. C. Russell, Demographic patterns in history. *Popul. Stud.*, 1:388–405, 1948.

only the genuine English entries. It is quite likely that Russell's figures for late childhood and early adolescence are also rather too low. His adult data agree well with those of Cowgill. The latter's figures may be a little too high on account of emigration, but the error is not cumulative in this case. Both sets of data give very much higher mortalities under 70 years of age than those of the United States today. Russell's figures, however, suggest low specific mortality rates at ages over 70 than now prevail. A like effect was observed by Macdonall in his work on the demography of ancient Rome. It may be due to the survival of more genetically rather short-lived individuals into the older age classes in modern populations than in those of ancient Rome and medieval England.

The examples given in figure 32 show that in man, under a fair range of different social and environmental conditions, high juvenile mortality is followed by a low and almost constant death rate in reproductive adults, with a progressive increase in mortality in postreproductive or senile individuals. Much of the rest of this chapter consists implicitly of a study of variations on this pattern. This, however, will not be explicitly apparent until a number of seemingly diverse situations have been examined.

Terminology of survivorship curves

In spite of the last few sentences, it is convenient, before we begin examining the information about survivorship in organisms other than man, to put our human preconceptions aside, and to consider certain simple situations which we might expect to be approached in actual cases. This consideration at least leads to a convenient terminology.[13]

We may reasonably assume that no organism is as such immortal, the so-called immortality of unicellular organisms involving reproduction and, in a steady-state system, equivalent death. The first simple possibility is that for any population there is a maximum life-span, more or less the same for all individuals of like genetic composition, and that the species under consideration is so well adapted to its environment, that so long as the latter remains unchanged all individuals live out their allotted life-span. The survivorship curve would thus run parallel to the abscissa and then drop suddenly to zero. Such an extreme curve is spoken of as a *negatively skewed rectangular curve* (figure 33, a, d). A curve of this sort is of course never realized in practice, though it remains the goal of everyone devoted to improvement in public health.

A second hypothetical but simple possibility might be that the rate at which the organisms composing a population die, is determined entirely by their interaction with the environment, so that there is a constant chance of death throughout any interval of life. This would give an exponential decline in $\mathbf{l}_x$, so that the survivorship curve would be a straight line running down diagonally with a slope given by the mortality rate when $\mathbf{l}_x$ is plotted logarithmically and time arithmetically (figure 33, b). Such a curve, which is quite often realized, at least in birds, is usually termed *diagonal.*

A third extreme possibility is logically less neat, though very important biologically. Experience suggests that the youngest animals are often the least prepared for life. The death rate is at first extremely high but declines rapidly, most mortality being juvenile mortality. When $\mathbf{l}_x$ is plotted arithmetically, this gives a curve that is the inverse of the negatively skewed rectangular type where mortality is concentrated at the end of the possible life-span. The survivorship curve of a population where death is mainly in the early stages is therefore called *positively skewed rectangular* (figure 33, c, f). In view of the frequent use of the three terms for the shapes of survivorship curves, it is important to note

13. See particularly R. Pearl, *Introduction to Medical Biometry and Statistics.* Philadelphia and London, W. B. Saunders, 2nd ed., 1930, 450 pp.

R. Pearl and J. R. Miner, Experimental studies on the duration of life: XIV. The comparative mortality of certain lower organisms. *Quart. Rev. Biol.*, 10:60–79, 1935.

E. S. Deevey, Life tables for natural populations of animals. *Quart. Rev. Biol.*, 22:283–314, 1947.

B. Kurtén, see note 7.

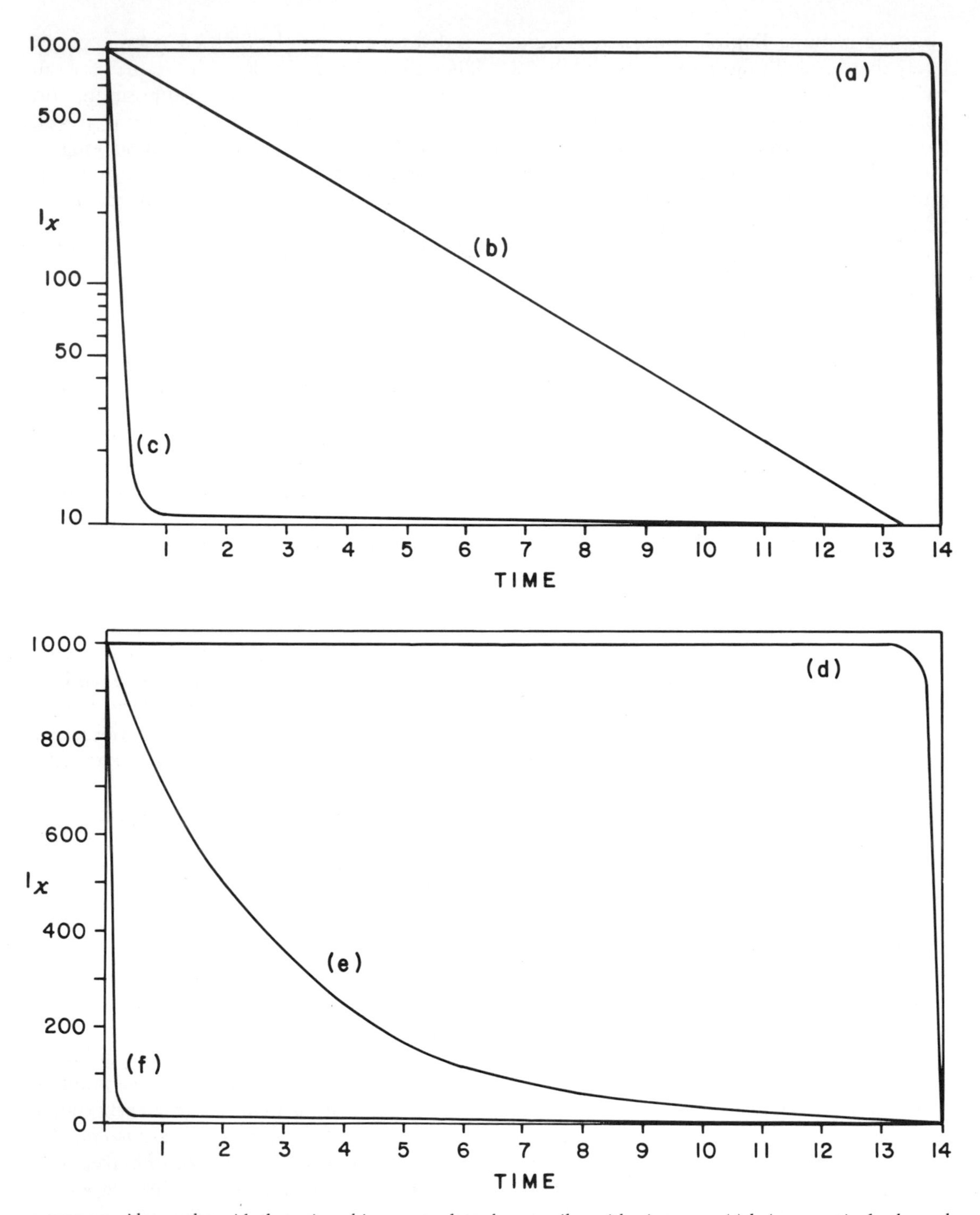

FIGURE 33. Above, three ideal survivorship curves plotted on semilogarithmic paper: (a) being negatively skewed rectangular, corresponding to a determined physiological life expectancy; (b) being diagonal, with a constant death rate; and (c) being positively skewed rectangular with immense juvenile mortality. Below (d, e, f) same on arithmetic scale.

that the diagonal curve in which specific mortality is constant throughout life is not diagonal when plotted arithmetically (figure 33, e), while very extreme positively skewed rectangular curves are rather less rectangular when plotted logarithmically. The simplest of such curves[14] is given by $N_t = N_o t^{-m}$, which gives a straight line plotted on double logarithmic paper. Most actual curves will be intermediates between the three types, though quite typical diagonal curves are found, after the first season, for birds. Kurtén's characterization of the curves which lie above the diagonal as *convex* and those that lie below as *concave* is convenient. Actually, as has already been hinted, it will become increasingly apparent that the most general situation is that of initial high juvenile and final high senile specific mortality, with a long intermediate period of lower and often fairly constant adult mortality. The observed form of the survivorship curves depends on the relative magnitudes of the mortalities of these three periods. Some divergence even from this very general pattern has, however, been observed.

In making comparisons between species of widely different mean longevities, we usually compute the mean longevity of a cohort, and then regraduate the age of each age class in terms of percentage difference from the mean. In using this procedure, which often shows up resemblances in the form of survivorship curves, it is important to remember that though two diagonal survivorship curves of quite different slopes may thus be superimposed, showing that in both cases the chance of death remains constant over a long period, the magnitude of that chance is very different in the two species and that difference is something of biological significance.

Life tables and survivorship curves of experimental populations

It is possible to produce artificially controlled survivorship curves that approach a negatively skewed rectangular form, though they probably are not very interesting. Rotifers which lay, parthenogenetically, large eggs one at a time, and so produce young that are not much smaller than adults, may have little infant mortality.[15] Since many cultures are clones, the genetic diversity can be minimal. With care in culture, so that adequate food is available, predators are absent, and the animals are not overcrowded, one might expect to approach a negatively skewed rectangular curve, and the expectation is in fact fulfilled (figure 34).

An even more artificial situation is exemplified by an early experiment of Pearl[16] in which a number of newly hatched female *Drosophila melanogaster* were put in a bottle without food. A negatively skewed rectangular survivorship curve merely indicates that the initial population was fairly homogeneous; one wonders if this experiment later gave rise to Wittgenstein's famous dictum[17]

14. The simplest situation is presumably that in which the specific mortality is inversely proportional to time. This gives $\ln N_t = \ln N_o - m \ln t$ where the specific mortality is $-m/t$. This may be taken as equivalent to $\mathbf{q}_x$ if the time units are short compared with the maximum life-span.

15. For *Proales decipiens* see:

B. Noyes, Experimental studies on the life history of a rotifer reproducing parthenogenetically (Proales decipiens). *J. Exp. Zool.*, 35:225–55, 1922.

R. Pearl and C. R. Doering, A comparison of the mortality of certain organisms with that of man. *Science*, 57:209–12, 1923.

For *Lecane inermis* see:

H. M. Miller, Alternation of generations in the rotifer Lecane inermis Bryce: I. Life histories of the sexual and non-sexual generations. *Biol. Bull.*, 60:345–81, 1931.

16. The data are tabulated in Pearl and Miner (see note 13, in which full references to Pearl's early works are given). The example of genetic differences, used in figure 35 is from R. Pearl, S. L. Parker, and B. M. Gonzalez, Experimental studies on the duration of life: VIII. The mendelian inheritance of duration of life in crosses of wild type and quintuple stocks of Drosophila melanogaster. *Amer. Natural.*, 57:153–92, 1923.

17. "309 What is your aim in philosophy? To shew the fly the way out of the fly-bottle." L. Wittgenstein, *Philosophical Investigations*, German, with English translation by G. E. M. Anscombe. New York, Macmillan, 1960, x, xe, 232, 232e pp. (originally Oxford, Blackwell, 1953). Wittgenstein was a friend of C. H. Waddington (personal communication) from whom he may have learned of the culturing of *Drosophila* in fly-bottles.

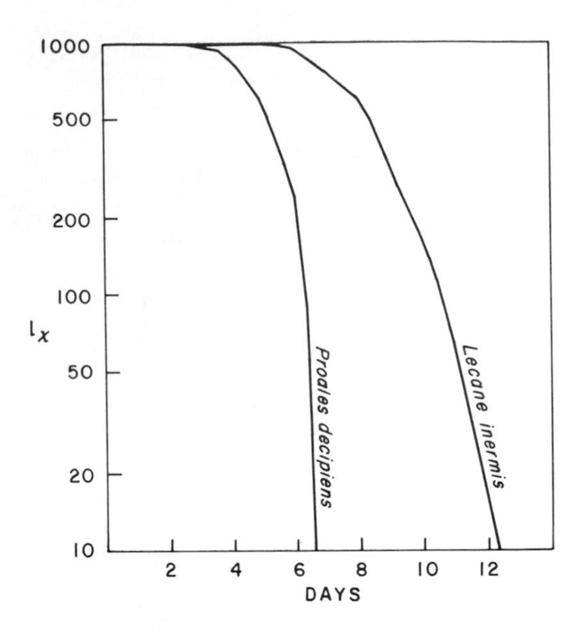

FIGURE 34. Survivorship curves of amictic females of two rotifers, *Lecane inermis* and *Proales decipiens*, in culture (data of Miller and Noyes).

that as a philosopher he showed the fly the way out of the fly-bottle.

Not all artificially controlled experimental situations lead to negatively skewed rectangular survivorship curves. In laboratory colonies of rodents,[18] mortality may be high in infancy, then drop to a minimum, and slowly rise throughout adult life until death has claimed that whole cohort (figure 35). Other animals show the same sort of phenomenon. In such cases it appears that the laboratory and natural mortalities may differ but run along parallel lines as the population ages.

The really interesting studies of mortality and survivorship in the laboratory are those in which cohorts of some taxon are compared under different demographic or environmental conditions, or in which cohorts of rather different genotypes are studied side by side. Various species of water fleas of the genus *Daphnia* and of pomace flies of the genus *Drosophila* have proved particularly useful in this work.

In both genera, and probably generally throughout the poikilothermous or so-called cold-blooded animals, expectation of life tends to be greater at low than at high temperatures within almost the entire range of temperatures that permit survival of the species studied.

Population density is clearly very important as a factor controlling longevity, but its effect is complicated and in no case fully understood. It was early observed that very small populations of *Drosophila* do less well than those maintained at an optimal density of 35 to 50 adults in a one-ounce bottle. This is probably due to excessive multiplication of wild yeasts in cultures containing too few individuals to consume the less desirable microorganisms just after the start of an experiment when few are present. The mechanical effects of burrowing larvae may also improve the culture medium as a place in which to live.

In *Daphnia* there are elaborate experi-

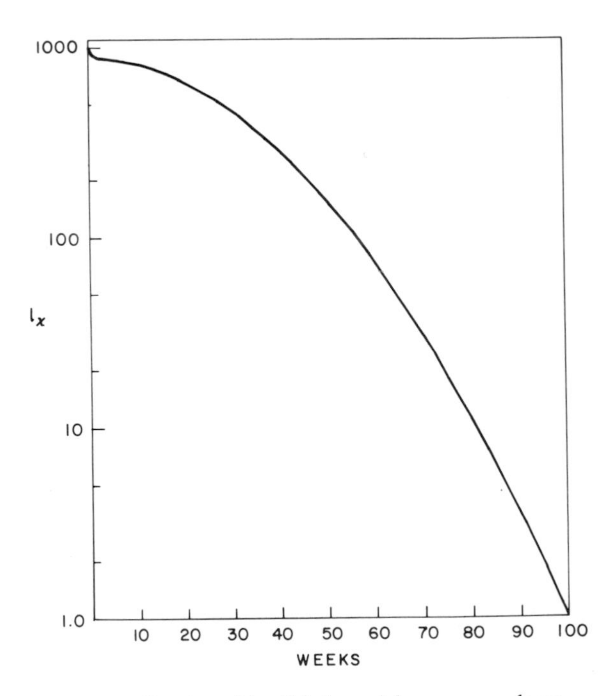

FIGURE 35. Survivorship (l_x) in a laboratory colony of the European field vole (*Microtus agrestis*) of British origin. From the data of Leslie and Ranson, as smoothed by them.

18. P. H. Leslie and R. M. Ranson, The mortality, fertility, and rate of natural increase in the vole (*Microtus agrestis*) as observed in the laboratory. *J. Anim. Ecol.*, 9:27–52, 1940.

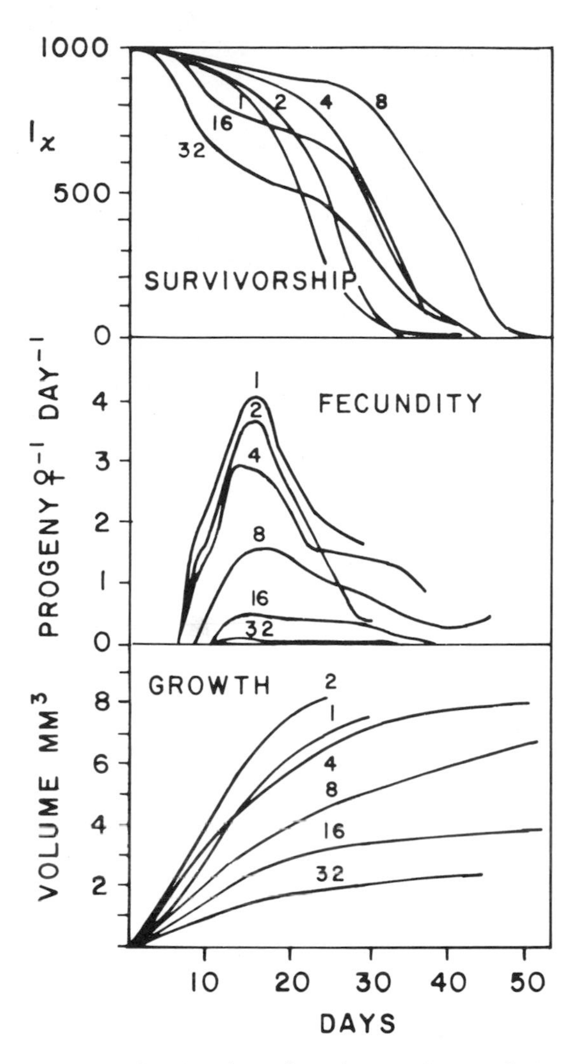

FIGURE 36. Survivorship, fecundity, and growth in *Daphnia pulex* reared at constant density by Frank, Boll, and Kelly. The figure over each curve gives the number of animals per milliliter maintained in the culture from which the curve is obtained. Note that fecundity is greatest at the lowest density, but growth rate at twice and survivorship at eight times this density.

ments[19] conducted by Pratt, by Frank and his associates, and by Smith, who were able to study cohorts maintained at approximately constant density. In Frank's experiments, both the size of the animals and their fecundity were studied as well as the longevity. None of these quantities was maximal at the same density (figure 36). Longevity was greatest at a density of 8 animals per milliliter, and least at a density of 1 animal per milliliter. Initially the populations at the very high densities of 16 and 32 animals per milliliter did less well than did the animals at 1 per milliliter, but later in the experiment the death rate in these two crowded cohorts fell off markedly, the survivorship curves being of a shape strikingly different from that observed at lower densities.

Fecundity, measured as average births per day, declined more or less in inverse proportionality to density. Growth rate was highest in the animals having 2 individuals per milliliter; the cohorts held at 1 per milliliter did less well, though a little better than those at 4 per milliliter. The more crowded populations showed marked inhibition of growth. Individuals kept at a density of 32 per milliliter, have after 40 days, when they have become postreproductive, a volume of about 2.5 mm^3, which volume would be characteristic of an animal about 8 days old, and just beginning to reproduce, in the cohort with 1 individual per milliliter.

It is quite obvious that the variation in all three quantities considered cannot be due solely to the more crowded populations making greater demands on the supply of food.

MacArthur and Bailey[20] had earlier found the rate of the heartbeat in *Daphnia* to be reduced by crowding; they believed that the rate indicated the metabolic activity of the animal, and supposed that within certain limits both crowding and lowering of temperature reduce metabolic rate and so increase

19. D. M. Pratt, Analysis of population development in Daphnia at different temperatures. *Biol. Bull.*, 85:116–40, 1943.

P. W. Frank, C. D. Boll, and R. W. Kelly, Vital statistics of laboratory cultures of *Daphnia pulex* DeGeer as related to density. *Physiol. Zool.*, 30:287–305, 1957.

P. W. Frank. Prediction of population growth found in *Daphnia pulex* cultures. *Amer. Natural.*, 94: 357–72, 1960.

F. E. Smith, Population dynamics in *Daphnia magna* and a new model for population growth. *Ecology*, 44:651–63, 1963.

20. J. W. MacArthur and W. H. T. Bailey, Metabolic activity and the duration of life: II. Metabolic rates and their relation to longevity in Daphnia. *J. Exp. Zool.*, 53:242–68, 1929. J. W. MacArthur was the father of R. H. MacArthur.

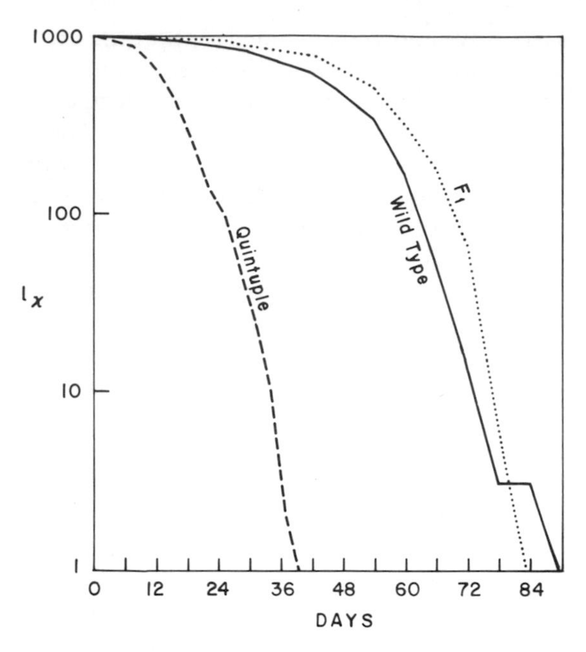

FIGURE 37. Survivorship of adult *Drosophila melanogaster* of two strains, one a wild type, the other (quintuple) with five recessive mutants including vestigial, and of the F_1 hybrid. Later generations show much diversity, but in general extracted vestigial stocks show the low survivorship of the original quintuple stock (Pearl, Parker, and Gonzalez).

longevity. It would, however, seem likely that in quite crowded cultures mutual mechanical stimulation would produce avoidance movements and so increase metabolism; the two effects could no doubt lead to an optimum density, though the evidence that the whole picture is controlled by general metabolic rate is extremely tenuous. The survivorship curves suggest that at high densities the adverse effects of crowding operate more strongly on small, young, rather than on large, older animals, even though in an old population the total biomass is greater than in a young one.

While there are certainly differences in longevity between different clones within a species of *Daphnia*, the most important work on genetic differences in life tables or survivorship curves is derived, as might be expected, from *Drosophila*. An early experiment from Pearl's laboratory is illustrated in figure 37.

In the same two species of Australasian *Drosophila* as were investigated by Ayala and

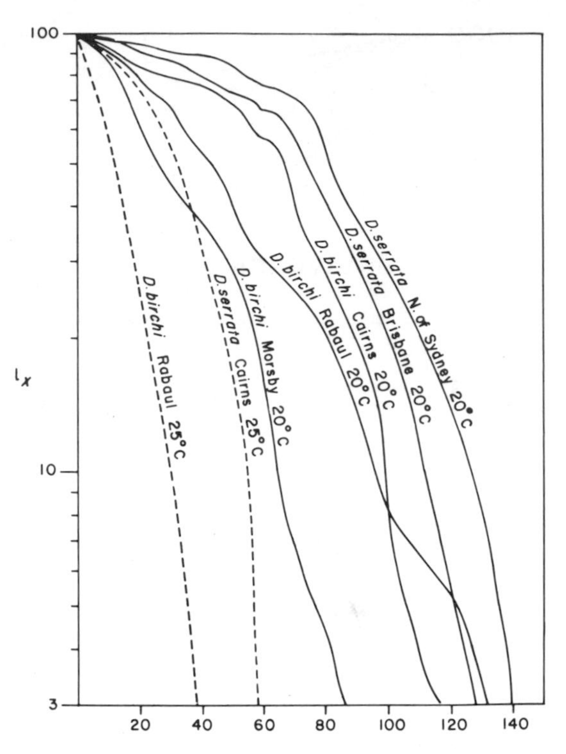

FIGURE 38. Survivorship curves for two strains of *Drosophila serrata* and three of *D. birchi*. Redrawn on a logarithmic scale from Birch, Dobzhansky, Elliott, and Lewontin. At 25°C (broken line) all but the Rabaul strain of *D. birchi* lie very close to the Cairns strain of that species. At 20°C all five strains differ according to geographical origin, the more southern temperate regions producing the longest-living flies.

discussed in chapter 1, Birch, Dobzhansky, Elliott, and Lewontin[21] studied survival very carefully, using two populations of *D. serrata* and three of *D. birchi* (figure 38). At 25° there was little difference in average longevity, except that paradoxically the most equatorial population from Rabaul in New Britain survived significantly less well than the others. At 20°C the survivorship curves indicate *D. serrata* as having a greater expectation of adult life than the more northern *D. birchi*; all populations give average survivorship curves that are arranged according to the latitude of origin and are usually well separated. Females live much longer than males, the ratios of life expectancy of the sexes in the various populations of the two

21. L. C. Birch, T. Dobzhansky, P. O. Elliott, and R. C. Lewontin, Relative fitness of geographical races of *Drosophila serrata*. *Evolution*, 17:72–83, 1963.

species varying irregularly around 1.3 at 20° and 1.5 at 25°C.

A series of experiments on the longevity of the adults of the mosquito *Aedes aegypti* by Crovello and Hacker[22] showed very great differences between populations from different localities, but these differences were not related to the feral or urban origins of the populations, as were simultaneously observed differences in reproductive rate.

Survivorship and mortality in natural populations
Two ways can be used to construct life tables and survivorship curves in natural populations. One way depends on plants and animals having structures to which recognizable additions are made yearly or sometimes more frequently. This is true of the annual rings of trees in humid temperate climates; of the recognizable annual increments of stems of mosses and Lycopodiaceae; of some skeletal structures in corals, molluscs, and sea urchins; of the scales and otoliths of fishes, the scutes of many turtles and tortoises; of horns and antlers of ungulates; the sites of corpora lutea formation in the ovaries of some female mammals; of overall size or size of particular structures such as teeth and, less exactly in a negative sense, of tooth wear in some mammals. Such indicators of age can be used in two ways. We may collect a sample of the organism to be studied, age the individuals, and so construct a time-specific life table. Alternatively we may search for and find the remains of dead organisms, and ultimately determine from them the age at death of a large number of randomly discovered individuals; from this array of ages we can construct a composite age-specific life table.

For animals that do not naturally reveal their ages to the observer we can use a system of marking, preferably at birth. The marked individuals now can provide a sample giving a time-specific table or curve, or efforts may be made to find remains of dead marked individuals from which an age-specific table can be constructed. Though all possible methods have probably been tried, the greatest amount of data most likaly concerns birds marked as nestlings or fledglings, the bands from which have been later recovered, usually on dead birds.

When individuals can be recovered alive, recognized by a band number, and then released, or where individuals can be marked by color bands, actual cohorts can be followed and life tables constructed, assuming negligible emigration from the marking site. This method has been used for a few birds and several bats.

All attempts to construct life tables for nonhuman populations suffer from various kinds of error. As we have noted earlier, the time-specific and age-specific methods of procedure are equivalent only when the population and age distributions are stable. Any method involving discovery of dead organisms in nature may be misleading when dead individuals of some age group are less easily found or less easily aged than those of other ages. Usually this leads to a serious deficiency of juveniles. In the case of marked animals, the tag may inhibit the growth and development of the animal bearing it, as seems to be the case in some fishes, or it may be lost or removed after a time, as is the case with long-lived and strong birds, so that no very old individuals appear in the record. Nevertheless a great deal of information has been obtained that is of sufficient quality to show what happens to populations under natural conditions.

It is most convenient to begin by considering the two classes of homoiothermal or warm-blooded vertebrates, the birds and the mammals, for much more is known about their vital statistics than has been discovered about organisms of other groups.

Survivorship and mortality in birds
The work on birds began in the 1930s with a study of the European starling (*Sturnus vulgaris*) in Holland by Kluijver,[23] followed

22. T. J. Crovello and C. S. Hacker, Evolutionary strategies in life-table characteristics among feral and urban strains of *Aedes aegypti* (L.). *Evolution*, 26:185–96, 1972.

23. H. N. Kluijver, Waarnemingen over de levenswijze van den spreeuw (*Sturnus v. vulgaris* L.) med behulp van geringde individuen. *Ardea*, 24:133–66, 1935.

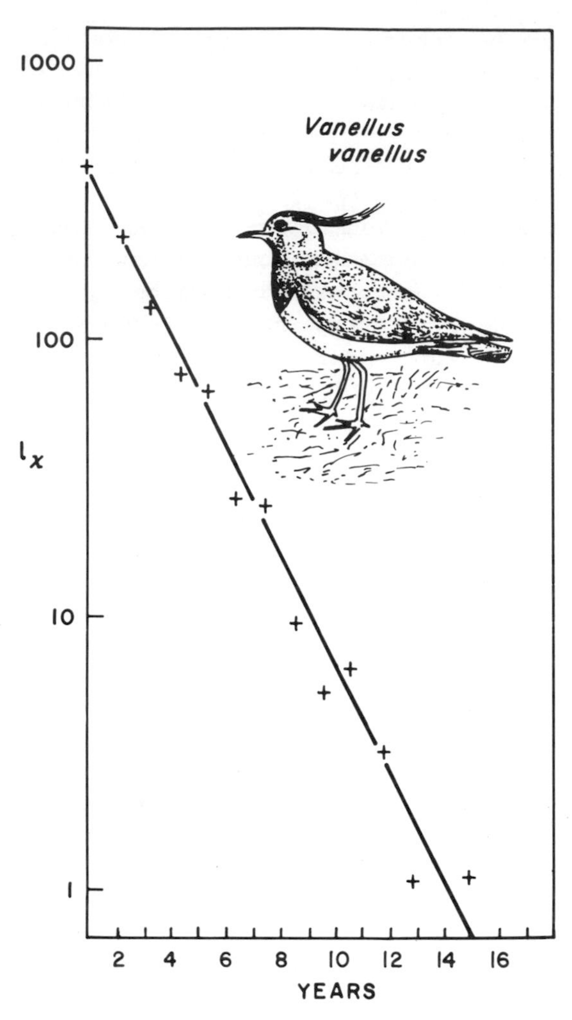

FIGURE 39. Composite age-specific survivorship curve for the lapwing, *Vanellus vanellus*, based on ringed birds found dead in Europe (data collected by Kraak, Rinkel, and Hoogenheide). Mortality is certainly higher in the first 6 months of life, but is not adequately estimated. The data are therefore presented from the end of the first year onward.

by the work of Nice[24] on the male song sparrow (*Melospiza melodia*) in North America. Both workers followed a cohort of birds banded with colored rings which permitted recognition throughout life. Soon afterward a group of ornithologists[25] in Holland (figure 39), and David Lack[26] in England employed the data that were accumulating on birds ringed as fledglings and later found dead. Such data permit composite age-specific life tables and survivorship curves to be constructed (figure 40). The classic papers of these early workers established quite clearly the usual avian pattern. There is a period of relatively high juvenile mortality affecting nestlings, fledglings, and quite young fully independent birds for roughly the first half year of life. After this the mortality rate, though it varies from species to species, is essentially constant for a number of years. It is reasonable to suspect high senile mortality, but experience with captive birds under good conditions indicates that they can live very much longer than the average wild bird actually does live. Botkin and Miller[27] think that it is possible that specific mortality continually but very slowly increases from a post-juvenile minimum, At least in early adult life, this increase must often be unappreciable.

Lack points out that the maximum recorded age of the European robin (*Erithracus rubecula*) is at least 11 years, though the average age achieved in nature is only 1.1 years after fledging, while a European blackbird (*Turdus merula*) has been known to live 20 years in captivity, though wild adults have an average life of 1.9 years. The senile death rate is therefore demographically of no direct importance. It is, however, obvious that there is an inherent factor regulating the average length of life, so that the rate of dying, though largely independent of age, is highly species-

24. M. M. Nice, Studies on the life history of the song sparrow: Vol. 1. A population study of the song sparrow. *Trans. Linn. Soc. N.Y.*, 4, vi, 247 pp., 1937.

25. W. K. Kraak, G. L. Rinkel, and J. Hoogenheide, Oecologische bewerking van de Europese ringgegevens van der Kievit (*Vanellus vanellus* (L.)). *Ardea*, 29:151–57, 1940.

A. Kortlandt, Levensloop, samenstelling en structuur der Nederlandse aalscholver bevolking. *Ardea*, 31:175–280, 1942.

26. D. Lack, *The Life of the Robin*. London, H. F. and G. Witherby, 1943 (revised ed., 1946).

D. Lack, The age of the blackbird. *Brit. Birds*, 36: 166–75, 1943.

D. Lack, The age of some more British birds. *Brit. Birds*, 36:193–97, 214–21, 1943.

D. Lack, *The Natural Regulation of Animal Numbers*. Oxford, Clarendon Press, 1954, viii, 343 pp. (reprinted 1967). This book is outstanding for its simplicity and clarity as well as for the vast amount of information that it contains.

27. D. B. Botkin and R. S. Miller, Mortality rates and survival of birds. *Amer. Natural.*, 108:81–192, 1974.

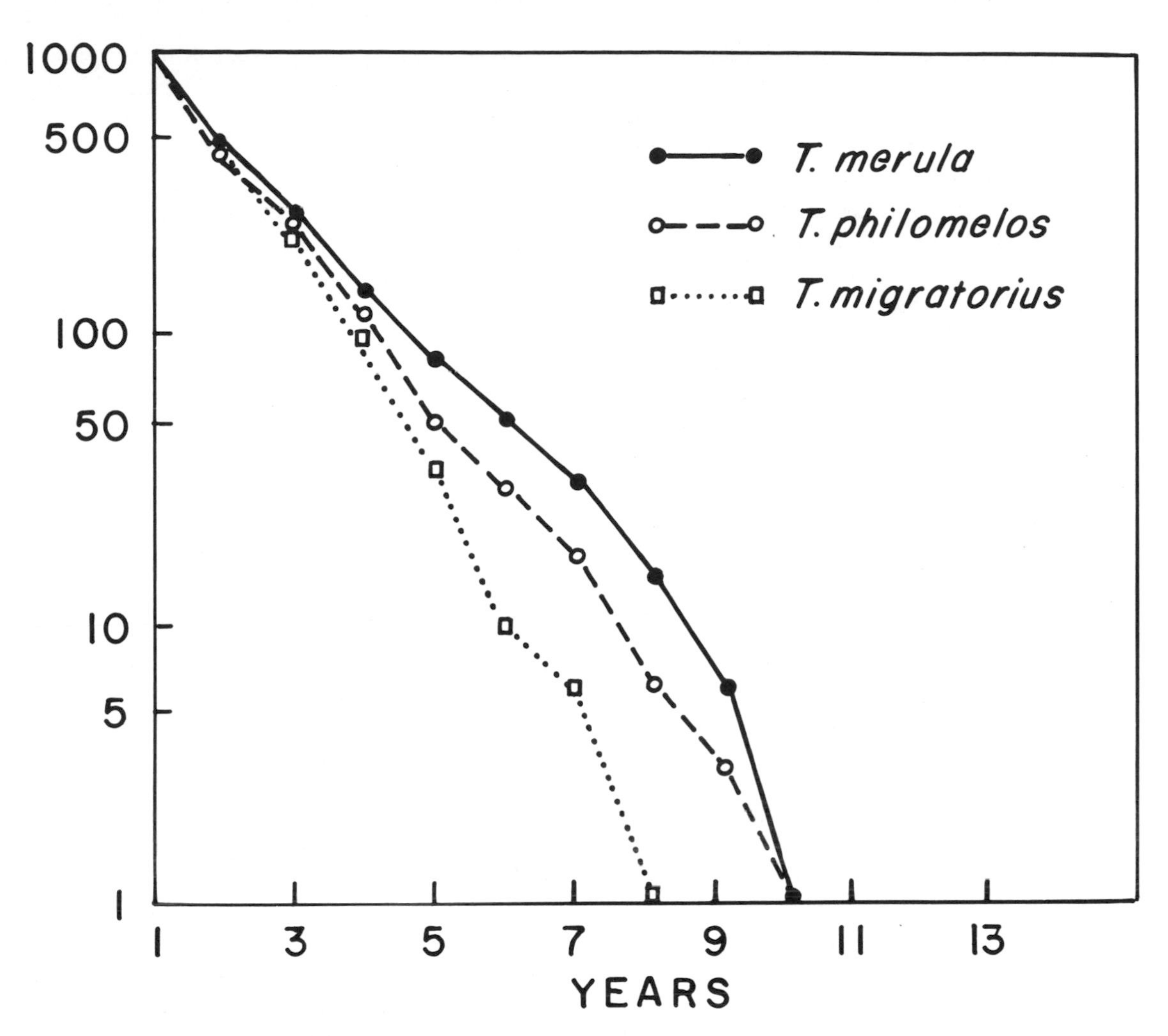

FIGURE 40. Composite age-specific survivorship curves for three species of *Turdus*, the European blackbird (*T. merula*), the song thrush (*T. philomelos*), and the American robin (*T. migratorius*). As before, initial juvenile mortality would be greater than implied by the later life indicated here.

specific. It would not be surprising to find that the actual survival is correlated with maximal possible longevity as well as dependent on the random causes of death occurring at a roughly constant rate at any time of life, the potentially longer-lived birds standing up to wear and tear better than the potentially short-lived species.

There is some correlation between mean longevity, as given by Lack for twenty-eight species of birds for which adequate data exist, and the sizes of these birds. There are, however, curious discrepancies. The swifts (*Apus* spp.) are much more long-lived than one would expect, having in nature an expectation of life four to five times than that of the ecologically similar European swallow (*Hirundo rustica*). As will appear later, the steady and by no means negligible adult death rate of the smaller, largely passerine, birds is probably less of a disadvantage than would be expected to any species in which learning is limited and experience of little importance in rearing young.

In the generally larger seabirds, a prereproductive period[28] of fairly high mortality

28. N. P. Ashmole, Sea bird ecology and the marine environment. In: *Avian Biology*, ed. D. S. Farner and T. R. King. New York, Academic Press, 1963, pp. 223–86.

lasting several years is usually followed by a very long period of quite low mortality in which the population declines slowly and regularly. Most of the earlier work on such birds involved primarily life tables for bird bands rather than birds. In the herring gull (*Larus a. argentatus* in Europe, *L. a. smithsonianus* in Atlantic America), which usually first breeds in its fourth year, about 30 percent of the individuals that have survived the juvenile mortality of their first 6 months of life will have died before breeding, even though during their first year they have achieved full adult size. Once such a bird becomes of breeding age the specific mortality is maintained for a number of years at a quite low figure, probably between 0.05 and 0.15 per individual per year in most large seabirds. Ashmole believes that in the prereproductive years the young bird gets a chance to learn enough about its environment to be able to risk any dangers of predation or starvation that might result from its first attempt at breeding. Thereafter there is a long period of low mortality when the bird, having learned how to live, can safely engage in reproduction. Tolonen has shown in a very careful study of feeding behavior of immature gulls that there is indeed conspicuous improvement in efficiency over the years. In potentially long-lived species such as the herring gull, which can live 40 years or more in captivity, the slow exponential decline of any cohort may reduce each element of the population to a very small number before detectable senile mortality begins. Actually, the pattern observed in long-lived seabirds is not very different from that exhibited by small passerine land birds, save that the time of both the period of early high and that of adult low mortality is greatly extended.

In penguins[29] the prereproductive period seems to be about 6 years, while in the albatrosses it is about 9. The period over which researches on these birds have been conducted is not yet long enough to indicate what sort of expectation of life they may have.

There is some indication that in bower birds,[30] which have an extraordinarily elaborate courtship display involving the construction of elaborate bowers, breeding begins much later than in other passerine birds.

Survivorship and mortality in mammals

The mammalian data are far less numerous than the avian, as in general mammals are less easy to study than are birds. The results of such studies as have been carried out are, however, far more diverse than those derived from birds. The literature has been critically reviewed by Caughley,[31] who accepts unequivocally only two investigations as adequately satisfactory at the time that he wrote. He admits, however, that a good deal can be learned from some of the less perfect studies. The two best cases are the composite age-specific table and curve produced from Murie's study of the Dall sheep (*Ovis dalli dalli*) in Alaska, and Caughley's own time-specific table and curve for the thar (*Hemitragus jemlahicus*), an ungulate belonging to the Caprinae, native to the Himalaya, but now introduced into New Zealand, where Caughley's work was done.

The raw data on the Dall sheep consist of records of recovery of the remains of dead animals that can be aged from the presence and development of horns. These records were given by Murie and analyzed by Deevey[32] in a classical paper on the life tables

J. A. Kadlec and W. H. Drury, Structure of the New England herring gull population. *Ecology*, 49: 644–76, 1968.

K. Tolonen, *Behavioral Ecology of* Larus argentatus *and* Larus marinus: *Age-Specific Differential in Feeding Efficiency, a Probable Factor in the Evolution of Delayed Breeding*. Thesis, Yale University, 1976.

29. P. Jouventin, Mortality parameters in emperor penguins *Aptenodytes forsteri*. In: *The Biology of Penguins*, ed. B. Stonehouse. Baltimore, London, and Tokyo, University Park Press, 1975, pp. 435–46.

The emperor penguin probably lives in a more rigorous environment than any other bird. Its juvenile mortality is only about 29 percent. Breeding usually begins at 6 years of age. A single egg is laid; in captivity the bird has lived at least 34 years.

30. A. J. Marshall, Display and bower-building in bower-birds. *Nature*, 153:685, 1944.

31. G. Caughley, Mortality patterns in mammals. *Ecology*, 47:906–18, 1966.

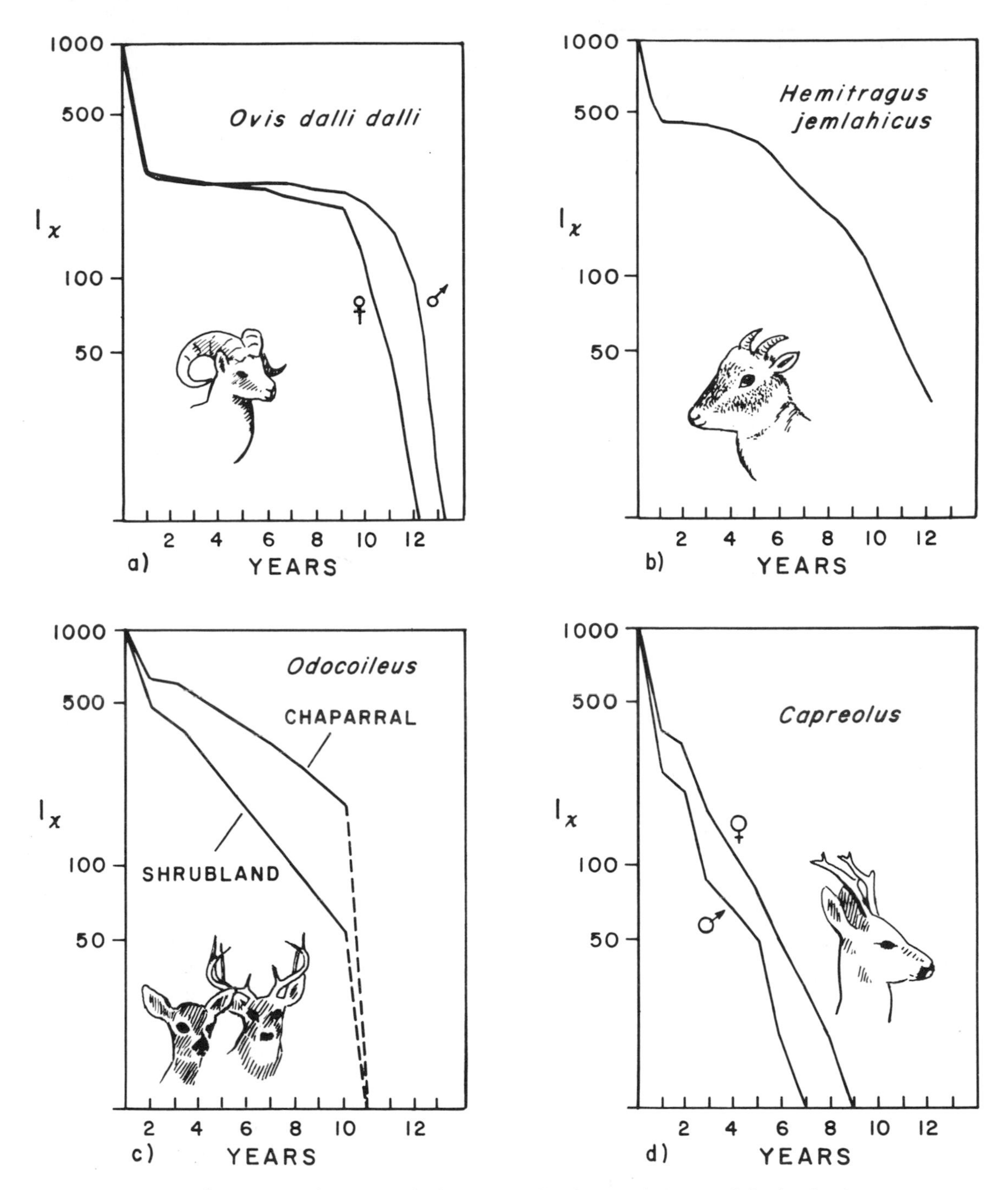

FIGURE 41. Survivorship curves of (a) the Dall sheep (*Ovis d. dalli*) in Alaska ♂ and ♀; (b) the that (*Hemitragus jemlahicus*), introduced into New Zealand; (c) the Columbian black-tailed deer (*Odocoileus hemionus columbianus*) in Lake County, California; and (d) the roebuck (*Capreolus capreolus*) in Denmark (data from Murie, Caughley, Taber and Dasmann, and Andersen).

32. A. Murie, *The Wolves of Mount McKinley*. U.S. Dept. Interior, Nat. Park Service, Washington, D.C. Fauna of the National Parks of the U.S. Fauna series no. 5, 1944, xx, 238 pp.
See also Deevey, note 13

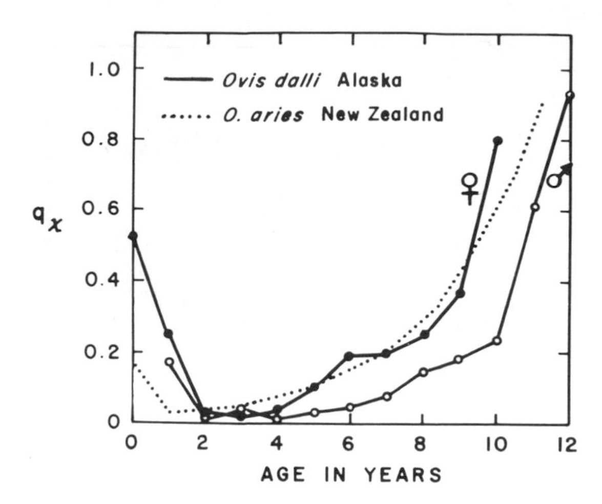

FIGURE 42. Specific mortality of Dall sheep (♂ and ♀, solid lines) compared with that of the domestic sheep (*Ovis aries*) (dotted line) in New Zealand (data from Murie and from Hickey, as analyzed by Caughley).

and survivorship curves of wild animals. Subsequently Taber and Dasmann[33] reexamined the original data (figure 41, a).

After a high initial mortality in the first 2 years, of the order of rather more than half the lambs produced, the specific mortality becomes very low (figure 42) for several years, rising at the 5th year in females and rather later in males. In the 10th year in females and the 12th in males, 80 percent or more of the survivors die. The resulting survivorship curve (figure 41, a) at first drops sharply, then becomes very flat, finally starting to fall again ever faster till the whole population disappears. The main cause of death of *Ovis d. dalli* in nature is predation by wolves. The predators can attack the lambs easily, particularly when they begin straying. By the opening of the 2nd year the sheep are fleet enough and experienced enough to avoid wolves, but as they get older they tend to become arthritic and so are more and more easily run down. In view of this highly characteristic cause of mortality, it is very curious to find that domestic sheep in New Zealand, where, whatever causes of death prevail, predation by wolves is not among them, show a remarkably similar variation of mortality with age. Presumably if an elderly Dall sheep were not eaten by wolves, it would soon die in some other way.

Though wolves have probably never been a demographically important cause of human death, and are certainly no more of any mortal significance to the human inhabitants of the United States, whose habitats and ways of life also differ greatly from those of sheep on New Zealand farms, the survivorship curve of modern *Homo sapiens* in North America, as elsewhere, is qualitatively very like that of the sheep, whether wild or domesticated. There is a fairly high juvenile and very high senile mortality, with little death in between. The Dall sheep differs conspicuously from contemporary Western man, however, in having longer-lived males than females.

Caughley's study of *Hemitragus* in New Zealand was based on a large sample of females culled from a relatively stationary population and aged by horn growth. His results indicate a fairly high juvenile mortality followed by a sharp minimum, after which the mortality rate increases almost linearly with age. The mortality curve is, however, derived from a smoothed time-specific survivorship curve. The actual points on the latter are rather irregular in the first 5 years of life, and the sharpness of the minimum may well be due to the function chosen for smoothing and have no biological meaning. The original data would be consistent with practically no deaths during the 2nd and 3rd years of life, as in *Ovis dalli*.

Since Caughley's paper was published a very important study of the African buffalo (*Syncerus caffer*) by Sinclair[34] has appeared. The survivorship and relative morality curves are comparable to those for the Dall sheep; there is, however, no detectable sexual

33. R. D. Taber and R. F. Dasmann, The dynamics of three natural populations of the deer *Odocoileus hemionus columbianus*. *Ecology*, 38: 233–46, 1958.

J. Andersen, Analysis of a Danish roe-deer population. *Danish Rev. Game Biol.*, 2:127–55, 1953.

34. A. R. E. Sinclair, The natural regulation of buffalo populations in East Africa: III. Population trends and mortality. *East Afr. Wildl. J.*, 12:185–200, 1974.

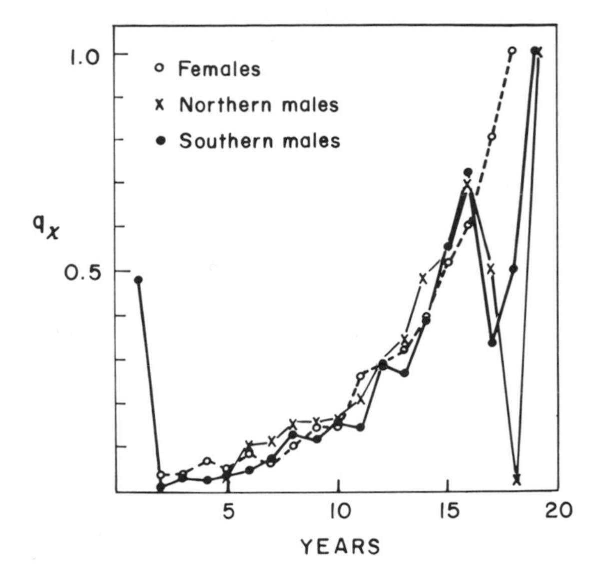

FIGURE 43. Specific mortality in three different groups of East African buffalo (*Syncerus caffer*) in the Serengeti region. Note the general resemblance of the pattern to that found in the Dall sheep or in man, except for an initial lack of sex differences followed by a curiously low male mortality in the 17th and 18th years after most females have died (Sinclair).

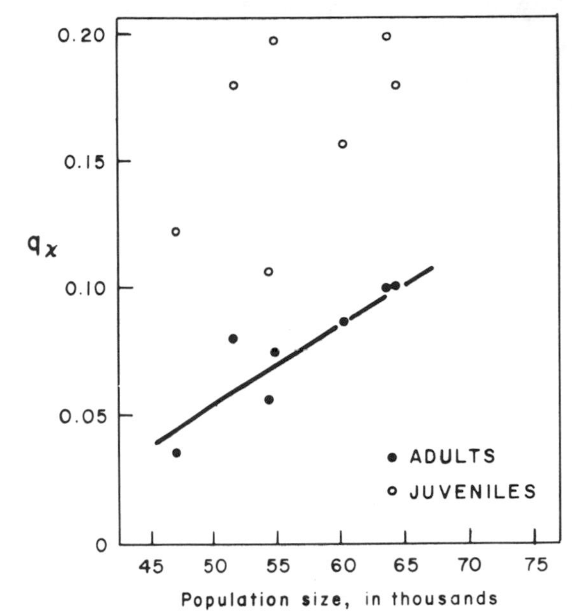

FIGURE 44. Specific mortality plotted against population size of *Syncerus caffer* in the Serengeti region. The adult mortality is clearly dependent on population size and so is regulatory. The higher juvenile mortality shows a comparable, but weaker, and statistically not significant trend (Sinclair).

difference until very late in life. Although all females had died by the end of the 18th year, in both subpopulations studied there was a pronounced drop in male mortality in the 17th and 18th years, so that 19-year-old males still existed (figure 43). This observation was confirmed on another herd.

The incidence of adult mortality (figure 44) is highly density-dependent and appears to be due at least in part to malnutrition, particularly in the dry season. Juvenile mortality, though higher than adult, is not significantly density-dependent on the basis of the available date. It is probably mainly due to disease and malnutrition with a relatively small amount of predation by hyaenas and lions. Predation by lions on adults occurs, but is demographically unimportant.

Jordan, Botkin, and Wolfe,[35] moreover, have given data for the population of the moose (*Alces alces andersoni*) on Isle Royale in Lake Superior in which the survivorship curve is not unlike that of the Dall sheep. Mortality is largely due to predation by wolves (*Canis lupus*). It is evident that in animals with the general intelligence of ungulates, comparable survivorship curves may be due either to starvation or to loss by predation.

In addition to these studies of ungulates, there are several less complete investigations of considerable interest. Kurtén,[36] using very large quantities of fossil material collected by Langrelius from the *Hipparion* beds of the Pliocene of China, was able to construct life tables for three species of ungulates and one carnivore. Two of the ungulates, *Plesiaddax depereti* and *Urmiatherium intermedium*, belong to extinct genera allied to the musk-ox, while the third is a gazelle. The carnivore *Ictitherium wongi* is allied to the modern hyaenas.

The skeletal remains of these animals are probably derived from individuals perishing in floods; the life tables based on the fossils are therefore time-specific. The ages at death were determined from tooth dimensions

35. P. A. Jordan, D. B. Botkin, and M. L. Wolfe, Biomass dynamics in a moose population. *Ecology*, 52:147–52, 1971.

36. B. Kurtén, see note 7.

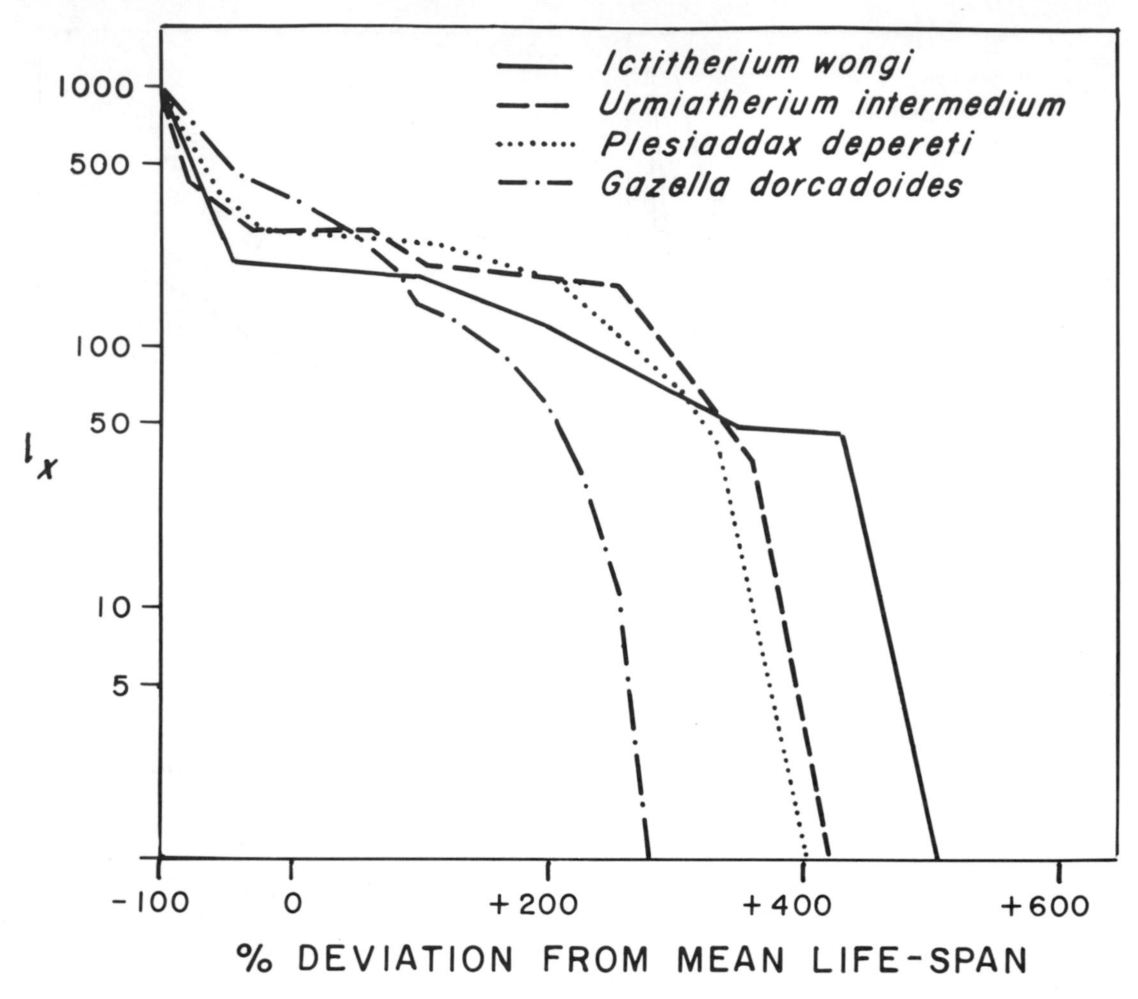

FIGURE 45. Survivorship curves constructed from fossil skeletal remains of three ungulate and one carnivore population from the *Hipparion* beds of the Pliocene of China. Time is given as deviation from mean life-span. Note that two of the ungulates, *Urmiatherium intermedium* and *Plesiaddax depereti*, exhibit curves strikingly like that of the Dall sheep (Kurtén).

which fell in discontinuous groups. There is clearly a deficiency of very young individuals, but some rough estimates of juvenile mortality are possible. The time-specific survivorship curves for the four species, incorporating the less reliable juvenile losses, are given in figure 45. The two members of the Ovibovinae or musk-ox subfamily exhibit survivorship curves very like that of the Dall sheep. The carnivorous *Ictitherium* has a slowly increasing mortality during most of its adult life. The fossil gazelle shows an initial more or less diagonal curve, but with a sudden terminal drop.

An extremely interesting situation has been analyzed by Taber and Dasmann[37] who studied the deer *Odocoileus hemionus columbianus* in two dissimilar habitats in California, one being chaparral, consisting of a mixture of shrubs and herbs, and the other shrubland in which the shrubs had been artificially encouraged (figure 41, c). In the chaparral there were about 10 deer per km^2, and the mature does produced 0.77 fawn per individual per year. In the artificially improved shrubland the population was about 34 deer per km^2, and the mature does produced a mean number of 1.65 fawns per individual

37. R. D. Taber and R. F. Dasmann, see note 33.

per year. However, the greater fecundity and the greater density were accompanied by a greater death rate in the females (males, being subject to hunting, were not considered in the study), so that in what would ordinarily be considered the better environment the survivorship was much poorer than in the unmodified chaparral. In such a case it must be remembered that replacement is more rapid in the less long-lived population in the more nutritious environment, so that in some cases at least evolution might proceed faster than in the leaner terrain supporting fewer longer-lived animals.

Taber and Dasmann attribute the difference between the survivorship curves of the black-tailed deer and the Dall sheep to the former being largely limited by food supply, the latter by predation. Hunger can affect an individual at any age, but predation primarily eliminates the young and inexperienced or the old and ill. This argument, however, cannot be applied to the buffalo studied in Serengeti by Sinclair, which shows a mortality curve very like that of the Dall sheep, though its population is controlled not by predation but by food supply.

In addition to the artiodactyls just discussed, Goddard[38] has published a study of survivorship in the black rhinoceros (*Diceros bicornis*) in Tsavo National Park, Kenya. There is moderate juvenile mortality, followed by a period of fairly low death rates of about 7 percent annually in young adults 5 to 10 years old. After this the mortality slowly increases. The expectation of life at birth is 8.4 years: the oldest individual recorded was 37 years old.

Though the typical distribution of mortality in the mammals so far discussed involves a relatively high and falling juvenile death rate, a period of greater or less length in early adult life when mortality is low, and a period of increasing late adult or postreproductive death, a certain number of very diagonal and therefore avian-looking survivorship curves have been published.

In the chamois, *Rupicapra rupicapra*, in the Alps, population structure has been studied in the male by examination of horns from hunted animals.[39] The main source of mortality may well be hunters who seem now to take any available male. The diagonal survivorship curve thus may be, at least partly, an artifact. The roe deer, *Capreolus capreolus*, a population living in an isolated tract of woodland in Denmark, exhibited a very striking diagonal curve (figure 41, d) certainly not due to human interference. In this species adult males appear intolerant of the young of previous years, so that an isolated population is always losing individuals through forced emigration. The apparent survivorship curve may thus formally be compared with Cowgill's sixteenth-century cohort of figure 30. A diagonal curve may be implied by data on the snowshoe hare, *Lepus americanus*, in Michigan. It is possible that such curves will be found more commonly in mammals when more small species have been studied.

So far such studies have mainly been made on bats, as marked individuals can be followed in bat colonies over a period of years.[40] The rates at which the numbers of such individuals decline indicate a relatively constant specific disappearance after the first year after banding. If it is assumed that death rather than failure to return to the original colony is the cause of disappearance after the first year, the rate of disappearance gives a specific mortality rate, implying a diagonal survivorship curve (figure 46, a). In a number of European species this varies from 0.200 per individual per year in Daubenton's bat

38. T. Goddard, Age criteria and vital statistics of a black rhinoceros population. *East Afr. Wildl. J.*, 8:105–21, 1970.

39. M. Couturier, *Le Chamois: Rupicapra rupicapra (L.)*, Grenoble, B. Arthaud, 1938, xi, 855 pp.

J. Andersen, see note 33.

As this book goes to press, essentially diagonal survivorship has been recorded in the surprisingly large population of the red fox inhabiting suburban London. See S. Harris, Distribution, habitat utilization and age structure of a suburban fox (*Vulpes vulpes*) population. *Mammal. Rev.*, 7:25–39, 1977.

40. J. J. Bezem, J. W. Sluiter, and P. F. van Heerdt, Population studies of five species of the bat genus *Myotis* and one of the genus *Rhinolophus*, hibernating in the caves of S. Limburg. *Arch. Néerl. Zool.*, 13: 511–19, 1960.

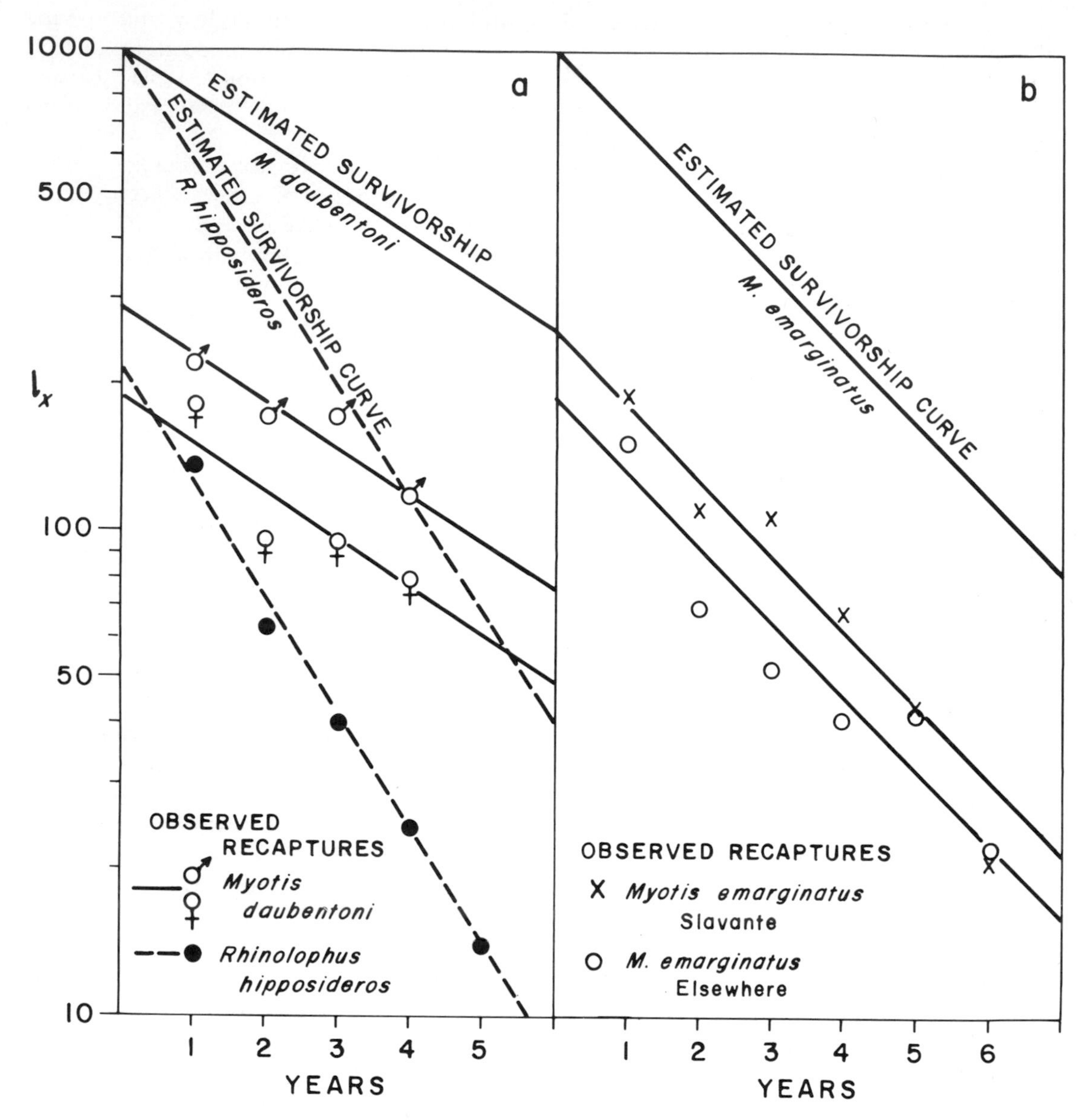

FIGURE 46. Survivorship of bats in caves in Holland; in all cases there was a much greater loss during the first year than later. If the subsequent years' losses are due to death, a fairly constant specific mortality seems to be implied. In (a) this is shown to be much less for both sexes of Daubenton's bat than for the lesser horseshoe bat. In (b) an intermediate death rate appears characteristic of more than one colony of another species of *Myotis* (Bezem, Sluiter, and van Heerdt).

(*Myotis daubentoni*) to 0.433 in the lesser horseshoe bat (*Rhinolophus hipposideros*). There is evidence that different colonies of the same species show similar mortality and survivorship, so that the species differences are probably real (figure 46, b). No sex differences in mortality have been discovered. The mean ages for the European species studied ranged from 2.3 years in *Rhinolophus hipposideros* to 5.0 years in *Myotis daubentoni*, the maximum known ages in these two species being 8 and 20 years respectively. The considerable potential life-spans of bats have already been noted; this is doubtless due to the very long

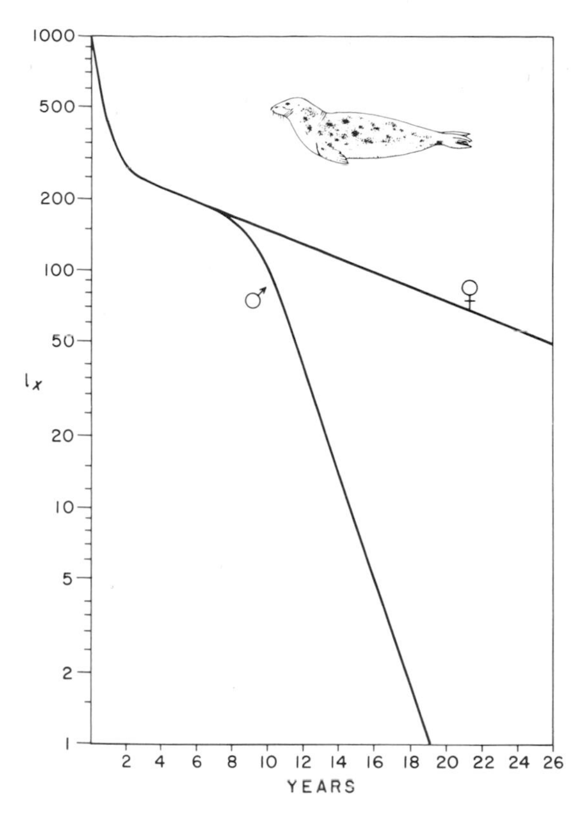

FIGURE 47. Somewhat idealized survivorship curves for the two sexes of the grey seal (*Halichoerus grypus*) living on the coasts of Britain, showing enormous mortality of males after they reach breeding age (Hewer).

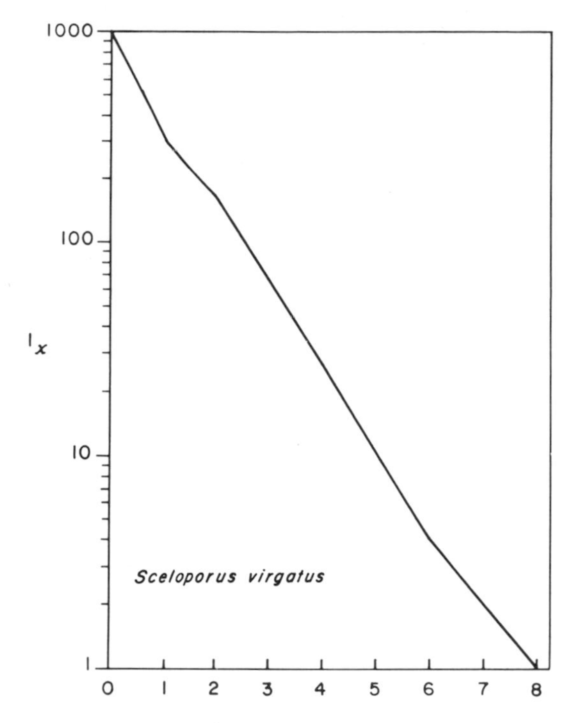

FIGURE 48. Diagonal survivorship curve of the lizard *Sceloporus virgatus* (Vincgar).

periods, both in the 24-hour and annual cycles, when bats are inactive.[41]

Finally, an extreme example may be given of a case[42] in which excessive male dominance and polygyny lead to so much death in the male part of the population that, as soon as breeding begins, the survivorship curves of the two sexes diverge enormously. The case is that of the grey seal (*Halichoerus grypus*) on the coasts of Britain. The animal can be aged from periodic structures layed down in the growth of the canine teeth. The curves (figure 47) are somewhat hypothetical, the constancy of the specific death rates after 10 years of age, of 0.067 per individual per year in females and 0.40 in males, being consistent with, but by no means certainly implied by, the data; great differences in the survivorship of the two sexes are, however, certain.

Survivorship and mortality in the lower vertebrates

There are a few data on reptiles coming mainly from turtles[43] and lizards.[44] As might be expected, such animals tend to have dia-

41. J. Drost, La longévité des chiroptères. *Mammalia*, 18:231–36, 1954.

My friend Alvin Novick tells me that infant mortality due to accidents in the nursing period and at the time of first flight must be very high. He believes that the capacity to lower body temperature during periods of inactivity is the fundamental cause of the long maximum longevity of bats.

42. H. R. Hewer, The determination of age, sexual maturity, longevity and a life-table in the grey seal (*Halichoerus grypus*). *Proc. Zool. Soc. Lond.*, 142:593–623, 1964.

43. For freshwater turtles see H. M. Wilbur, The evolutionary and mathematical demography of the turtle *Chrysemys picta*. *Ecology*, 56:64–77, 1975.

The extraordinary population, of at least 100,000 animals, of the giant tortoise, *Geochelone gigantea*, on Aldabra Island in the Indian Ocean is now under study. Thirty-five percent of the animals are under 20 years of age, and only about 20 percent are sexually

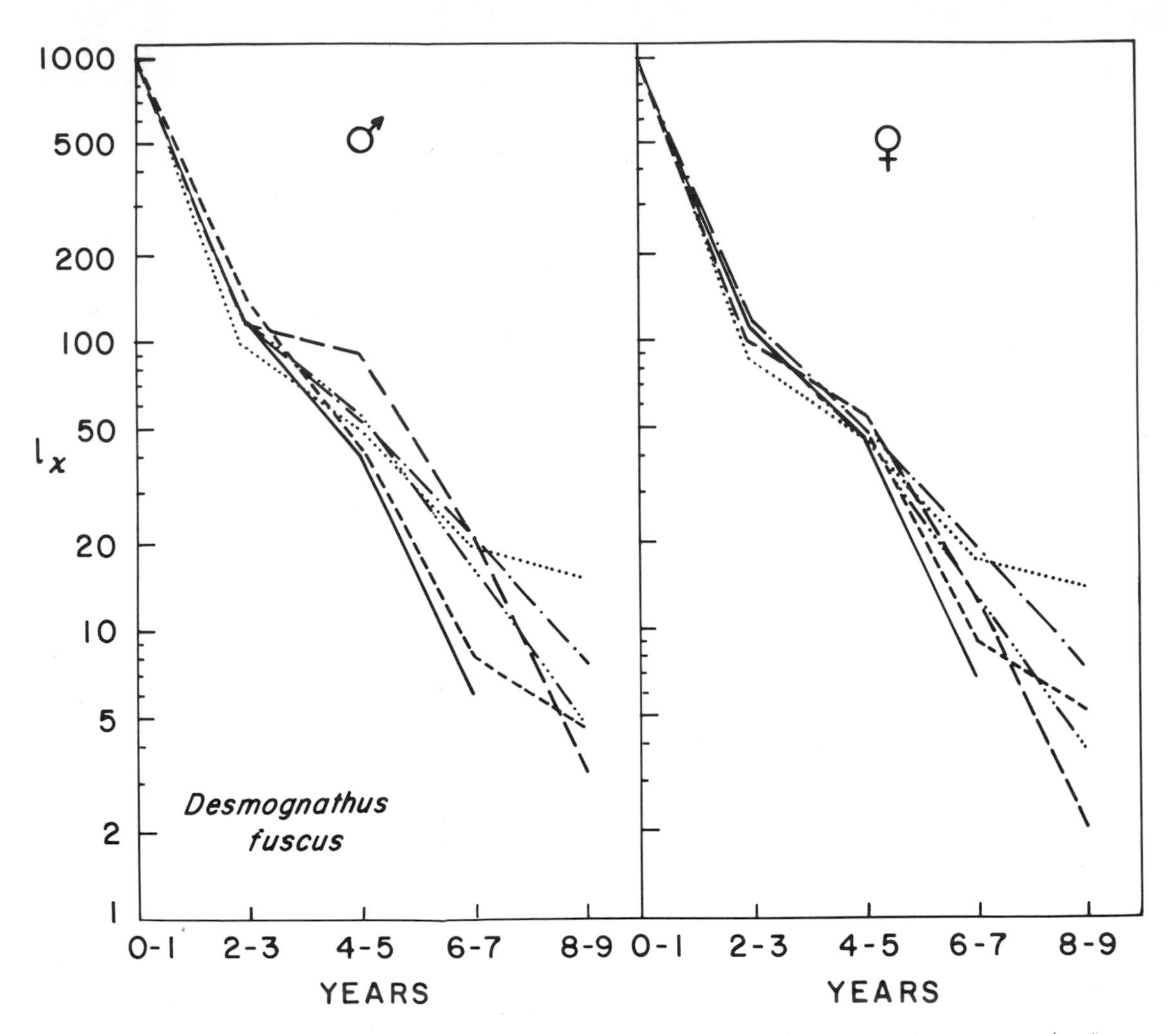

FIGURE 49. Survivorship curves for six male and female local populations of the salamander, *Desmognathus fuscus*, in Maryland, showing a greater amount of variability between such populations in males than in young or middle-aged females (Danstedt, slightly modiffed).

gonal adult survivorship curves (figure 48), though the initial juvenile mortality may be relatively much greater than in birds. Some amphibia[45] have also been examined, and also are found to exhibit constant and by no means insignificant mortality at least after metamorphosis, so that diagonal survivorship curves are produced. Danstedt has made an elaborate study of the demography of the plethodontid salamander, *Desmognathus fuscus*. It is possible to age these animals by size and condition of the reproductive organs. Though there are certain difficulties in producing survivorship curves, which inevitably

mature. Some animals may be over 50 years old. This age distribution is perhaps due to a rather rapid increase in the population in recent decades. For further details see P. Grubb, The growth, ecology, and population structure of giant tortoises on Aldabra Island. *Phil. Trans. Roy. Soc. Lond.*, 260B:327–72, 1971.

44. M. B. Vinegar, Demography of the striped plateau lizard, *Sceloporus virgatus. Ecology*, 56:172–82, 1975.

B. Barbault, Population dynamics and reproductive patterns of three African skinks. *Copeia*, 1976:483–90 (a very interesting paper that became available too late to use).

45. R. T. Danstedt, Local geographical variation in demographic parameters and body size of *Desmognathus fuscus* (Amphibia: Plethodontidae). *Ecology*, 56:1054–67, 1975.

depend on various kinds of indirect estimation, it seems that when comparison is made between populations living in different localities, males show greater variation in adult mortality than do females (figure 49). Danstedt believes this is due to selection operating more rigorously on females for whom the cost of reproduction is greater than for males. Actually, if one population is discounted, the effect is quite small, but is interesting enough to warrant further study. The variations observable in male and older female subpopulations are of course quite likely to be determined environmentally. Another amphibian case will be mentioned later in a more general context (pp. 86–87).

Survivorship and mortality in fishes

Two difficulties are apparent as soon as we consider the fishes. First, in most species the number of eggs laid is considerably greater than among the terrestrial vertebrates, and the size of the eggs is correspondingly smaller. This difference is exemplified to an extreme degree in pelagic marine fishes such as the herring, mackerel, anchovy, and the like. When there are very numerous small eggs, sampling of the fry, or in many cases of morphologically distinguishable larvae, involves quite different methods from those used in the study of older stages. It is therefore never easy to put a whole life table together.

The second difficulty arises from the fact that with practically every fish that can be caught in sufficient quantity to provide suitable samples, fishing has previously been practiced intensively to provide human food. Ricker[46] wrote in 1949 that "studies of unexploited populations are prized rarities in the fishery literature." The known number of such rarities has increased a little since he wrote, but not very much. There seems to be no case in which such an unexploited population has been studied from egg to oldest surviving adult.

Where an enormous number of small eggs are produced, up to the order of one million per female fish, it is obvious that in any species with an equilibrium population, the mortality must also be enormous.

In a celebrated study of the Atlantic mackerel, *Scomber scombrus*, Sette[47] found that in 1932, which was hydrographically unfavorable for the development of the plankton on which the young fish feed, the mortality in the first 70 days of life was 0.999994 per egg laid. The survivorship curve (figure 50) apparently consisted of two segments in which mortality was constant, joined by a short segment in which a much higher mortality prevailed. This period of high mortality corresponded to the rapid development of fins as the larva metamorphosed into a postlarval stage. Whatever happens in the later history of such a cohort, it is obvious that the survivorship curve is about as near to a positively skewed rectangular curve as one is likely to find. In most years no effective recruitment to the adult stock will have taken place. This indeed was the case in 1932 in the Atlantic mackerel.

The few survivorship curves that exist for unexploited populations[48] suggest that adult populations usually die at a fairly constant rate, giving a diagonal survivorship curve (figure 51). In at least one population of a white fish, *Coregonus clupeaformis*, from a lake on Shakespeare Island in Lake Nipigon,

46. W. E. Ricker, Mortality rates in some little exploited populations of freshwater fishes. *Trans. Amer. Fisher. Soc.*, 77:114–28, 1949.

See also J. H. Beverton and S. J. Holt, On the dynamics of exploited fish populations. *Fisheries Investigations*, ser. 20, vol. 19. London, H. M. Stationery Office, 1958, 533 pp.

47. D. E. Sette, Biology of the Atlantic mackerel (*Scomber scombrus*) of North America: Part 1. Early life history including the growth, drift, and mortality of the egg and larval populations. *Bull. Fisher. Wildl. Serv.* (U.S. Dept. Interior), 50 (*Fisher. Bull.* 38):109–237, 1943.

48. W. E. Ricker, Natural mortality among Indiana bluegill sunfish. *Ecology*, 26:111–21, 1952. See also Ricker, note 46.

W. P. Williams, The growth and mortality of four species of fish in the River Thames. *J. Anim. Ecol.*, 36:695–720, 1967.

R. H. K. Mann, Observations on the age, growth, reproduction, and food of dace *Leuciscus leuciscus* (L.), in two rivers in southern England. *J. Fish Biol.*, 6:237–53, 1974.

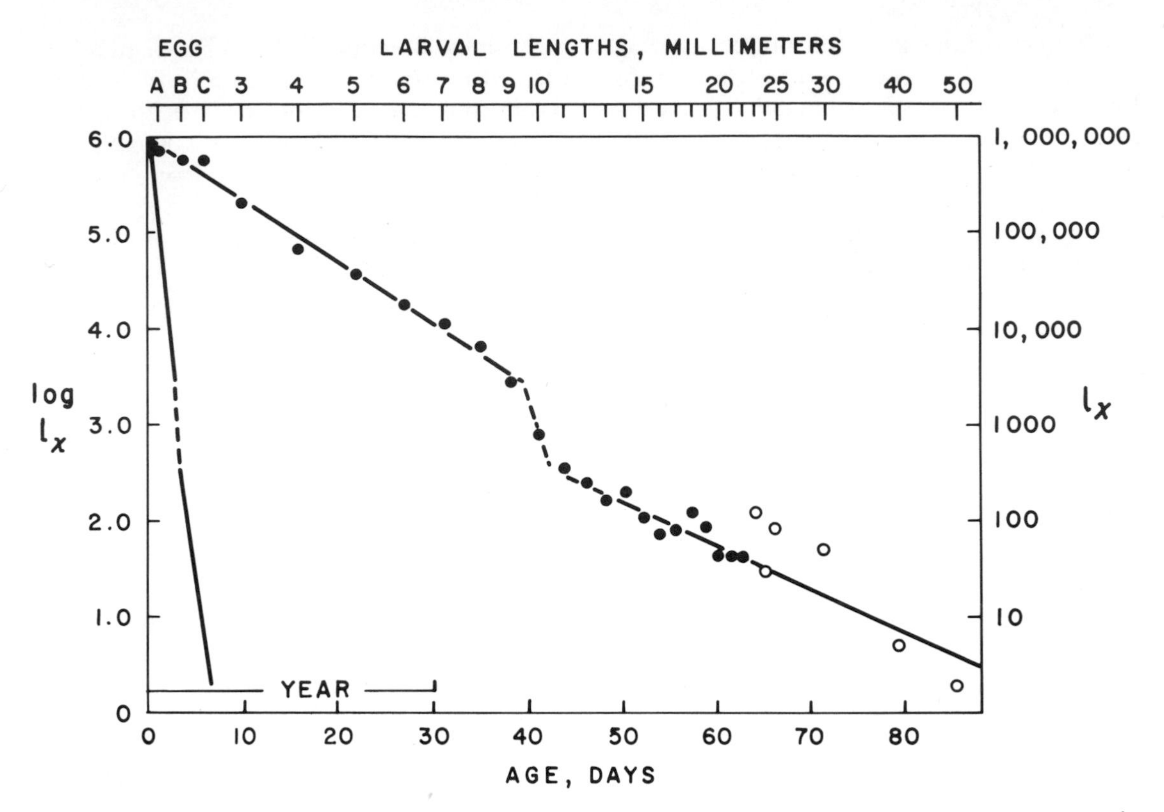

FIGURE 50. Survivorship curve of eggs and larvae of the Atlantic mackerel, *Scomber scombrus*, for the 3-month period after breeding in 1932. The two solid lines are fitted by least squares and are joined by a broken line fitted by eye. The changes in the death rate between 35 and 40 days correspond to the development of fins and metamorphosis into the postlarval stage. Open circles represent less reliable data not used in fitting a regression line. The curve is roughly reproduced on the left on a scale more comparable to that used in the graphs of other species of relatively long-lived vertebrates (Sette, modified).

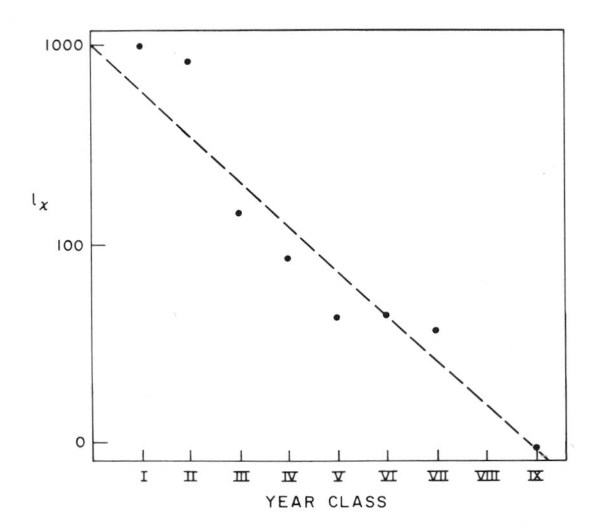

FIGURE 51. Survivorship of the dace, *Leuciscus leuciscus* in the River Stour, Dorset, England (Mann, modified).

in which specimens of the fish may survive into their 26th year, there is apparently a very slight increase in mortality (figure 52) until the 19th year, after which death seems to occur increasingly faster. Alternatively, the observations could be due to a systematic increase in the size of the population, though at a decreasing rate, over the quarter of a century prior to sampling. In other unfished Canadian lakes, in which the growth of the fish is apparently more rapid, the specific mortality in the later half of life is essentially constant, but much higher than in the Shakespeare Island lake, varying from 0.57 to 0.81 per individual per year.

Survivorship and mortality in Crustacea

Two cases are of interest, one relating to a freshwater copepod with five generations in

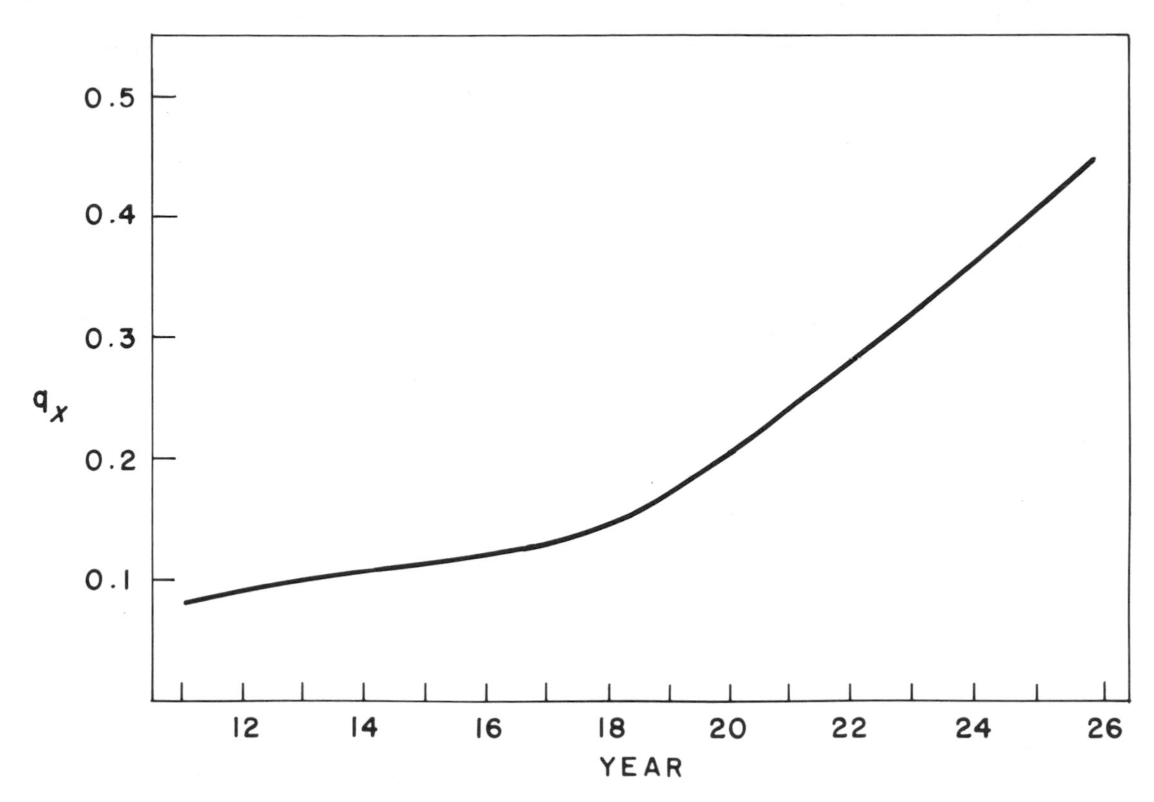

FIGURE 52. Apparent specific mortality in an old population (12–26 years) of the whitefish (*Coregonus clupeaformis*) in a lake on Shakespeare Island in Lake Nipigon (Ricker).

a year, the other to a prawn living for several years. Both species have very small planktonic larvae; this habit persists in the copepod, while the prawn becomes nectobenthonic, swimming when not resting on the bottom.

The work on the copepod,[49] *Diaptomus (Aglaiodiaptomus) clavipes*, was done in a pond in Oklahoma and was supplemented by laboratory studies. Numerous periodic collections were made in the pond and from them the appearance and disappearance of the various instars or stages separated by molts could be determined, as well as the changes in numbers of the instars as the seasons progressed. From these data it is possible, by some mathematical manipulation, to construct a composite age-specific life table and survivorship curve. This can be compared with a directly observed laboratory population (figure 53).

The laboratory population showed lower mortality at all larval stages than did the various generations studied in nature. This was particularly true of the juvenile mortality during the egg and first three naupliar stages, during the first 2 days of life.

Adult mortality appears very low until the maximum adult life-span of about 40 days is reached. Gehrs and Robertson write as if the adult survivorship curve is essentially negatively skewed rectangular, though it is clearly somewhat hypothetical.

Juvenile mortality in nature was found to increase steadily throughout the five generations observed, from about 0.83 per individual for the egg to N_3 stage at the beginning of the season, to 0.97 at the end. Later larval mortality, however, showed very little variation between the generations. Such high mortality is, however, not necessary in cope-

49. C. W. Gerhs and A. Robertson, Use of life tables in analyzing the dynamics of copepod populations. *Ecology*, 56:665–72, 1975.

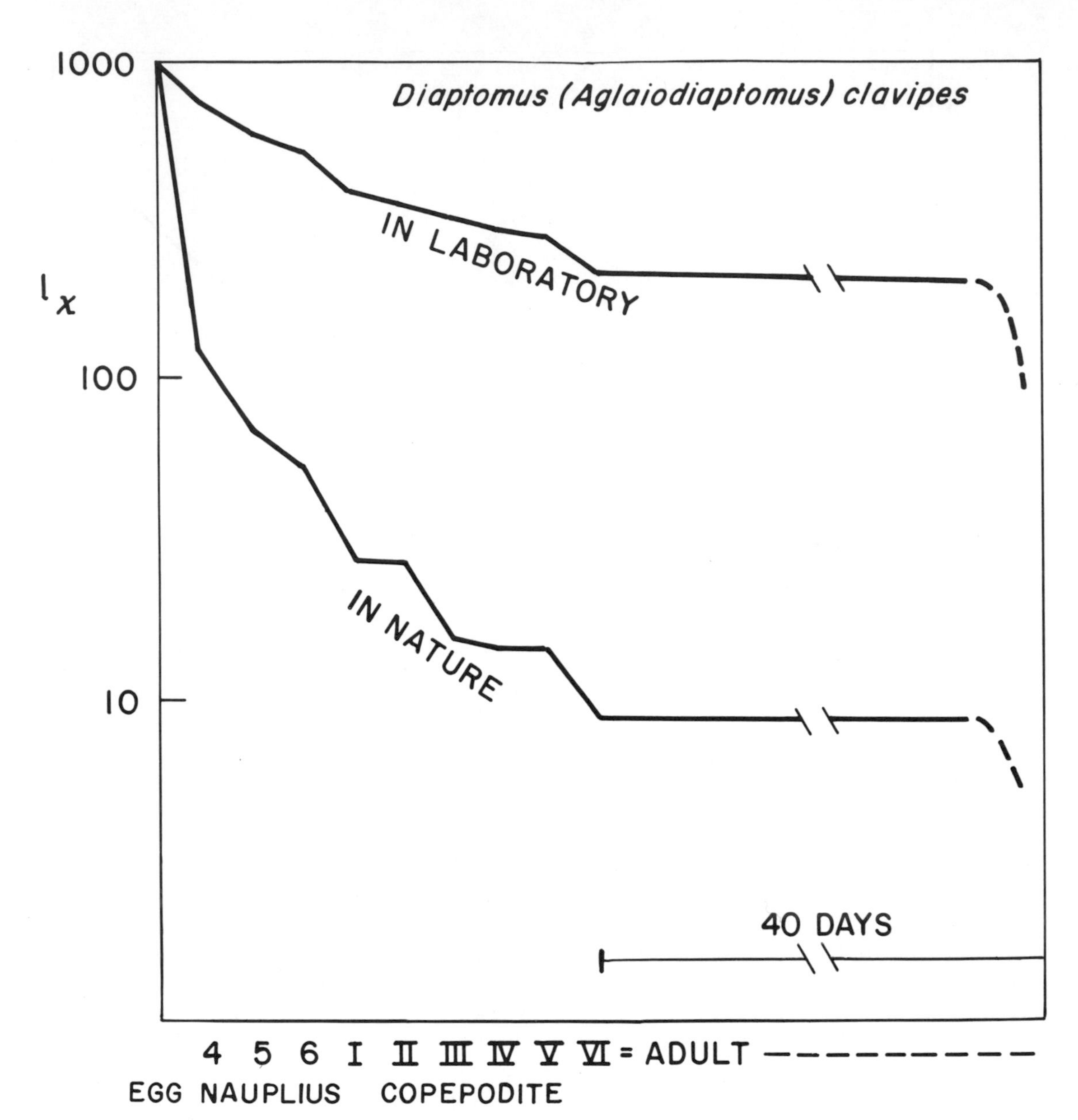

FIGURE 53. Survivorship curves for *Diaptomus (Aglaiodiaptomus) clavipes* for the entire life history from egg to the beginning of the adult (CVI) instar reared in the laboratory, and in the average population of an unspecified pond presumably in Oklahoma. Adult survival was believed to have been very high for about 40 days after the beginning of the adult instar, followed by intense senile mortality.

pod populations in nature, for Tonolli[50] found that 6 out of 11 eggs laid per litre per year by *Eudiaptomus padanus* in Lago Maggiore reached the fifth copepodite stage.

Kurtén's estimates[51] for the relatively long-lived prawn, *Leander squilla*, derived from the data of Höglund, indicate enormous mortality in the 1st year (figure 54) when

50. V. Tonolli, Studio sulla dinamica de popolamento di un copepode (*Eudiaptomus vulgaris* Schmeid. *Mem Ist. Ital. Idrobiol.*, 13:179–202, 1961. The species is now referred to *E. padanus* (Burckhardt) Kiefer.

51. B. Kurtén, see note 7.

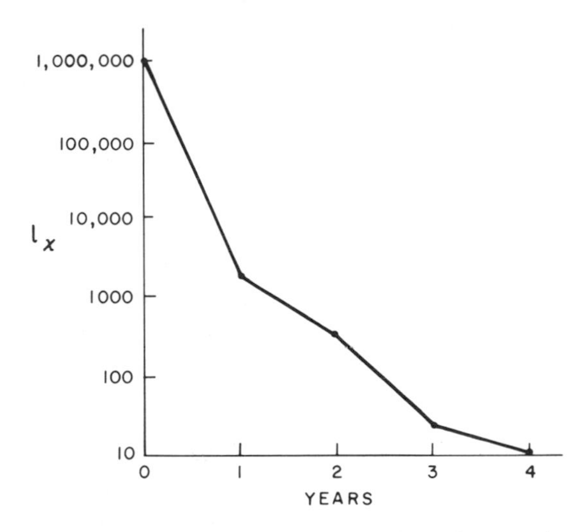

FIGURE 54. Survivorship curve calculated by Kurtén for a Swedish population of the prawn, *Leander squilla*. There is some uncertainty about the figures after the 2 year; they are here given as averages for two succeeding years, 1940 and 1941. The enormous decrease in the 1 year is reminiscent of what may happen in pelagic marine fishes, such as the mackerel.

about 99.8 percent of the eggs and early stages die, followed by 40–90 percent mortality in the 2nd and 3rd years. The situation here, as far as the data go, is probably comparable with what often happens in fishes.

In addition to these cases, a good deal of work has been done on barnacles. Here much intraspecific competition may occur after a population has grown more or less logistically and consists of a maximum number of small individuals, which by growth may displace each other. The results are often very irregular.[52]

Survivorship and mortality in insects

A great deal of work has recently been done on the life tables of insects; such investigations are often very useful in applied entomology. Comparison of survivorship and relative mortality in two populations, one of which has been subjected to some insecticide, is obviously a very good way of learning in detail about the efficacy of control.

Price,[53] in an excellent book on the ecology of insects, has given data about a number of insect life tables. Considering nineteen species of phytophagous insects belonging to the Homoptera, Coleoptera, and Lepidoptera, he found that in eleven species (group B) 70 percent of the mortality occurred before the mid-larval stage and in seven species (group A) only 40 percent of the mortality fell in this period. One species was intermediate. When the survivorship curves (figure 55) of these species are plotted semilogarithmically, the first group are seen to have more or less diagonal and the second group convex curves.

Whether the intervening region in which this intermediate species fell really separates two largely discontinuous groups of curves will only be apparent as more data accumulate.

A very careful study of populations of two Psocoptera of the genus *Mesopsocus*, insects living on the bark of larch trees, made in Britain by Broadhead and Wapshere,[54] suggests little difference between two closely allied species, the deaths in both of which seem largely dependent on predation by insects and birds (figure 56).

Detailed study of single species, such as the sawfly *Neodiprion swainei* on jackpine in Quebec (figure 57) shows how the minor features of survivorship depend in holometabolous insects on the sequence of stages.[55] There is evidently good survival in the egg, but considerable mortality after hatching;

52. H. Hatton, Essais de bionomie explicative sur quelques espèces intercotidales d'algues et d'animaux. *Ann. Inst. Oceanogr., Monaco*, 17:241–348, 1938. Hatton's work has been discussed in detail by Deevey (note 7).

J. H. Connell, The influence of interspecific competition and other factors on the distribution of the barnacle *Chthamalus stellatus*. *Ecology*, 42:710–23, 1961.

53. P. W. Price, *Insect Ecology*. New York, London, Sydney, Toronto, Wiley-Interscience, 1975, xii, 514 pp.

54. E. Broadhead and A. J. Wapshere, *Mesopsocus* populations on larch in England: The distribution and dynamics of two closely-related co-existing species of Psocoptera sharing the same food resource. *Ecol. Monogr.*, 36:327–88, 1966.

55. J. M. McLeod, The Swain jackpine sawfly, *Neodiprion swainei*, life system: Evaluating the long-term effects of insecticide applications in Quebec. *Environm. Entom.*, 1:371–81, 1972.

P. M. Price, Strategies for egg production. *Evolution*, 28:76–84, 1974.

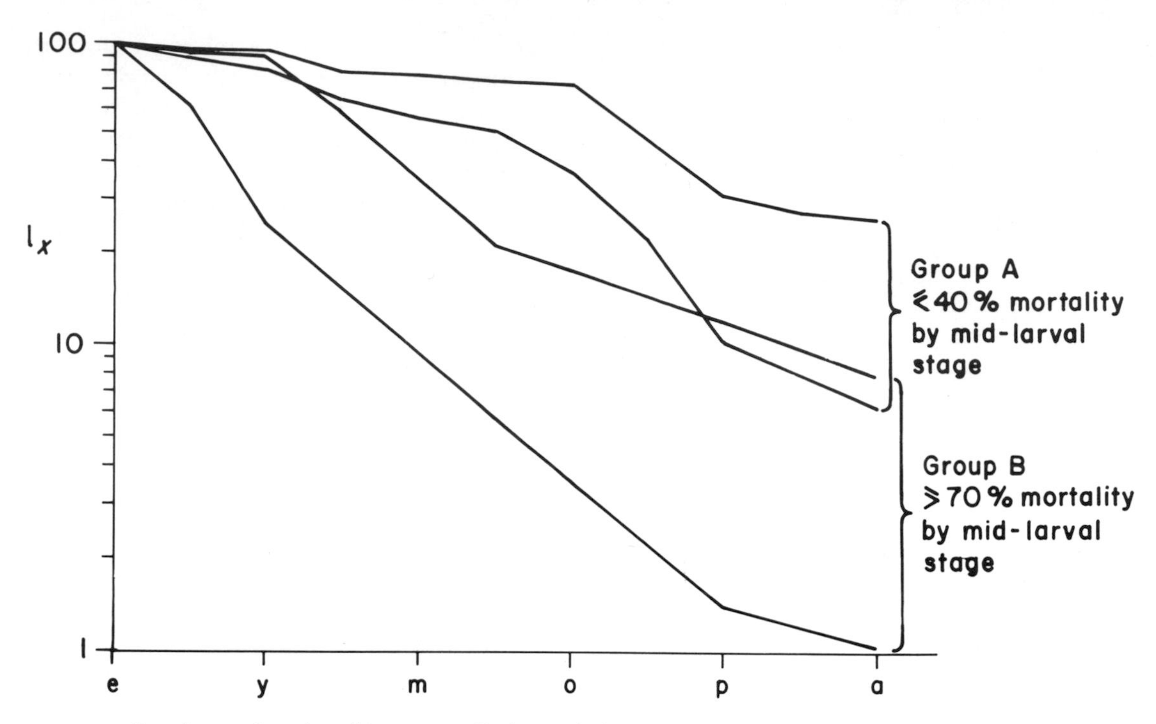

FIGURE 55. Envelopes of survivorship curves of holometabolous phytophagous insects showing division into seven (group A) with 40 percent of the mortality before the mid-larval period, and nine (group B) with 70 percent of the mortality before this time; e, egg; y, young larva; m, middle-aged larva; o, old larva; p, pupa; a, adult (Price, modified).

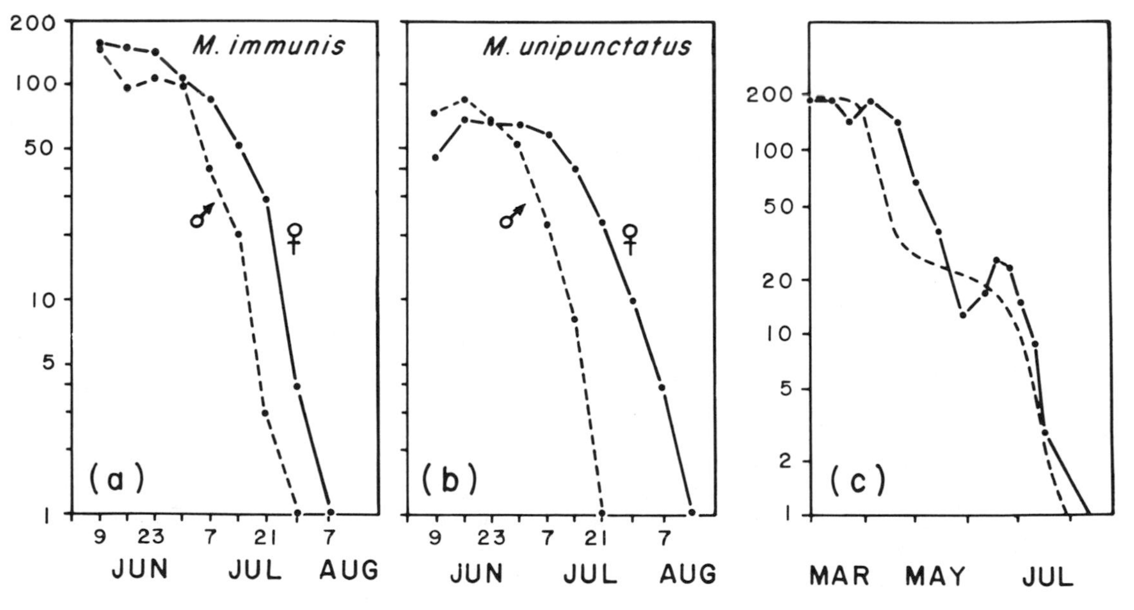

FIGURE 56. (a) Declines of populations, giving a close approximation to survivorship curves, of males and females of *Mesopsocus immunis* on larch trees at Harrogate in the sumer of 1958; (b) the same for *M. unipunctatus*; (c) combined population of both species (solid line) and a theoretical declining population (broken line) obtained by integrating the calculated mortality rate based on the local abundance of insectivorous birds and of the larvae of the lacewing fly *Wesmaelius*.

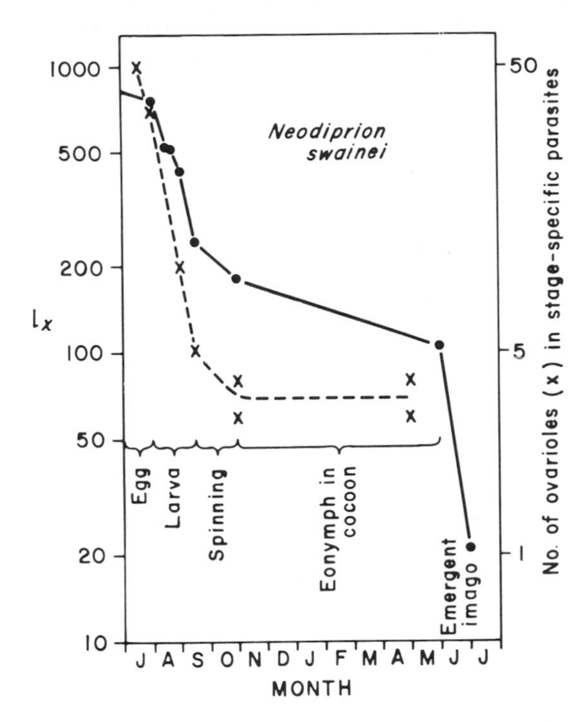

FIGURE 57. Survivorship curve (solid line) of the sawfly *Neodiprion swainei*, with an indication of the number of ovarioles (x and broken line) in the hymenopterous parasites of this species, those infecting early instars being much more fecund than those parasitizing later stages (McLeod, and Price).

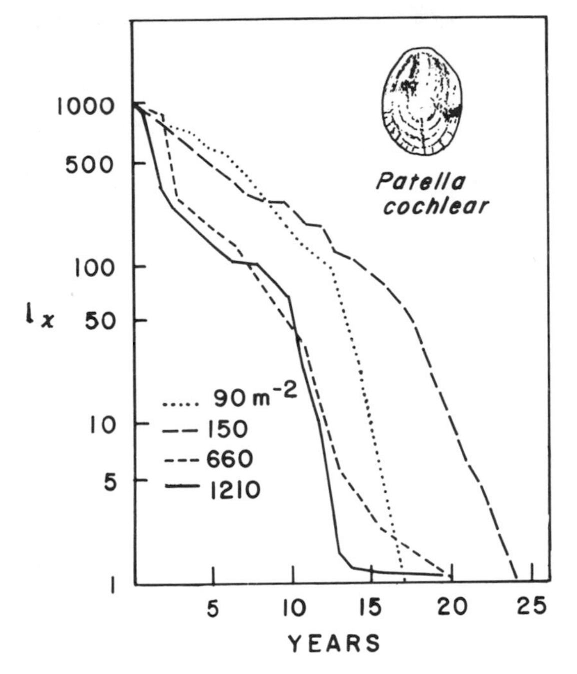

FIGURE 58. Survivorship curves of the South African limpet, *Patella cochlear*, from different environments differing greatly in the density of their populations.

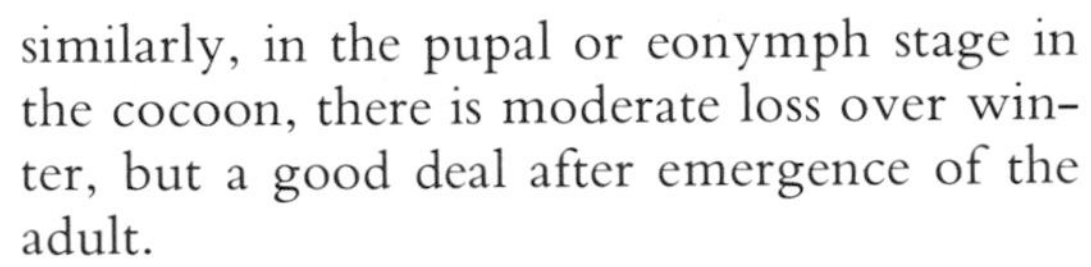

similarly, in the pupal or eonymph stage in the cocoon, there is moderate loss over winter, but a good deal after emergence of the adult.

It is interesting to find that the characteristic hymenopterous parasites of this insect have ovaries consisting of very varying numbers of ovarioles, the species parasitizing the early stages of *N. swainei* being much more fecund than those living on the rarer though larger later stages.

Some very interesting work by Grüm[56] on the mortality of carabid beetles in Poland appeared just as the manuscript of this book was completed. Mortality is usually greater in active stages, namely, the larva and the nonhibernating imago, than in the egg, pupa, or hibernating adult. Species breeding in spring usually have a shorter total but longer adult life-span than those breeding in autumn. Variations in mortality with activity are attributed mainly to variations in the intensity of predation on the beetles and their larvae.

Survivorship and mortality in molluscs

Some molluscs live for a number of years, can be aged by shell structure or by size, and are either attached or sufficiently territorial to permit recognition throughout life as individuals. A very good case is provided by the South African limpet, *Patella cochlear*, recently studied by Branch.[57] No direct evidence exists about larval mortality before settling. When settlement has taken place, juvenile mortality depends largely on the number of young individuals that have had

56. L. Grüm, Mortality patterns in carabid populations. *Ekologia polska*, 23:649–65, 1975. It should be noted that the relationship of mobility to risk of death by predation is a complex one, now under intense investigation by R. Strickler.

57. G. M. Branch, Interspecific competition in *Patella cochlear* Born. *J. Anim. Ecol.*, 44:263–81, 1975.

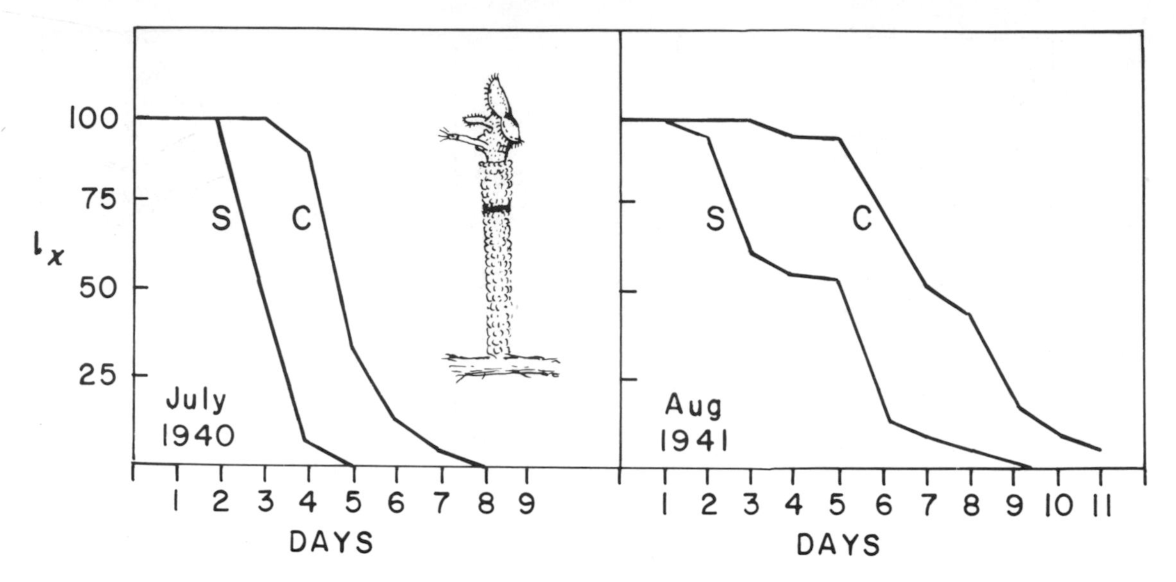

FIGURE 59. Survivorship of sessile rotifer, *Floscularia conifera* after setting, in two successive years, near New Haven, Conn. The individuals are classified as to whether they are solitary (s) attached directly to *Utricularia* plants, or are colonial (c) associated with other *Floscularia*. The life expectancy of the latter group was, in both years, greater than that of the former (Edmondson).

to settle on the shells of older specimens. These young tend to move to the rock surface as it becomes available. This process is apparently particularly dangerous, so leading to high death rates in very dense (660 per m^2) populations. After the younger limpets have taken their place on rock and have produced a scar on which they sit, the mortality becomes relatively independent of density (figure 58). Growth and gonad maturation are restricted in dense populations, which, however, appear to exhibit higher senile mortality than do less dense groups.

Survivorship and mortality in sessile rotifers

In a classic series of experiments, Edmondson[58] studied survivorship in *Floscularia conifera* (figure 59), a species that builds a tube of fecal pellets as it grows. By letting India ink or a carmine suspension sink through a colony on a *Utricularia* plant in nature, it is possible to produce daily rings on the tube. Collecting the colony after a suitable interval allows one to obtain a time-specific life table from the ages of the marked individuals. The survivorship curves tend to be convex. In both years in which experiments were run, the individuals that had settled on the tubes of older animals had a greater mean life-span than those settling on the *Utricularia*. There was a suggestion of a slight inhibition of growth in young colonial individuals, but for most of their lives the growth rates of solitary and colonial individuals were comparable. Colonial animals probably produce rather more eggs in unit time than do solitary ones. It is therefore clear that colonial animals have a greater rate of increase. There would be strong selection in favor of any genetic mechanism favoring settling on tubes rather than *Utricularia* leaves.

Survivorship and mortality in reef-building organisms

A considerable field certainly awaits exploration. Grigg[59] has recently examined some aspects of the demography of a gorgonian or sea fan, *Muricea californica*, on the Cali-

58. W. T. Edmondson, Ecological studies of sessile Rotatoria: II. Dynamics of populations and social structures. *Ecol. Monogr.*, 15:141–72, 1945.

59. R. W. Grigg, Age structure of a longevous coral, a relative index of habitat suitability and stability. *Amer. Natural.*, 109:647–57, 1975.

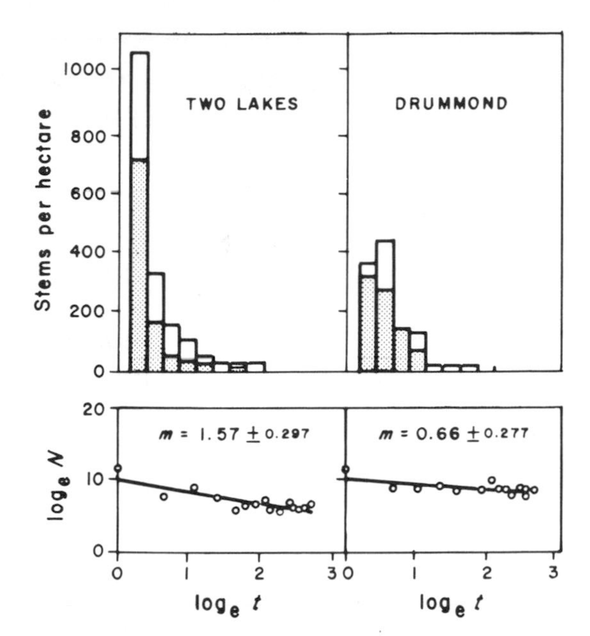

FIGURE 60. Upper panels, distribution by size classes (0–10.2, 10.2–20.3, 20.3–30.5 cms, etc. breast height diameter) of sugar maple (stippled) and all other trees (open) in two stands studied in Wisconsin. Lower panels, $\log_e N$ against $\log_e t$. The number N is given as stems per 0.405 hectare, t is in years (Hett and Loucks).

fornian coast. The individuals were aged by size; growth rates are known at two of the stations and so permit conversion of the measurement of height to one of age. This can be controlled to some extent by study of annual growth rings. Sexual reproduction begins when an individual is about 10 years old; the maximum age recorded is about 50 years. The distribution of age classes is very uneven. Grigg assumed a constant specific death rate and used the variance of the observed data about the best monotonic regression line as a measure of the unpredictability of the environments in which the sea fan lives.

Survivorship and mortality in plants

The same general approach that has been developed from actuarial practice as a method of studying animal populations can of course also be applied to plants.

It has long been known that the survivorship curves of forest trees are likely to be of an extreme positively skewed rectangular type. Anyone who has cultivated a garden containing a maple or an oak tree will appreciate this intuitively.

The simplest kind of extreme concave curve is given by[60]

$$N_t = N_0 t^{-m}$$

or

$$\ln N_t = \ln N_0 - m \ln t. \qquad (2.1)$$

The relative mortality $\mathbf{q}_x$ or $-m/t$ decreases with time. At least in the case of the sugar maple, *Acer saccharum*, this type of relationship holds for the first 15 years of life. In the two examples from Hett and Loucks's[61] work in Wisconsin (figure 60), it is probable that the difference in the slope of the survivorship curves is due to differences in the vertical structure of the stands, so that less light and direct rain reach the lower levels in the stand with the higher mortality. Foresters tend to regard the later survivorship of trees as approximately conforming to a diagonal curve, the relative mortality being constant. It is, however, probable that over the period of chief interest to them, the decline is small enough for significant departures from either an exponential or a power law to be inappreciable in practice.

The demography of herbaceous plants has been extensively studied by Harper[62] and his associates, who not only have collected a great deal of new data, but have reanalyzed

60. See note 14.

61. J. M. Hett and O. L. Loucks, Sugar maple (*Acer saccharum* Marsh.) seedling mortality. *J. Ecol.*, 59:507–20, 1971.

62. Harper has done more than anyone to combine the approaches of plant and animal ecologists to produce a satisfactory demography of plants. Two general summaries are of particular importance:

J. L. Harper, A Darwinian approach to plant ecology. *J. Ecol.*, 55:247–70, 1967.

J. L. Harper and J. White, The demography of plants. *Ann. Rev. Ecol. Syst.*, 5:419–63, 1974.

The reanalysis of much early work, followed by one of the most intensive demographic studies of any group of nonhuman organisms made in nature, will be found in:

J. Sarukhán and J. L. Harper, Studies on plant demography: *Ranunculus repens* L. and *R. acris* L. I. Population flux and survivorship. *J. Ecol.*, 61:675–716, 1973.

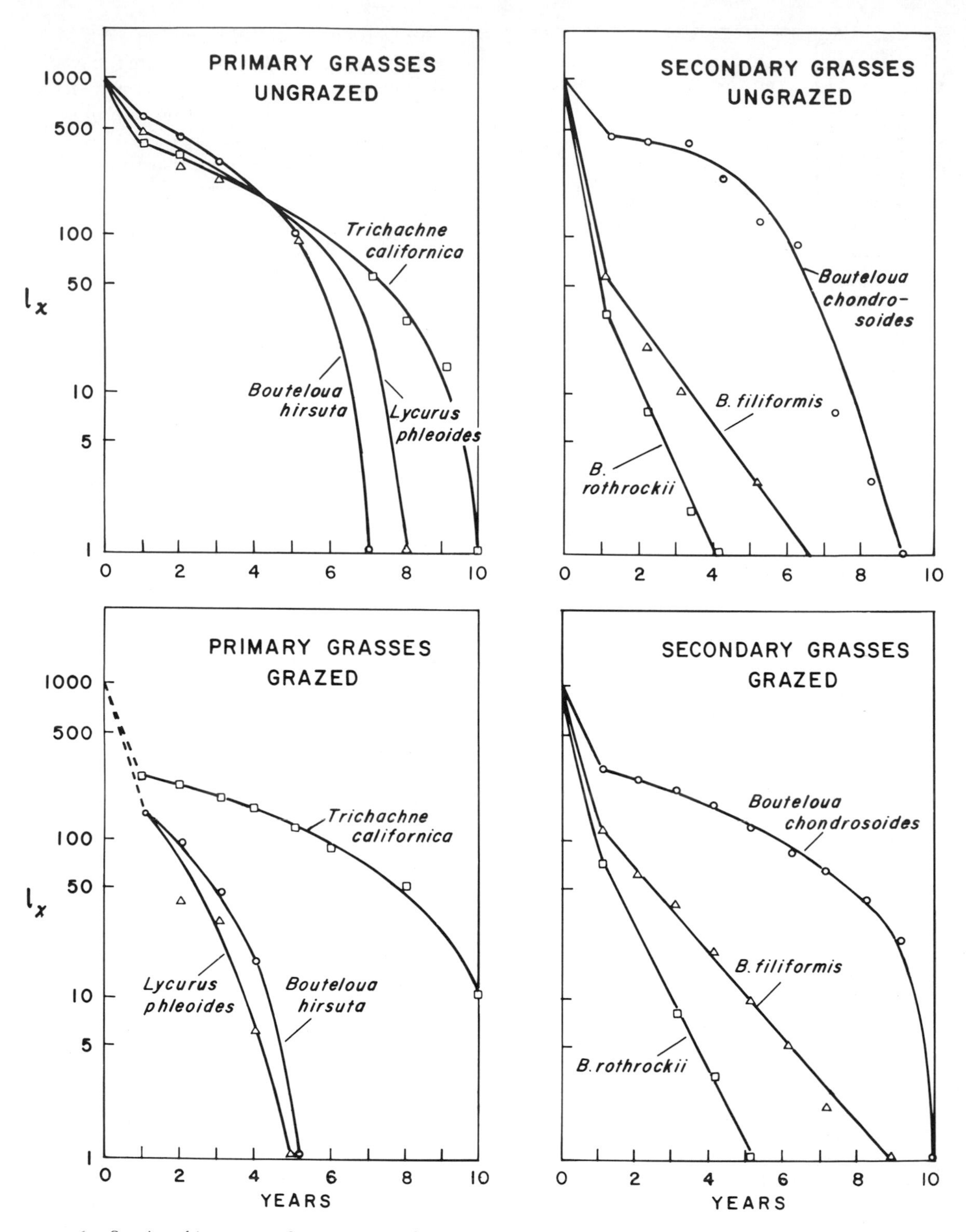

FIGURE 61. Survivorship curves of grasses in southern Arizona. Primary grasses are those predominant on land protected from grazing or maintained in good condition by proper management, while secondary grasses occur abundantly on grazed ranges in poor condition (data of Canfield, analyzed by Sarukhán and Harper).

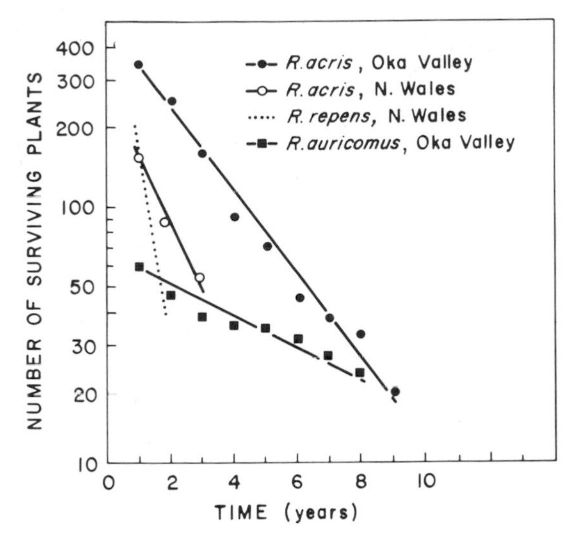

FIGURE 62. Survivorship curves of three species of European buttercup of the genus *Ranunculus*. The line for *R. acris* in North Wales is one of two that are parallel, while that for *R. repens* is based on a very large number of points along one of four parallel lines. Note the striking specific diversity, but the general comparability of the Russian and Welsh populations of *R. acris*, though those from the latter country are slightly less long-lived (redrawn from Sarukhán and Harper, the Russian data from Rabotnov).

much of the work of their predecessors. They have also encouraged investigations in many areas far from North Wales, so that at the present time more demographic work is probably being done on flowering plants than on any other group of organisms; only a sample of the results of this activity is given here. The fact that the smallest individual plant can be entered on a large-scale map by the use of a pantograph, directly copying the distribution in the field, makes the study of plant demography far more accurate than is possible in most studies of animals, provided that the investigator has almost limitless patience.

After a fall due to juvenile mortality which varies greatly with the amount of seed set, the survivorship curves of perennial herbs vary from markedly convex to strictly diagonal. The variation is evidently comparable to that found in mammals.

Practically the entire range can be found in a single genus such as the grass *Bouteloua* in Arizona, studied by Canfield, whose data were plotted as survivorship curves by Sarukhán and Harper. Congeneric species may react quite differently to environmental factors. Grazing reduces the life expectancy of *Bouteloua hirsuta*, has little effect on *B. chondrosoides*, but actually increases longevity somewhat in *B. filiformis* and *B. rothrockii*. Comparable differences can be observed when other, less closely allied grasses are compared. The cases where grazing promotes the expectation of life are presumably due to elimination of more easily eaten, but otherwise more successful competitors (figure 61).

The buttercups of the genus *Ranunculus* have been studied both in Russia and North Wales. When survivorship is considered over yearly intervals, the survivorship curves are diagonal, but with different slopes in different species and in widely spaced localities. Two populations of *R. acris* in North Wales gave almost identical slopes for the diagonal survivorship curves, but one from the Oka valley in Russia died rather less rapidly. *R. auricomus* in Russia had a mean life of twice or more than that of *R. acris*, which was longer-lived than *R. repens* (figure 62). The seasons of dying in this genus are restricted and characteristic of the species, as is shown by comparing specific mortality in *R. repens* and *R. bulbosus* (figure 63.) The period of low mortality in the middle of the growing season corresponds to the period of flowering. At least in the vegetative propagules of *R. repens*, there is a strong indication of mortality being density-dependent (figure 64). In this species there is some evidence of seedlings showing a higher initial mortality than do newly formed vegetative propagules or ramets. If this is substantiated, it is obviously very interesting; unfortunately so few seedlings could be studied at any one time that

J. Sarukhán, Studies on plant demography: *Ranunculus repens* L., *R. bulbosus* L., and *R. acris* L. II. Reproductive strategies and seed population dynamics. *J. Ecol.*, 62:151–77, 1974.

The data on grasses, reanalyzed by Sarukhán and J. L. Harper, are given in R. H. Canfield, Reproductive and life span of some perennial grasses of southern Arizona. *J. Range Mgmt*, 10:199–203, 1957.

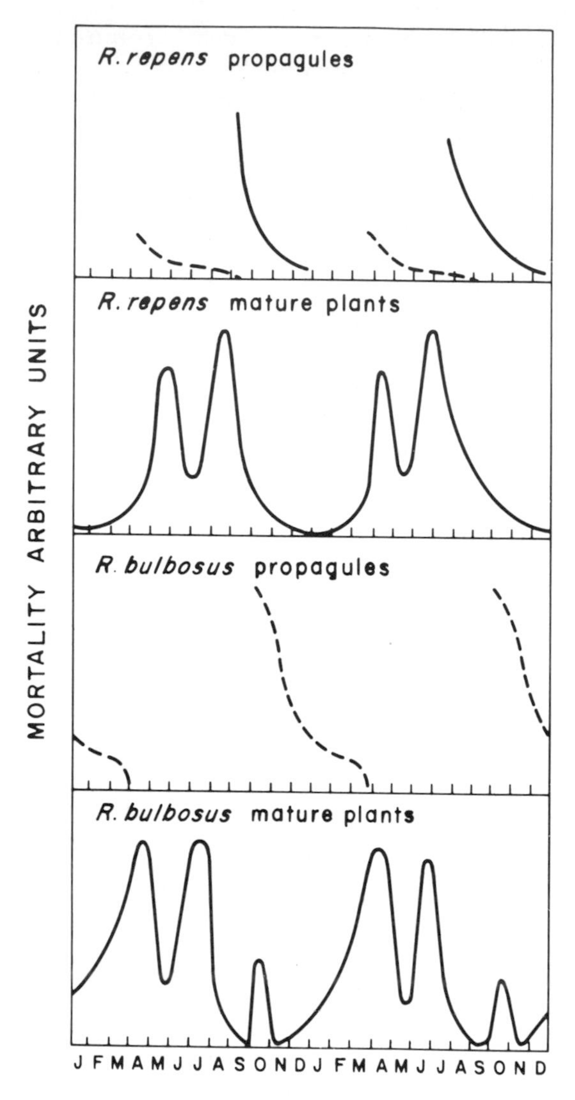

FIGURE 63. Ideal representation of temporal variation in specific mortality of *Ranunculus repens* (above) and *R. bulbosus* (below). The solid lines in the lower panels for each species represent adult mortality, with a minimum at midsummer during flowering. The solid lines in the upper panels for each species represent seedling mortality, the broken line for *R. repens* mortality of vegetative propagules or ramets (Sarukhán and Harper).

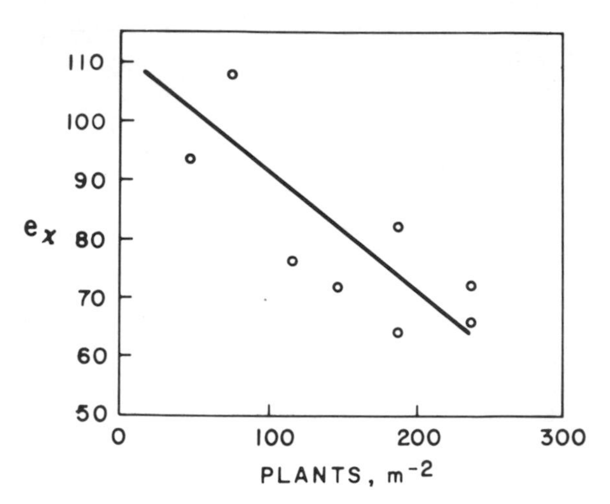

FIGURE 64. Specific mortality of asexually produced ramets of *Ranunculus repens* as a function of density (Sarukhán and Harper).

it is not possible to exclude completely the possibility of the high seedling death rate being due to environmental factors independent of the age of the plant.

Recently very extensive work has been done by Watson[63] on mosses of the family Polytrichaceae in North America (figures 65, 66). The data refer to the individual ramets of which a clonal clump of moss is composed. The leaves produced during the growing season vary in length, the longest being produced at the height of the season. It is thus possible to recognize growing seasons on any ramet and ascertain its age. Six species of star mosses were studied, three referable to *Polytrichum* and three to *Polytrichastrum*, all growing on Mount Washington, New Hampshire. With one exception, there was extremely little difference in the survivorship curves of the six species. There is, moreover, essentially no difference between a mean survivorship curve for all species growing singly and a mean curve for all species when growing in mixed clumps. Furthermore, when the pure and mixed stands are compared, the individual species only show differences of an insignificant kind, mainly when the cohorts are reduced to very small numbers. If competition were significant between the co-occurring species, we should expect a reduction in life expectancy of one competitor, but there is no reason to suppose that this would be balanced by an improvement in life expectancy of the more successful species

63. M. A. Watson, *The Population Biology of Six Species of Closely Related Bryophytes* (Musci). Yale Ph. D. thesis, 1975, 239 pp., *Diss. Abstr.*, 36 B1:78–79 (order no. 5-15385).

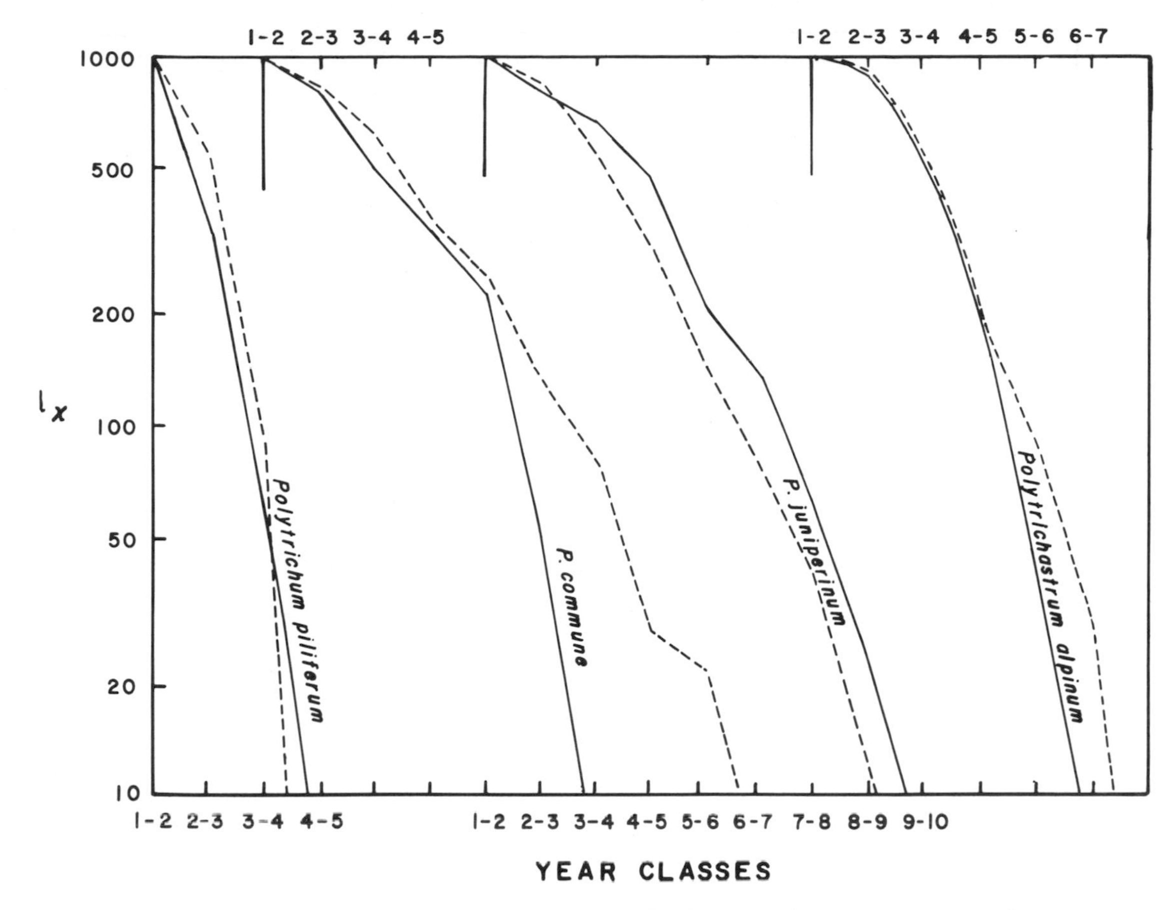

FIGURE 65. Survivorship curves of four species of moss of the family Polytrichaceae on Mount Washington. Note the much less long life of *Polytrichum piliferum* than that of the other species. In *P. juniperinum* and *P. commune* the effects of association seem to work in opposite directions, but are probably not significant (Watson).

over its performance by itself. This, however, would be needed to produce the observed similarities if significant competition took place. These observations therefore suggest that the factors determining life-span are of little or no importance in permitting or preventing allied species from co-occurring. In this, the Polytrichaceae seem to be very different from the grasses and probably also from the buttercups just studied.

Key factor analysis

A simple method of studying variations in mortality, when life tables are available for a species over a period of years, was introduced by Varley and Gradwell,[64] following an initial suggestion by Haldane. The specific mortality is determined either over each stage of the life history, or for each major cause of death. If the logarithms of each specific mortality be designated $k_1, k_2, k_3, \ldots$, the sum

$$K = k_1 + k_2 + k_3, \ldots,$$

is the logarithm of the specific mortality over the whole life-span. We then plot K, k_1, $k_2, \ldots$, over a period of years. It is often

64. G. C. Varley and G. R. Gradwell, Key factors in population studies. *J. Anim. Ecol.*, 29:399–401, 1960.

J. B. S. Haldane, Disease and evolution. In: Simposio sui fattori ecologici e genetici della speciazione negli animali. *La Ricerca Scientifica*, 19 (suppl.): 3–11, 1949.

J. R. Krebs, Regulation of numbers in the great tit (Aves: Passeriformes). *J. Zool. Lond.*, 162:317–33, 1970.

A further development of the method is given in H. Podoler and D. Rogers, A new method for the identification of key factors from life table data. *J. Anim. Ecol.*, 44:85–114, 1975.

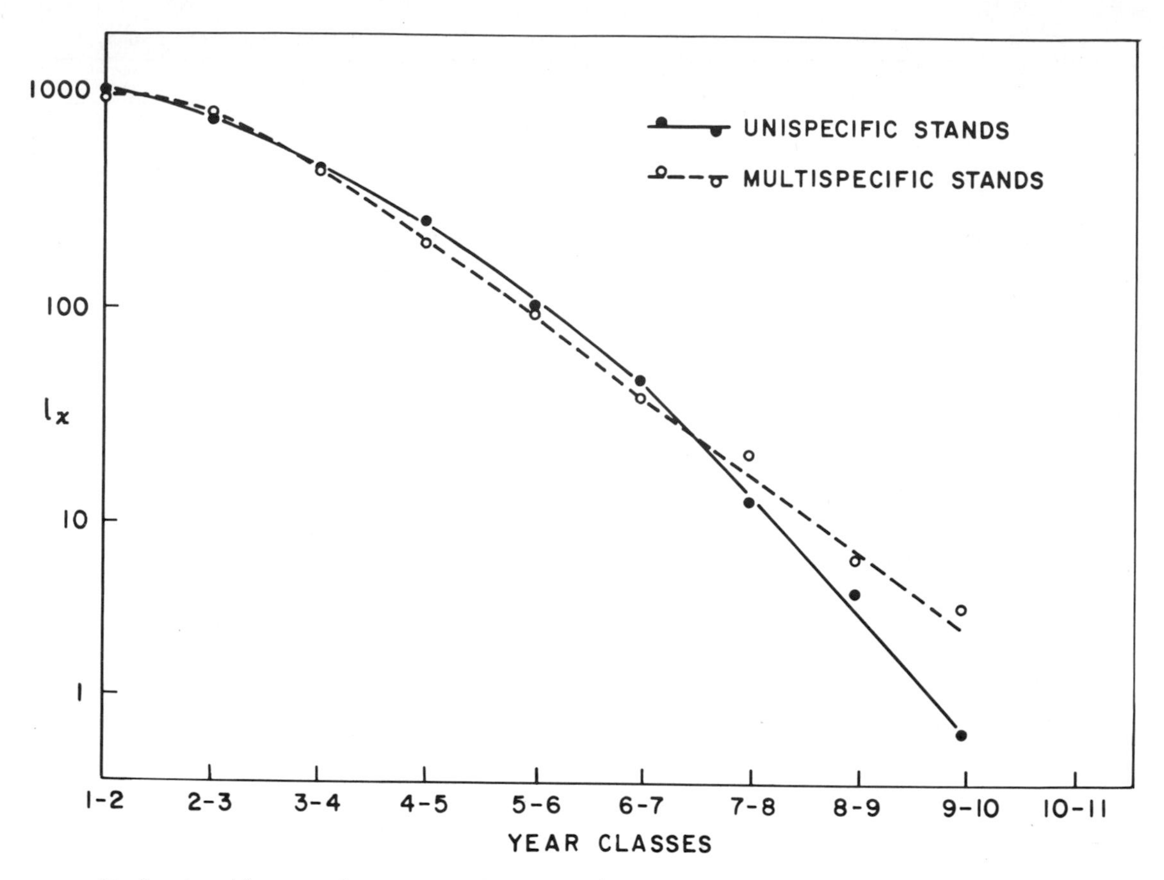

FIGURE 66. Survivorship curves for mosses on Mount Washington for all 6 species of Polytrichaceae growing in unispecific stands (solid line) or in multispecific, intrafamilial, association (broken line) (Watson).

apparent, without any further analysis, which of the *k* values may be regarded as key factors, contributing significantly to the variation in the total mortality. Thus in the entomological example in figure 67, the variations from year to year are primarily due to winter disappearance, damped a little by pupal predation; parasites and disease are unimportant.

In the example from *Parus major* (figure 68) variation in postjuvenile is far more important than in juvenile mortality. It must, however, be born in mind that the key factors are primarily concerned with the variance of the population and not with the regulatory adjustment of the mean size about which it oscillates. In the example of the great tit, the mean population size is certainly primarily controlled by density-dependent reproductive success. Similarly, in the populations of buffalo studied by Sinclair,[65] though the irregular variation in population size is largely due to juvenile mortality, the mean population is likely to be determined by the much more density-dependent adult mortality.

Some general ideas about survival and mortality

If we look over all the cases that have been discussed, which constitute a very moderate but reasonably representative sample of what is known, we get an impression of a fundamental pattern rather different from the classical rectangular and oblique types that are usually invoked.

If we omit organisms with a resting stage of some sort intercalated in the middle of the life cycle, it might be reasonable to suggest that all survivorship curves are based on variations in the slope of the three components, corresponding to the juvenile, adult, and senile segments of the life-span. The

65. A. R. E. Sinclair, see note 34.

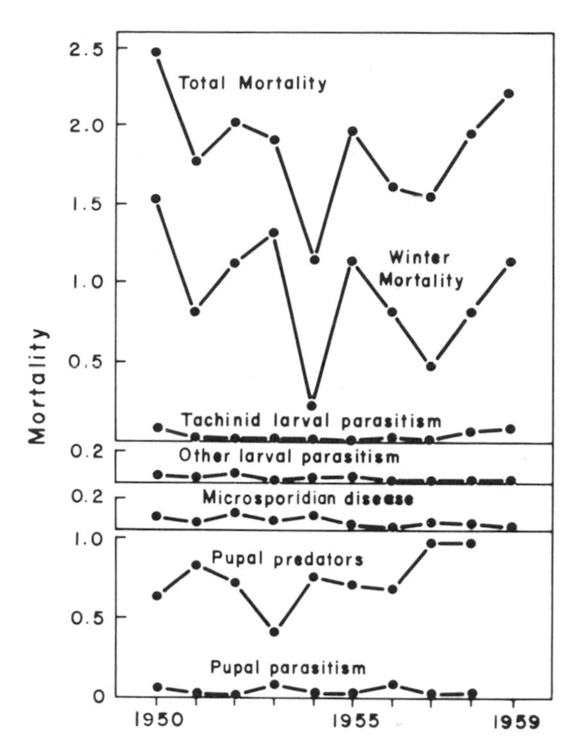

FIGURE 67. Key factor analysis of mortality of the moth, *Operophtera brumata*, showing that the various sources of mortality that can be identified quite specifically are not those most significant in determining the variation from year to year.

mortality rates prevalent in the adult portion will typically be lower than in the other two. When the mortality rate of the adult segment is very low, we get curves like those of modern man, the African buffalo, or the grass *Bouteloua chondrosioides*. When the mortality in the middle segment is fairly high, there are very few individuals left to exhibit senile mortality and for the greater part of the life-span a constant adult specific mortality is observed. When juvenile mortality is much greater than either adult or senile mortality, as in the mackerel, many parasites and most benthic marine invertebrates and forest trees, we have the extreme concave type of curve.

It is evident from the studies that have been stimulated by Fisher's concept of reproductive value, to be discussed in the next chapter,[66] that the form of the survivorship curve is under the control of natural selection. There is always an antithesis, corresponding essentially to that between *r*- and *K*-selection, between producing numerous young as early as possible before much adult death has occurred, and producing and rearing young efficiently but more slowly while the reproducing population is dying. The way that this antithesis is most effectively handled by a species of any given set of biological properties will determine its pattern of survivorship.

The main variation on a tripartite scheme of characteristic juvenile, adult, and senile mortalities is due to the intercalation of resting stages or other major metamorphic interruptions in the life history.

Istock[67] has pointed out, in an elaborate theoretical study, that in any organism with a complex life history, in which resource utilization occurs in two quite different ways in the two stages of a life history, the demographic characteristics of the two stages are controlled by largely independent events. In such a case a maximum equilibrium popu-

66. R. A. Fisher, *The Genetical Theory of Natural Selection*. Oxford, Clarendon Press, 1930, xiv, 272 pp. See also chapter 3, note 1.

67. C. A. Istock, The evolution of complex life cycle phenomena: An ecological perspective. *Evolution*, 21:592–604, 1967.

The evolutionary processes of development and abolition of metamorphosis may have had some unexpected results. W. H. R. Rivers, in *Instinct and the Unconscious* (Cambridge University Press, 1920, viii, 252 pp.), and again in Psychological dissociation as a biological process (*Scientia*, 35:331–38, 1924), suggested that the mechanism of what would now be called repression was originally evolved to provide a means of inhibiting responses appropriate in water as those appropriate on land were developed in the course of the ontogeny of an ancestral amphibian.

Shortly after this, without giving any indication that he knew of Rivers, S. S. Flower published a note, Loss of memory accompanying metamorphosis (*Proc. Zool. Soc. Lond.*, 1927 (1): 155–56), pointing out that *Salamandra salamandra*, trained to feed from the hand of the investigator, lost tameness at metamorphosis and had to be trained *de novo*. See also G. E. Hutchinson, Psychological dissociation as a biological process. *Nature*, 120:695, 1927.

In amphibian metamorphosis, any such mechanism is presumably activated by the thyroid, but might lose its endocrine control in subsequent evolution. As far as I am aware, no one has followed this possibly fruitful lead.

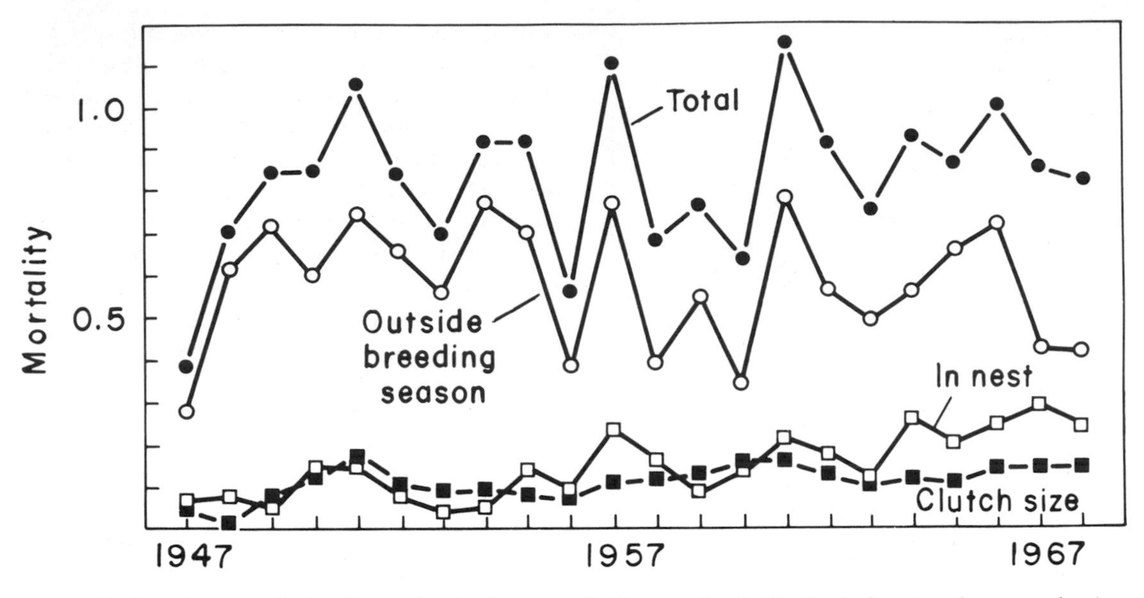

FIGURE 68. Key factor analysis of mortality in the great tit, *Parus major*, in England, showing that reproductive mortality is not the most significant fraction of the total in determining mortality from year to year.

lation will be one in which there are enough larvae to saturate the adult habitat on metamorphosis and enough adults with a sufficient birthrate to saturate the larval habitat with newly produced larvae. Since the properties of the two habitats will not necessarily vary concurrently and any undersaturation in one will be favorable to a competitor, the population of such a metamorphosing species is inherently less stable than if all resources were gathered throughout life in one habitat.

Though holometabolous insects are extremely successful, as are those hemimetabolous species that are initially aquatic and later terrestrial or aerial, there is a persistent tendency to suppress feeding in the imagines, notably in many stoneflies, caddisflies, moths, and two-winged flies, and of course most spectacularly in the mayflies. A less striking trend to abolish the larval stage can be seen in the viviparous tsetse flies, producing larvae ready to pupate, and in some other Diptera. Istock further points out that the whole history of the coelenterates toward either medusae or polyps from some ancestor that presumably showed elaborate metagenesis may be an example of the same sort of thing. The amphibia provide further examples, and at least one line went on to become amniote.

The most striking case of an interruption in the life cycle is birth itself, which nearly always seems to be followed by a period of increased mortality. Sometimes, as in the salamander, *Ambystoma tigrinum* (figure 69),[68] the process is multiple, both laying of eggs and hatching being followed by high mortalities, separated by a rather lower value. In organisms such as the mackerel, most tapeworms, many marine invertebrates, and most forest trees, in which the wastage of eggs or seeds is enormous, each egg or seed is comparatively inexpensive, and the advantages of wide dispersal, which may be the only way of insuring that a few zygotes find potential environments for development, evidently are sufficient to justify the fantastic reproductive losses that such species ordinarily endure. If we turn to the other end of the spectrum, we find in nearly all birds and all mammals, in a number of fish, many insects, some other invertebrates, and a few plants such as mangroves, elaborate means are taken to reduce juvenile mortality at the cost of producing relatively few young. There is nothing surprising in these two methods,

68. J. D. Anderson, D. D. Hassinger, and G. H. Dalrymple, Natural mortality of eggs and larvae of *Ambystoma t. tigrinum*. *Ecology*, 52:1107–12, 1971.

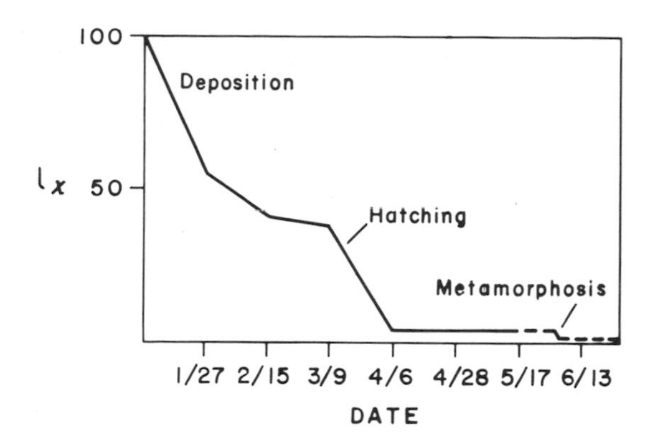

FIGURE 69. Survivorship curve of the salamander, *Ambystoma tigrinum*, showing the mortality of eggs after laying and of larvae after hatching (Anderson, Hassinger, and Dalrymple).

respectively prodigal and frugal, of insuring an adequate replacement of parents by young. It is, however, rather curious that in spite of elaborate evolutionary effort, to say nothing of public health schemes in man, infant mortality seems inevitably to exist. Hamilton[69] has even supposed that under some circumstances, a high juvenile mortality is the expression of an adaptation; the phenomenon, however, seems too widespread to be fully explained in this way. Some such mortality must indeed be inevitable, due to a genetic load of recessive genes which on meeting produce nonviable young. It is reasonable to suppose that such lethal genes are most likely to be expressed when the organism has used up its yolk or has had to give up complete physical dependence on its mother. It seems, however, hardly likely that this is the whole story. We may be faced with the possibility that the evolutionary expense of providing really adequate infantile care is just too great. This of course does not mean that we cannot and should not try to improve what natural selection has been able to do rather imperfectly. Scientifically it is quite likely that there is a real problem of considerable interest here.

SUMMARY

A *life table* is a table which indicates how many of a *cohort* of say 1,000 individuals, starting at time zero, are alive at later times. If time is measured in years, the number surviving to year x is usually termed $\mathbf{l}_x$. The *specific death rate* $\mathbf{q}_x$ over any time interval x to $x + 1$ is the number of individuals dying during the interval (i.e., $\mathbf{l}_x - \mathbf{l}_{x+1}$) divided by $\mathbf{l}_x$. If we plot $\mathbf{l}_x$ against x, we obtain a *survivorship curve*. If the scale, ordinarily the ordinate, of $\mathbf{l}_x$ is graduated logarithmically, x being arithmetic, the slope of the survivorship curve gives instantaneous values of $\mathbf{q}_x$. If $\mathbf{q}_x$ is constant, then the survivorship curve on semilogarithmic paper is a diagonal line. Two other simple-minded situations may be born in mind. If no members of the cohort die until they are senile and they then all die of old age, the survivorship curve will be rectangular, following the top and right-hand edges of the paper. This is what public health aims at, but never achieves completely. Such a survivorship curve is *negatively skewed rectangular*, or, more simply, in its less perfect manifestations, *convex*. If we imagine the opposite situation in which almost all mortality is infantile, we have a *positively skewed rectangular* or *concave* curve. The simplest concave curve is given by $N_t = N_0 t^{-m}$. This gives a straight line plotted on double-logarithmic paper.

The life table obtained by following a cohort is said to be *age-specific*, all individuals surviving having the same age. If in a population that is stationary in numbers and age structure, a count is made of all individuals of each age class, such a count gives an equivalent *time-specific* life table.

Studies of human life tables and survivorship curves show great variations in time, largely due to changes in public health procedures. A time-specific curve is therefore not equivalent to an age-specific one.

Where organisms in a sample of a population can be aged by periodic structures or by study of growth in size, time-specific tables and survivorship curves can be made, though it is seldom possible to tell how valid an estimate of the age-specific table or curve they give.

Age-specific curves can be obtained where

69. W. D. Hamilton, The moulding of senescence by natural selection. *J. Theor. Biol.*, 12:12–45, 1966.

enough animals found naturally dead can be aged, either by inherent structures or because the individuals, such as banded birds, were marked at birth. In general, some confusion between time- and age-specific procedures has probably not produced illusionary results in studying nature.

Laboratory studies of survivorship and mortality have been numerous. Under extremely artificial conditions such as a collection of unfed flies of identical age, convex rectangular curves can be obtained. Since the risk of death is usually less in a laboratory experiment than in nature if adequate cultural techniques exist, convex curves are obtained in most cases. They are primarily of interest when environmental or genetic differences exist between different populations. High temperatures regularly decrease life expectancy, within the limits tolerated by the organism, in cold-blooded or poikilothermal organisms. The density of a population often influences the longevity of its members. High densities usually produce high mortality rates, but there is often an optimum density, at which the longevity is greatest and below which the mortality rate again increases. Experiments, notably with the water flea *Daphnia*, indicate that the existence of an optimum density for longevity does not imply that this density is optimal for growth in length or for reproduction, the various demographic parameters being to some extent independent.

Considerable genetic differences in the longevity of various laboratory strains of *Drosophila melanogaster* and of various wild populations of other species of that genus are known.

Among wild populations, birds have been particularly well studied. After an initial high, but declining, juvenile mortality lasting for a few months in small passerine birds to at least 4 years and probably more in some seabirds, specific mortality becomes essentially constant, so that the chance of, say, a 2-year-old thrush dying within a year are about the same as those of a 5-year-old. The average length of life of such birds is but a small fraction, of the order of 10 percent, of the known maximum life. Practically no birds survive to give evidences of senility.

A very different pattern is shown by various ungulates and some other mammals including man. There is a relatively short period of fairly high infantile mortality, then a period of very low constant specific mortality in adult life, and finally a rather rapid decline of the population as it becomes senile and the specific mortality mounts. This pattern has been most critically demonstrated in the Dall sheep in Alaska, and in the African buffalo in the Serengeti Park. There is evidence of a comparable pattern in some fossil ungulates. A number of mammalian survivorship curves are intermediate between this and the diagonal type; the latter is likely to be found in small mammals and is certainly exhibited by bats. Sexual differences in mortality are known in mammals, but neither sex is consistently the long-lived one.

A very interesting and seemingly paradoxical situation can arise when two populations in different environments are compared, for the richer and more productive environment may be inhabited by a larger population, but one composed of animals with a lower expectation of life, mortality being highly density-dependent.

The lower vertebrates are characterized by moderate to very high infantile mortalities. In an unfavorable year, 99.9994 percent of the eggs of a mackerel are fated to die or to hatch to larvae that disappear in 70 days; none may survive to adult life. Populations of such fish usually consist of a limited number of year classes, produced only when conditions were peculiarly favorable; in many years no effective reproduction takes place.

Among invertebrates, a good deal of work has been done on insects in which a variety of convex and oblique curves are recorded. Whenever complicated life histories occur, specific mortality is likely to vary greatly from stage to stage. Some work has been done in aquatic invertebrates; the clearly density-dependent effects on attached animals like limpets and barnacles may be noted. In the rotifers, in at least one sessile tube-living species, longevity appears much greater when a larva settles on a preexisting tube than on the plants to which such tubes are fixed.

In plants, the same actuarial approach as has been adapted to the study of animals can be very successful, though less work has been done so far. Forest trees tend to have a huge juvenile mortality, but the specific mortality decreases with time rather regularly.

Herbaceous plants show a great variety of survivorship curves, even within a single genus. There is much evidence of both species-specific and environmental effects. The only case well studied in mosses, however, shows very little variation among six species of Polytrichaceae.

When we have life tables for a single species over a series of years, large variations between them may often be observed. If information about the incidence of mortality with time is available, simple graphical methods permit the isolation of key factors on which the great variations depend. The mean size of the population around which variation occurs is usually set by the less variable demographic parameters, which are density-dependent.

Life tables and survivorship curves may be partly understood by thinking of the life-span as composed of three segments. In the first, or *juvenile segment*, mortality is high though declining; in the second, or *adult segment*, mortality is lower and fairly constant; in the third, or *senile segment*, mortality rises. If the adult mortality is moderate, it may be great enough to prevent any individual reaching senility, as in most birds; if it is very low, a striking convex pattern is produced, as in the Dall sheep or man. The differences to be observed in survivorship are reasonably attributed to various solutions of the problem raised by the antithesis between an organism reproducing early before much mortality has occurred or later when experience of living has been acquired but by a population that is already significantly declining. The antithesis is of course presented primarily by animals with well-developed nervous systems.

In cases with complex life histories, periods of high or low mortality may be intercalated into the life history. Life histories in which quite different resources are used at different stages inevitably introduce some instability. Though such life histories are very common, there is a strong suggestion that the resulting instability has been important in the evolution of insects, coelenterates, and amphibians. Possibly any sudden metamorphosis is followed by a temporary increase in mortality. The first change, that of birth itself, seems inevitably followed by a higher death rate than later, even when there is evidence of great parental care. Part of this mortality may be adaptive and part be due to a genetic load of unfavorable recessives, which become apparent when parental dependence ceases. It is possible that in evolutionary, if not in moral terms, complete abolition of infant mortality is impossibly expensive.

Chapter Three

Why Do They Have So Many Children?

IF a population is not to decline and finally become extinct, its mean birthrate over a period of years must obviously be at least as great as its mean death rate. Ordinarily, as we have seen in chapter 1, ideal populations, at least of equilibrium species, tend to oscillate about an equilibrium population K, which represents the number of individuals that can be supported by the environment in the area studied. In a population following the Verhulst logistic growth equation, the rate of increase per individual

$$\frac{1}{N}\frac{dN}{dt} = r\frac{K - N}{K}$$

will decline steadily throughout time as the population grows. Since r may be thought of as a birthrate minus a death rate (**b** − **d**), the decline could be due either to a decrease in **b** or to an increase in **d** as N increases. Such changes are said to be *density-dependent*, while death caused by a drought or a storm, which affects individuals equally whether they belong to large or small populations, is said to be *density-independent*. We have already noted various cases of death rate and a few of birthrate (figure 36) being density-dependent. In this chapter a more detailed analysis of the ways in which **b** may vary as a population grows will be made.

It is necessary before proceeding to such an analysis to point out one aspect of natality that is obvious, but has sometimes been forgotten. In the cases of populations not consisting of a single clone, variation in fecundity will often be observable, and in such cases as have been investigated part of this variation is of genetic origin. This means that unless the offspring produced by more fecund parents have a lower life expectancy than those from less fecund parents, the population will tend to consist of a rather greater proportion of individuals descended from the more fecund genotypes than were present in any previous generation. This will be true until an optimal fecundity is reached, whether the population is increasing, decreasing, or stable. Since there will always be selection in favor of the more fecund, or more accurately, the optimally fecund genotypes, which leave the greatest number of descendants, regulation of a population is more likely to occur by way of density-dependent death rates than by density-dependent birthrates. Density-dependent regulation of birthrates does of course occur, but usually it is simply due either to large numbers competing for limited food inevitably producing fewer eggs per individual than would be formed if more food were available, or to the rather elaborate mechanisms which ensure optimal fecundity in species in which production of a supra-optimal clutch or brood does not end in the survival of more individuals than if the brood had been smaller. Marked limitation of overall fecundity probably occurs in all territorial species when the populations are large and some specimens fail to take suitable territories. Whether a limitation is ever developed strictly as an adaptive mechanism against depletion of the environment has been hotly debated. Neither the inevitable loss of fecundity through semistarvation nor the adaptive development of an optimal brood implies any modification of natural selection continually working in favor of increased viable reproductive output. The development of adaptations against environmental depletion does at least *prima facie* imply such a modification if the entire population is considered and involves what has been called *group selection*. The controversy as to whether anything that might be compared to family limitation in man has evolved in other animals will be considered in a later section, after the importance of territoriality has been discussed.

From a human point of view regulation by density-dependent death rates removing the offspring in excess of an equilibrium number seems wasteful and immoral. Medical science has developed largely as a means of eliminating this immoral waste, but without any compensatory regulation such elimination can cause disaster. In the most developed countries a new regulatory mechanism involving foresight and contraception is replacing density-dependent death, but in most societies this replacement is not being achieved fast enough, if at all.

Prodigal and prudent reproduction

Two extreme situations linked by every possible intermediate clearly exist. In pelagic fishes and many benthic marine invertebrates, enormous numbers of eggs or larvae are discharged into the water by every reproducing female, and in many parasites vast numbers of minute eggs pass out with the feces of the host and are easily dispersed by water or wind so that they contaminate the food of potential hosts or otherwise enter their tissues. In striking contrast to these *prodigal* species, we have those in which survival is assured, not by dispersing enormous numbers of eggs so that a few may be expected to develop in suitable environments, but by careful nurturing of a few eggs or newborn individuals, so that each offspring has a probability of survival to an adult of perhaps from 1 in 100 to over 1 in 10, rather than 1 in 1,000,000 to 1 in 100,000, as might be true of a pelagic fish or a tapeworm. Such reproduction may be termed *prudent*, though the two terms used are really only picturesque metaphors, the prodigal species being just as likely to succeed in leaving descendants as is the prudent. Usually, the extremely prodigal animals will exhibit very concave survivorship curves, while the extremely prudent species, such as contemporary Western man, who can still increase even with an average completed family of three, will have very convex curves.

In general, organisms having prodigal reproduction will be under the influence of r-selection, those having prudent reproduction will be more subject to K-selection.

Fisher's concept of reproductive value

In any population the probability of an individual of age x producing offspring over the short element of age dx is $\mathbf{l}_x \mathbf{b}_x\, dx$ and the probability over the whole life giving the total expectation of offspring is

$$\int_0^\infty \mathbf{l}_x \mathbf{b}_x\, dx.$$

If the age and specific fecundity distributions remain unchanged, the rate of increase r is also constant, so that

$$e^{rx} = \int_0^\infty \mathbf{l}_x \mathbf{b}_x\, dx.$$

or

$$\int e^{-rx} \mathbf{l}_x \mathbf{b}_x\, dx = 1. \qquad (3.1)$$

It is then possible to define the reproductive value of an individual of age x as

$$\mathbf{v}_x = \frac{e^{rx}}{\mathbf{l}_x} \int_x^\infty e^{-rt} \mathbf{l}_t \mathbf{b}_t\, dt. \qquad (3.2)$$

This concept of R. A. Fisher (figure 70) has been greatly developed by a number of

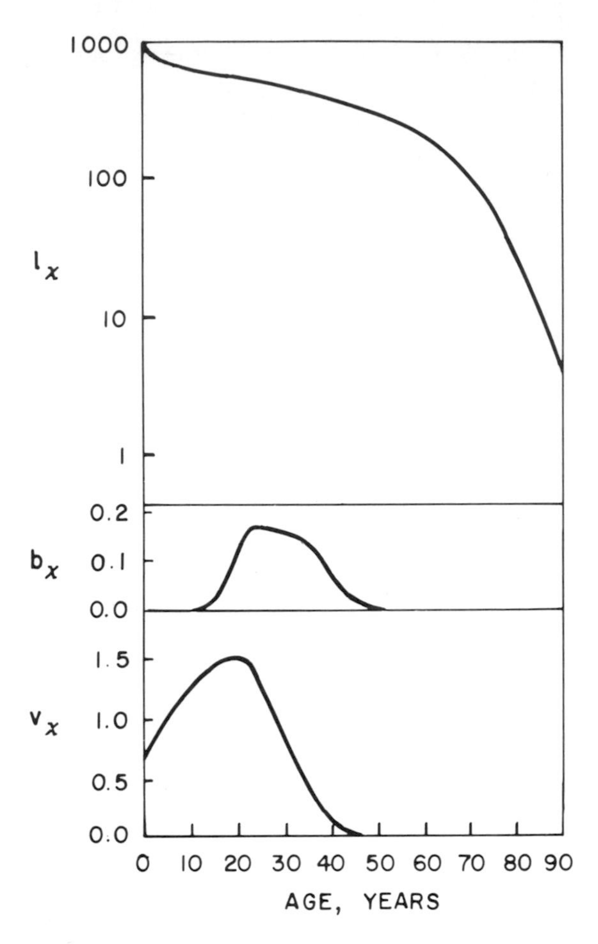

FIGURE 70. Survivorship ($\mathbf{l}_x$), natality ($\mathbf{b}_x$), and reproductive value ($\mathbf{v}_x$), plotted against age in years, for the human female population of Taiwan, about 1906 (Hamilton, modified).

investigators, notably G. C. Williams and W. D. Hamilton.[1]

There is an infinite number of ways of satisfying equation 3.2, depending on the forms of the survivorship and natality distributions $\mathbf{l}_t$ and $\mathbf{b}_t$, which will at least in part depend on the biological possibilities of the species under consideration. There is a choice between the initial prodigal production of young, which may lead to maximum fecundity, followed by the parents, or at least the female parent, becoming literally effete, in a state of senescence, or the prudent spacing of the young over a long period of time. This may mean that each offspring, if it is born, has a very good start in life, but since the number of parents is inevitably declining, a number of potential offspring never is born.

It is apparent from the work of Williams and of Hamilton that there must be a tendency for any improvement in $\mathbf{l}_t$ or $\mathbf{b}_t$ early in life to be reflected in a decline in values later. If either function increases early, the reproductive rate increases and the population becomes more crowded, so that early senility and death are the price payed for early successful reproduction.

In general, a tendency for selection to act more effectively during the period of high reproductive value will thus mold the life history; and, as Slobodkin puts it, one kind of mortality will appear to attract another because both will be most strongly selected against at the same time of life. Slobodkin points out that this makes any predator at least somewhat prudent, for the stages with lowest reproductive value will tend to develop mechanisms against being eaten less rapidly than will those with high values of $\mathbf{v}_x$. The evolutionary advantage of the latter

1. R. A. Fisher, *The Genetical Theory of Natural Selection*. Oxford, Clarendon Press, 1930, xiv, 272 pp.

The classic developments are in G. C. Williams, Pleiotropy, natural selection and the evolution of senescence. *Evolution*, 11:398–411, 1957.

W. D. Hamilton, The moulding of senescence by natural selection, *J. Theort. Biol.*, 12:23–45, 1966.

More modern developments may be found in B. Charlesworth and J. A. León, The relation of reproductive effort to age. *Amer. Natural.*, 110:449–59, 1976, and in papers referred to therein.

Slobodkin's prudent predator is discussed in L. B. Slobodkin, How to be a predator. *Amer. Zool.*, 8: 43–51, 1968.

L. B. Slobodkin, Prudent predation does not require group selection. *Amer. Natural.*, 108:665–78, 1974. However, see also the following which suggests the theory is incomplete.

V. C. Maiorana, Reproductive value, prudent predators and group selection. *Amer. Natural.*, 110: 486–89, 1976.

D. B. Mertz and M. J. Wade, The prudent prey and the prudent predator. *Amer. Natural.*, 110:489–96, 1976.

For sibling replacement in marsupials and its evolutionary significance, see P. Parker, An evolutionary comparison of marsupial and placental patterns of reproduction. In: *The Biology of the Marsupials*, ed. R. Stonehouse and D. P. Gilmore. London, Macmillan, 1977, pp. 273–286.

stages may in fact in some cases imply a disadvantage that must be tolerated in the reproductively less valuable age classes.

Usually there will be no selection acting to maintain, and often selection will indirectly eliminate, postreproductive stages. An important exception will be in those cases in which postreproductive individuals care for young and so can increase the fitness of their descendants. Hamilton, however, points out that occasionally, as in some aphids, a postreproductive period, fully as long relative to the whole life-span as that in man, may occur in animals where such a period can hardly be functional. Further investigation of this type of phenomenon may be interesting.

Hamilton has also considered possible cases of adaptive juvenile mortality. He examines the case of a pauciparous species in which, when one offspring dies, it is easily replaced, though reproduction would have been delayed if it had lived. Such *sibling replacement* is certainly characteristic of human reproductive behavior. If we imagine an infectious disease, the outcome of which is either immunization or death, it will be advantageous to get the disease over with as early as possible, while young can be replaced, for though the juvenile mortality will be increased, so will the ultimate production of viable young mature individuals. It is possible that part of the infantile mortality of man is due to Hamilton's process, though the detailed nature of such a selective mechanism is not obvious. The available data on juvenile mortality in the field seem to suggest, moreover, that the phenomenon is too widespread in animals for it to be usually adaptive.

It may be noted that in marsupials, there is an elaborate mechanism by which sibling replacement can occur very easily. A female kangaroo may have a joey at heel, one in the pouch, and a diapausing blastocyst in the uterus waiting to mature rapidly as soon as the pouch is vacant. The free young, however, receive far less maternal care than would equivalent young placentals, and the whole process is much more easily terminated than in the latter. The greater investment that the placental mother clearly makes probably leads, as Parker has recently pointed out, to mother-offspring bond that is far more conducive to the evolution of intelligence than is likely to occur in the marsupials. If this proves to be the case, it would provide an extraordinary example of a virtually unpredictable result arising from an initially small modification of reproductive behavior.

Semelparous and iteroparous species

In zoology species that produce a single clutch of eggs or brood of young are called *semelparous*, those that have a long life producing several broods or clutches are termed *iteroparous*. The terms, which very roughly correspond to *annual* and *perennial* among flowering plants, are not quite precise enough to cover all possible situations. The simplest contrast is between a species which has an annual cycle, passing the cold or dry season in diapause of some kind, and producing a single brood, and a species which, living a number of years, produces repeated broods, often at a specific time of year. This contrast is exemplified by the comparison of most insects with long-lived terrestrial vertebrates in temperate climates. There are certainly intermediate conditions between these formally well-defined possibilities. Most small passerine birds, though they can live many years, seldom do so, as we have already noted, but it is usual even for these functional annuals to breed more than once in a single season. The same situation certainly may be expected in many small mammals.

In some tropical iteroparous species, including man, breeding can occur at any time of year, producing a rather different kind of iteroparity from that usual in temperate regions with marked seasonal changes in temperature reflected in the season of birth.

Cole[2] pointed out, in a classic paper

2. L. C. Cole, The population consequences of life history phenomena. *Quart. Rev. Biol.*, 29:103–37, 1954.

G. Bell, On breeding more than once. *Amer. Natural.*, 110:57–77, 1975. This paper gives references to earlier works based on Cole's ideas.

S. C. Stearns, Life-history tactics: A review of the ideas (*Quart. Rev. Biol.*, 51:3–65, 1976) is an excellent review, which did not become available in time for use in this book.

published in 1954, that if the specific mortality rate did not change from the moment of birth up to the end of reproduction, the reproductive effort of an iteroparous species is equivalent to producing one extra member of the brood of a semelparous species, for all an iteroparous species does is for each breeding member to carry over to the next year along with its brood of offspring. This statement applies in its boldest form to a parthenogenetic animal, but can be suitably modified for an ordinary bisexual species. The conclusion struck Cole, and doubtless everyone else encountering it for the first time, as distinctly odd, even though it is obviously a correct deduction. The oddness is due to the fact that the initial assumptions about survivorship do not correspond to a probable biological situation. We may consider a mackerel in a bad year producing about 400,000 eggs to which the surviving iteroparous female is to be added. If all the eggs or the larvae that are produced perish, so the effective brood size is zero, the adults alone carry over to the next year but obviously permit the survival of the species.

In general, semelparity will be favored when the extra cost to the parents of staying alive between broods, as for instance over the winter, is greater than the cost of producing enough additional members of the brood to offset differences between juvenile and adult survival during that winter. Such young might pass the unfavorable season in the egg in diapause with a quite high probability of surviving till the next season favorable for growth.

Whenever survival of young is insured by processes that depend on adult experience, so breeding is delayed while the young adults are still being educated, as in seabirds, ungulates, or man, interoparity will obviously be favored.

Bell points out that since there is always a chance of death prior to reproduction, the sooner the latter occurs the more likely the reproducing animal is to leave descendants. Producing eggs or young as discrete broods may therefore be less advantageous than producing eggs one at a time, but beginning earlier than would be the case if a full clutch is to be produced. This type of iteroparity is characteristic of internal parasites which produce a continuous stream of eggs.

Density-dependent natality

We have already noted cases in laboratory populations where density-dependent natality has been observed. This may have various causes. In *Daphnia* it is quite likely due simply to the additional food being shared by an increasing number of females as the population grows. In *Drosophila* the disturbance of the females by other members of the population was believed by Pearl[3] to be a major factor in reducing the birthrate in crowded cultures. In laboratory mammals such as the domesticated mouse, the effects of crowding are probably more recondite.

The simplest kind of density-dependent effect on natality is certainly that due to many individuals eating more than do a few. Since all eggs or newborn young are produced by the ingestion, assimilation, and resynthesis of this food, each female will have less material to work with when the population is dense. This is probably an everyday occurrence throughout the animal kingdom, but the extent to which it is apparent is very variable.

In a case such as that of Kluijver's study (figure 71) of the great tit (*Parus m. major*) in Holland,[4] in which the clutch size and frequency of second broods clearly depend on density, it is difficult not to avoid the suspicion that food is partly involved, particularly as it is known that British populations (*P. major newtoni*) of this species may increase as food becomes more abundant. Kluijver himself, however, believed that frequent meetings, resulting in quarrels, may have a directly unfavorable influence on fecundity.

Wynne-Edwards[5] has collected from the literature a number of cases in which varia-

3. R. Pearl, The influence of density of population upon egg production in *Drosophila melanogaster*. *J. Exp. Zool.*, 63:57–84, 1932.

4. H. N. Kluijver, The population ecology of the great tit *Parus m. major* L. *Ardea*, 39:1–135, 1935.

5. V. C. Wynne-Edwards, *Animal Dispersion in Relation to Social Behaviour*. Edinburgh, Oliver and Boyd; New York, Hafner, 1962, xi, 653 pp.

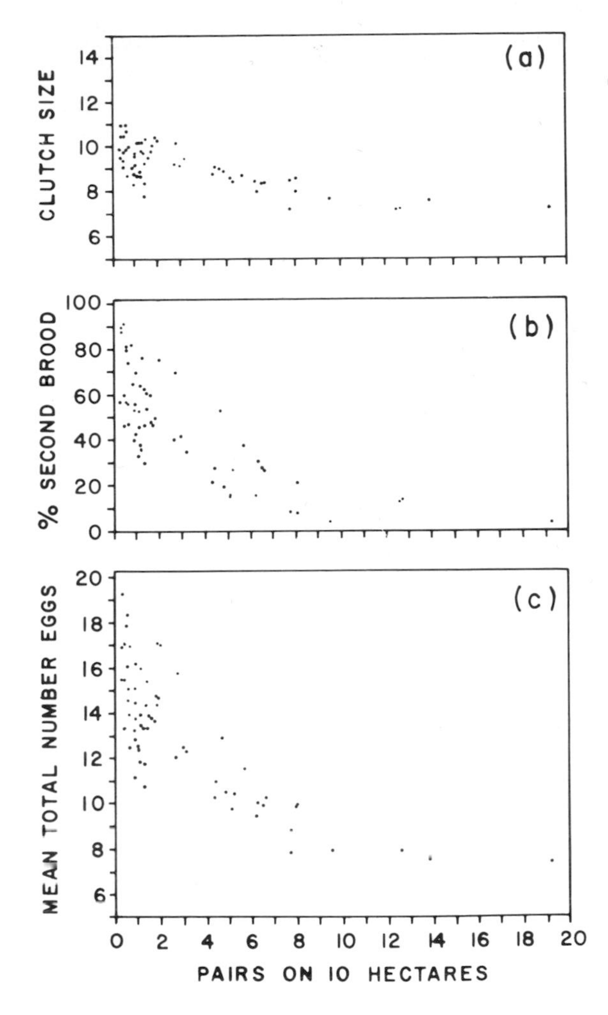

FIGURE 71. Relation of density to (a) clutch size in first broad; (b) percentage of pairs having a second brood; (c) total fecundity per pair, in the European great tit, *Parus m. major*, in Holland (Kluijver).

tions in food supply have a considerable effect on reproduction. Lions are said not to breed when game is scarce, and buzzards (*Buteo buteo*) in Britain may produce no offspring in seasons when rabbits (*Oryctolagus cuniculus*) are uncommon. Sheep farmers in Scotland have traditionally increased the nutrition of ewes about three weeks before mating in the not forlorn hope of producing twins or triplets. Snowy owls are known to lay more eggs in years of lemming maxima than in the intervening minima.

Darwin[6] wrote, "a large number of eggs is of some importance to those species that depend upon a fluctuating amount of food, for it allows them rapidly to increase in numbers." Nearly a century later MacArthur[7] pointed out that the opportunistic species of American warblers (Parulidae), which depend largely on spruce-budworm outbreaks for food, can indeed lay more eggs than their less opportunistic allies. *Vermivora peregrina* the Tennessee, *Dendroica tigrina* the Cape May, and *D. castanea* the bay-breasted warbler, all may lay up to 6 or 7 eggs, though no other species of the family have a clutch of more than 5 eggs, and that very exceptionally. In the case of the bay-breasted warbler, there is clear evidence (figure 72) of the dependence of large clutches on the incidence of spruce-budworm outbreaks. In the case of the less opportunistic species, the variation is very much smaller.

In mammals the case of the African buffalo, *Syncerus caffer*, studied by Sinclair,[8] is most interesting because here there seems to be no effect of density on natality, though adult mortality is clearly density-dependent and apparently controlled by the available supply of food. Moreover, in man, though fecundity is doubtless depressed by malnutrition, and juvenile mortality greatly increased, the relationship is much less neat than that which limits reproduction in *Daphnia* as food supply declines. One may suspect that at least in some mammals, there has been strong selection in favor of maintaining fecundity in the face of partial starvation.

Apart from territoriality, discussed in detail later, two more complicated situations are developed in some animals. In the first, the production of eggs or young in clutches or broods which receive parental care has resulted in the brood size being adjusted

6. C. R. Darwin, *On the Origin of Species by Means of Natural Selection; or The Preservation of Favoured Races in the Struggle for Life*. London, J. Murray, 1859, ix, 502 pp.; see chap. 3.

7. R. H. MacArthur, Population ecology of some warblers of northeastern coniferous forests. *Ecology*, 39:599–619, 1958.

8. A. R. E. Sinclair, The natural regulation of buffalo populations in East Africa: III. Population trends and mortality. *East Afr. Wildl. J.*, 12:185–200, 1974.

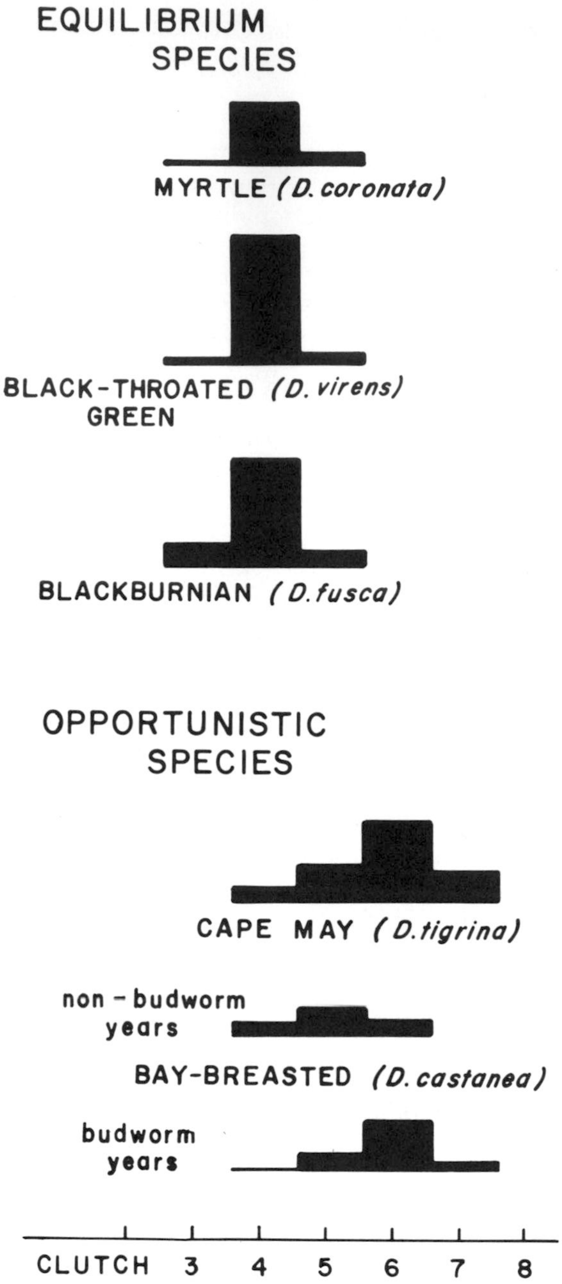

EQUILIBRIUM SPECIES

MYRTLE (D. coronata)

BLACK-THROATED GREEN (D. virens)

BLACKBURNIAN (D. fusca)

OPPORTUNISTIC SPECIES

CAPE MAY (D. tigrina)

non-budworm years

BAY-BREASTED (D. castanea)

budworm years

CLUTCH 3 4 5 6 7 8

FIGURE 72. Distribution of clutch sizes in five species of American warblers of the genus *Dendroica*. Above, three strictly determined equilibrium species almost always laying 4 eggs; below two opportunistic species, of which at least the bay-breasted warbler has a clutch depending on the food supply (data from MacArthur).

either environmentally or genetically to the resources available to the parent, so that there are not so many mouths to feed that all suffer. In the second, the inevitable decline in egg number with declining resources is somewhat mitigated by the eggs at times of low food supply and consequent low fecundity being not merely reduced in numbers, but increased in size. Both types of adaptation exhibit various complicating variations. The first pattern of adaptation is found in birds, the second in certain freshwater animals, notably copepods, and in a rather different way in fishes.

Clutch size in birds

Birds lay their eggs, in practically all cases, in cavities or nests that they have constructed in an enormous variety of ways and, except in the megapodes, brood the eggs until they hatch. On hatching the young may be quite helpless, remaining in the nest until they have developed more or less perfectly a plumage of feathers, during which time they are fed by one or both parents. Such young and the birds that produce them are said to be *nidicolous* or *altricidal*. In contrast to this, the young may be downy, able to leave the nest, here ordinarily on the ground, and in many cases to feed themselves. Such birds are *nidifugous* or *precocial*.

In general, nidifugous birds lay larger eggs than nidicolous ones; this is reasonable as the young are hatched at what is essentially a later stage in their life history.[9]

The number of eggs in a clutch is very variable throughout the class of birds, though there are regularities within the lesser taxa.

The four families (albatrosses, shearwaters, storm petrels, and diving petrels) of Procellariiformes, all oceanic birds that feed far offshore, lay clutches of but 1 egg, and this is also true of tropic birds, frigate birds, and some boobies, penguins, and auks. In the families of land birds in which clutches of 2 eggs are usual, a moderate minority lay only 1.

9. All the general information on the nests, eggs, and rearing of young in birds in derived from:

D. Lack, *The Natural Regulation of Animal Numbers*. Oxford, Clarendon Press, 1954, viii, 343 pp. (reprinted 1967).

D. Lack, *Population Studies of birds*. Oxford, Clarendon Press, 1966, v, 341 pp.

D. Lack, *Ecological Adaptations for Breeding in Birds*. London, Methuen, 1968, xii, 409 pp.

Occasionally no nest of any sort is constructed. The

Among such birds are a few quite small passerines such as the lemon-bellied crombec (*Sylvietta denti*), a wrenlike African warbler, and at the other end of the size range of birds, the condors (Cathartidae).

Many birds lay clutches of 2 eggs. Lack suspects that this may well be the modal number since most tropical passerine species belong here, as do all humming birds and most pigeons and nightjars.

At the other end of the scale, the blue tit (*Parus caeruleus obscurus*), one of the smallest Palaearctic birds, has been known in England to lay a clutch of 19; from 1 of 18 eggs, all hatching, every nestling was fledged. Of the nidifugous nonpasserine species the European partridge (*Perdix perdix*) may lay up to 17. Certain cases of much greater numbers are probably due to two females laying in the same nest.

Apart from obvious specific differences in clutch size, which presumably imply genetic determination, Perrins and Jones,[10] through the study of a banded population of great tits (*Parus major newtoni*) in England, showed that part of the variation in number is due to genetic factors for which the population is polymorphic, though environmental factors operating differently in different years also play a part in the determination of clutch size in this and in many other species.

It is probable that in most birds the clutch size is *determinate*, or more or less fixed by environmental and genetic factors independent of the sensory stimuli received by the bird from the tactile and visual information provided by a full clutch. In some birds, however, such stimuli are clearly needed to inhibit egg production. If an egg is removed, another is laid to replace it. In such *indeterminate* species, astonishing numbers of such replacements have sometimes been recorded.[11]

The cases in which birds have rapidly replaced eggs that have been removed indicate that in some species more eggs can be laid than normally constitute a clutch. In the parasitic cuckoos[12] in which eggs are laid for series of days followed by a sterile period, each run of eggs presumably corresponds to a clutch, but is larger than the clutch of nonparasitic species.

The initial work suggesting that the size of the clutch is related to the number of nestlings that can be reared, at least in nidicolous birds, was done by Moreau in Central Africa.[13] A little later Lack started his intensive investigations,[14] and a number of other workers have made possible the extension of the work. All these studies show that though the presence of more young to feed stimulates

fairy tern (*Gygis alba*) lays its egg on the flatter parts of a horizontal branch of a tree, murres or guillemots lay on flat rock ledges, and the two penguins of the genus *Aptenodytes*, the king penguin (*A. patagonicus*) and the emperor (*A. forsteri*), incubate their eggs on their feet, covered by a fold of abdominal skin (see also G. G. Simpson, *Penguins Past and Present, Here and There*. New Haven and London, Yale University Press, 1976, xi, 150 pp.).

The only birds that do not incubate their eggs are the Megapodiiade or brush turkeys of Australia northward and westward to the Philippine and Nicobar Islands; their eggs are laid in mounds and are kept warm by heat from the sun, decaying vegetation, or volcanic sources. The young hatch from very large eggs at a later stage than do other birds and can fly the day after hatching.

10. C. M. Perrins and P. J. Jones, The inheritance of clutch size in the great tit (*Parus major* L.). *Condor*, 76:225–29, 1974.

11. The greatest number recorded appear to be 71 in the yellow-shafted flicker (C. L. Phillips, Egg-laying extraordinary in *Colaptes auratus*. *Auk*, 4:346, 1887). Though the majority of birds are probably more or less determinate in egg number, the subject is not well explored. D. E. Davis, Determinate laying in barn swallows and black-billed magpies. (*Condor*, 57: 81–87, 1955) gives a summary. Early observations are given in:

E. C. Raven, *John Ray, Naturalist: His Life and Works*. Cambridge University Press, 1942, xix, 502 pp.; see pp. 478–80.

12. R. B. Payne, The evolution of clutch size and reproductive rates in parasitic cuckoos. *Evolution*, 28:169–81, 1974.

13. R. E. Moreau, Clutch-size: A comparative study with special reference to African birds. *Ibis*, 86:286–347, 1944.

R. E. Moreau, Relation between number in brood, feeding-rate and nestling period in nine species of birds in Tanganyika Territory. *J. Anim. Ecol.*, 16: 205–09, 1947.

14. D. Lack, Clutch and brood size in the robin. *Brit. Birds*, 39:98–109, 130–35; 1946.

D. Lack, The significance of clutch size: I, II. *Ibis*, 89:302–52, 1947.

D. Lack, The significance of clutch size: III. *Ibis*, 90:25–45, 1948.

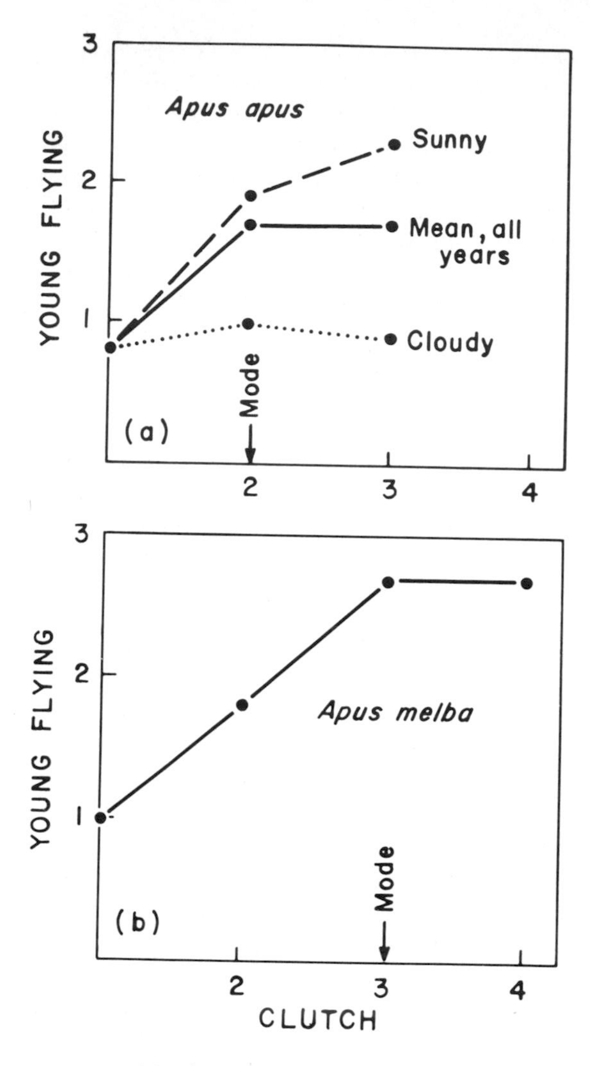

FIGURE 73. Number of young leaving nest as a function of clutch size in swifts; (a) *Apus apus* in Oxford, England, the success with more than one egg clearly dependent on the weather; (b) *A. melba* in Switzerland (data of Lack; Lack and Arn).

the parents into making more frequent visits to the nest, the increase is not proportional to the increment in brood size. This means that the average nestling in a large brood receives less food than that in a small brood. The consequences of this, however, are more complicated than might be supposed.

In the simplest cases, typified by the two European species of swift, the common swift (*Apus apus*) and the alpine swift (*A. melba*), there is clear evidence (figure 73) that the mortality rate of nestlings rises with the size of the brood, so that a point is soon reached (3 eggs for *A. apus*, 4 for *A. melba*) at which the addition of an extra egg makes no difference to the size of the brood that fledges.[15] Moreover, in *A. apus* this point is clearly dependent on the weather during the breeding season; in 1946–48 when there was less sunshine than usual at Oxford, 2-egg clutches produced a mean fledging brood of 1, and 3-egg clutches of 0.9, while in the succeeding four years which had average or more than average sunshine, 2-egg clutches produced 1.9 fledging young, 3-egg clutches, 2.3 young. The polymorphism that must exist in the population is presumably dependent on frequent changes in the optimal number of eggs.

More usually it appears that the nestling mortality is independent of the size of the brood. Fledglings from the larger clutches, which may be lighter[16] and less strong, may well have a higher mortality in their first few months of independent life. Data bearing on this are extremely hard to collect, involving ringing enormous numbers of chicks and finding a sufficient number that have died after leaving the nest to indicate juvenile survival or death rates. The most practical method is to tabulate all birds found to have died after 3 months of age. The number of such birds should depend on survival during the first 3 months. The available data for the European starling, some of which is plotted in figure 74, indicate that in spite of lack of differential mortality in the nest, birds from the larger clutches survive less well in their first 3 months of independent life than do those from clutches of not more than 5 eggs.[17] There is, moreover, a little direct evidence to indicate increased mortality from broods of

15. D. Lack and H. Arn, Die Bedeutung der Gelegegrösse beim Alpensegler. *Ornith. Beob.*, 44:188–210, 1947.

D. and E. Lack, The breeding biology of the swift *Apus apus*. *Ibis*, 93:501–46, 1951.

D. Lack, *Swifts in a Tower*. London, Methuen, 1956, 239 pp.

D. Lack, Further notes on the breeding biology of the swift *Apus apus*. *Ibis*, 98:606–19, 1956.

16. J. Gibb, Feeding rates of great tits. *Brit Birds*, 48:49–58, 1955.

17. D. Lack, Natural selection and family size in the starling. *Evolution*, 2:95–110, 1948.

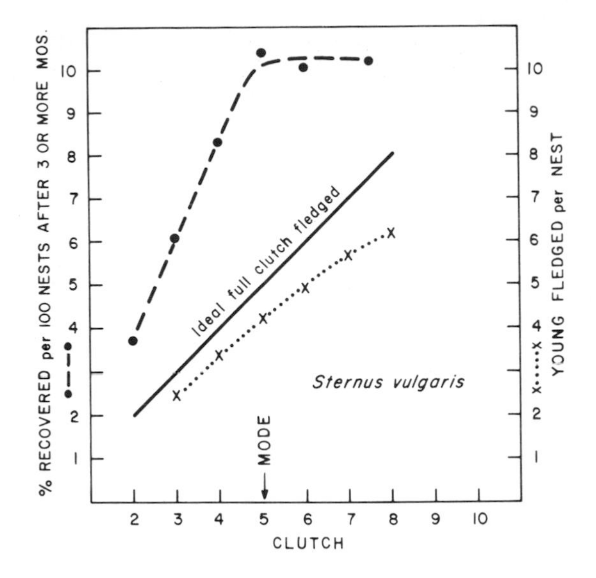

FIGURE 74. Relation of fledging success (dotted line) and of survival when at least 3 months old (broken line) in the starling (*Sturnus vulgaris*) as a function of clutch size. Note that although the chicks fledged always are rather fewer than the ideal number (diagonal line) determined by the number of eggs laid, increasing the size of the clutch up to 8 always increases the number of surviving young per brood at fledging. However, above the modal number of 5 eggs, such increase is not reflected in the number of juveniles more than 3 months old. Fledging data from Holland, survival times from Switzerland, but comparable data exist for other areas (data of Lack).

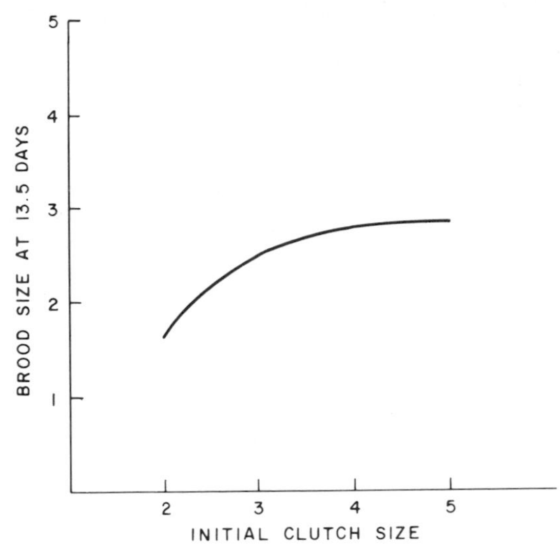

FIGURE 75. Survival rate of chicks or number fledged per brood in the house sparrow (*Passer domesticus*) at Oxford, England; clutches of 1 and 6 omitted (data of Seel).

6 to 8 young compared with those from smaller broods. It is probable that the pattern observed in the starling is the common one in passerine birds in temperate regions.

In the house sparrow,[18] a markedly colonial species, Seel has found that the number of visits by the parents increases as the brood is increased from 1 to 3, but not further. The modal number of eggs in a clutch is 4. The survival of chicks is rather low, and is shown in figure 75, omitting the 1- and 6-egg clutches, which are clearly too infrequent to give reliable results. The survival rate is greatest per chick for 3-egg clutches, as would be expected, but the surviving brood is slightly greater for 4-egg clutches. Adding an extra egg leads to no further increase in the brood produced. Seel suspects that in this case what sets the upper limit is the conspicuousness of frequently repeated visits, which might suggest to other sparrows in the colony that the birds were being very successful and so indicate good, but only partly exploited, sources of food.

A remarkable complication may occur when the energetics of small birds breeding in cool temperate climates is considered.[19] Heat loss may be a very important item in the energy budgets of the nestlings of such a bird as *Parus major*. When we have a large brood huddled together, the heat loss per chick will be considerably less than the heat loss of a single chick, so compensating for the smaller supply of food that can be put into each mouth. Where the parents continue feeding the young after they have left the nest, it is possible that the advantage to the individual of being a member of a small brood comes into play. There is indeed evidence of lower survival in the first 3 months of birds from larger clutches (10–13) than smaller ones in *P. m. major* in Holland.

18. D. C. Seel, Clutch-size, incubation, and hatching success in the house sparrow and tree sparrow *Passer*. spp. at Oxford. *Ibis*, 110:270–82, 1968.

D. C. Seel, Nestling survival and weight in *Passer*. spp. *Ibis*, 112:1–14, 1970.

19. T. Royama, Factors governing feeding rate, food requirement, and brood size of nestling great tits *Parus major*. *Ibis*, 108:31–47, 1966.

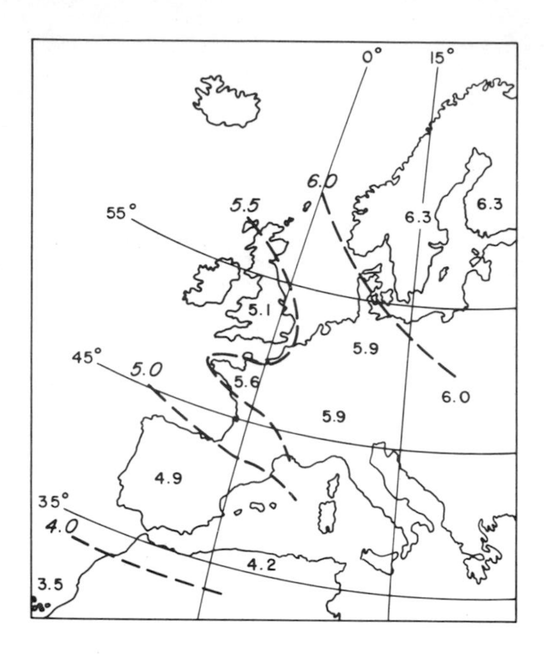

FIGURE 76. Distribution of clutch sizes of the European robin (*Erithacus rubecula*); data from Lack used to construct rough contours showing increase going from south to north and west to east.

Marked geographical variation in clutch size has long been known in such birds as do not lay a rigidly constant number of eggs, as do pigeons or Procellariiformes.[20] In Europe the usual pattern is an increase in number of eggs as one goes northward and a decrease as one goes westward. In North America the same trends are found, but the westward decrease is not observed until the climatically more oceanic area bordering the Pacific is reached. In general, clutch size increases as one goes poleward from the equator (figure 76) and decreases as one passes from more continental to more oceanic climates; there also is a clear tendency for insular populations to have a smaller clutch size than those of the adjacent mainland. These tendencies are exhibited by both nidicolous and nidifugous birds. There has been much discussion as to the meaning of the observed trends.

Lack supposed that the increase in clutch size in going poleward depended on an increase in daylight hours, permitting a longer working day for the parents. He also believed that an increased possibility of adult feeding permits the nidifugous species, notably ducks, to lay more when living in high than when in low latitudes.

The crossbill (*Loxia curvirostra*) and the raven (*Corvus corax*), which breed before the spring equinox in northern Europe, show a reverse trend to other birds, their clutch size increasing on going south. However, this nutritional theory cannot account for the whole story, for a number of tropical species nesting before the vernal equinox nevertheless shows an increasing clutch size on going north from the equator. Skutch[21] suggested that the low number of eggs laid at one time in the tropics is primarily due to predator pressure. The smaller the number of conspicuous visits by parents and the faster the young become independent, the smaller is the chance of their being destroyed by snakes, mammals, and other predators. There is no reason why this interpretation should exclude the trophic one favored by Lack.

Cody[22] has developed a comprehensive general theory of clutch size, presented in a geometrical form, though a verbal qualitative argument can give its main features. Cody accepts both Lack's and Skutch's hypotheses as partial explanations. He postulates, however, that in addition to food supply and predation, the effect of competition must be

20. See D. Lack, The significance of clutch size (*Ibis*, 89:302–52, 1947; and 90:25–45, 1948) for a review of the older data. Dr. G. Svärdson once told me that an investigator studying the quantitative aspects of the feeding of nestlings by passerine birds north of the Arctic Circle was defeated by the fact that the minimum sleep he needed was greater than that of the birds under a midnight sun.

21. A. F. Skutch, Do tropical birds rear as many young as they can nourish? *Ibis*, 91:430–55, 1949.

22. M. Cody, A general theory of clutch size. *Evolution*, 20:174–84, 1966. Cody is doubtless generally correct that the more stable an environment is for a species, the greater is the incidence of selection for K in populations of that species, and the more inter- and intraspecific competition will occupy the time and energy of the species with a consequent reduction in clutch size. The application of this idea to the New Zealand islands is not immediately self-evident; presumably great intraspecific competition is involved here.

considered, for in the presence of many competitors the realized niche (see chap. 5) of the species under consideration is reduced, and consequently some food will not be available that would have been if fewer competitors were present. It is reasonable to suppose that in stable environments with a diversified flora and fauna, more competitors are likely to be present than in more variable environments. We therefore usually will have, in passing from the tropics poleward, an increase in daylight hours available for foraging in late spring and summer, a decrease in the risk of predation, and a decrease in the intensity of competition, all three of which will promote selection for larger clutches as the distance from the equator increases. If we go from oceanic to continental climates, only stability is likely to decrease, predation and foraging time being constant; a small increase in clutch size is to be expected. Temperate islands will probably differ from the mainland of the same latitude in a greater meteorological constancy; predation will be low on both island and mainland, and clutch size should be lower on account of the increase in stability. This is confirmed by the islands off New Zealand, the mean clutch size being 4.5 in the birds of the mainland, while the mean clutch size for the vicariant endemic insular subspecies is 3.2. In tropical islands the possible increase in stability may well be offset by a lack of predators. In the Caribbean, seven species studied by Cody have a mean clutch size of 2.2 on both small islands and the mainland.

Certain other explanations may have to be entertained for special cases. In nidifugous birds, though the parents may not feed the young, they may protect them by distraction displays while they are moving about feeding. The demand made on such activities will depend on the amount of movement that is needed to find food. Safriel[23] has found that in the semipalmated sandpiper, *Calidris pusilla*, at Point Barrow, Alaska, ordinary clutches of 4 eggs produce an average of 1.74 fledged birds, but if the clutch is artificially raised to 5, the number of birds fledged falls to 1.0. There is no significant difference in growth rates, so that both subpopulations of young birds are getting adequate food. The reduction in average success in fledging is primarily due to more whole broods being predated when there were 5 rather than 4 eggs in the clutch. The main predators were jaegers (*Stercorarius*). It is not easy to exclude Lack's hypothesis that clutch size in nidifugous birds, which generally have large eggs, is regulated by the food available to the female, but at least in the case of the semipalmated sandpiper Safriel's explanation is adequate.

On account of the discreteness of production in clutches, their visibility, and other characteristics that make them easy to study, birds' eggs have provided more analyzable data about the ecology of natality in nature than have been obtained from any other group of organisms. The summary just given is a bare one; many cases, of interest primarily to the specialist, are omitted: these details may be found in the writings listed in note 9 of this chapter.[24]

There is a little information about mam-

23. U. N. Safriel, On the significance of clutch size in nidifugous birds. *Ecology*, 56:703–08, 1975.

24. A difficulty may arise from the fact that birds can lay only integral numbers of eggs, though the amount of food that can be brought to the nest during a given time could be enough to feed a fractional number of chicks.

R. E. Ricklefs (On the limitation of brood size in passerine birds by the ability of adults to nourish their young. *Proc. Nat. Acad. Sci. U.S.A.*, 61:847–51, 1968) thinks that in such a case the bird will lay the integral number next greater to the fractional number that could be fed, and that the slightly longer nestling period compensating for this would not be deleterious.

S. D. Fretwell, D. E. Bowers, and H. A. Hespenheide, Growth rates of young passerines and the flexibility of clutch size. *Ecology*, 55:907–09, 1974.

A quantitative theoretical treatment of the optimum relationship between number and size of young is given by C. C. Smith and S. D. Fretwell, The optimal balance between size and number of offspring. *Amer. Natural.*, 108:499–506, 1974. The mathematically minded reader will find this paper profitable.

The very curious cases in penguins of variable egg size are discussed by D. Lack, *Ecological Adaptations for Breeding in Birds* (see note 9), pp. 252–56.

The biological problems raised by clutch size were doubtless apparent in a vague way to preevolutionary naturalists, as is clear from the following quotation from James Grahame's *Birds of Scotland with other Poems* (Edinburgh, 1806, pp. 42–43), in his discussion of the wren, *Troglodytes troglodytes:*

mals.[25] Lord found in North American rodents and insectivores that nonfossorial species which did not hibernate had larger litters at higher latitudes, but that this was not true of pocket gophers and ground squirrels. In the Carnivora, the mustelids and felids show no such variation; the evidence for foxes is inadequate. Cody concludes that the mammalian effect is largely due to intensity of predation, which would have little or no effect on weasels and cats.

Egg number and egg size in char and whitefish

In a study of the alpine char (*Salmo alpinus*) in the lakes of the Faxälven drainage in Sweden, Määr[26] found that the uppermost and lowermost lakes containing the species supported sparse populations of relatively large fish, producing both absolutely and relative to the size of the ovigerous female more eggs than did the smaller fish composing the large populations of the more central lakes. At least in Leipikvattnet, the uppermost lake of the series, the eggs were much smaller (mean weight 40.3 mg) than in the other lakes; in Ströms Vattudal, the lowest locality for the species in the valley, the eggs were slightly smaller (mean weight 57.2 mg) than in any other population (mean weights 58.0–66.4 mg) studied. Svärdson produced a theoretical treatment, strongly influenced by Lack, to account for these observations. His conclusions were that where there was little intraspecific competition, natural selection would favor the individuals that produced most young; while where there is much intraspecific competition, the most efficient young would be favored, even though they were more expensive to produce.

A very interesting, but quite comparable case in the lake whitefish, *Coregonus clupeaformis*, has recently been described from two lakes in Alberta, Canada, namely, Pigeon Lake and Buck Lake.[27] In Pigeon Lake reduction of predators and competitors in the 18 years prior to the study has allowed the development of a large population of whitefish, but intraspecific competition, as is usual in such circumstances, has depressed the growth rate of these fish. No such reduction of predators or competitors has taken place in Buck Lake.

In both localities spawning nearly always begins when a fish has completed its 4 year; this means maturity at about 250 mm in Pigeon, and about 410 mm in Buck Lake. The egg number increases relative to length much faster in the Buck Lake than in the Pigeon Lake population (figure 77). The diameter of the Buck Lake eggs in 1977 was 1.77 mm, while that of the Pigeon Lake eggs was 1.87 mm; these differences imply that the egg volume is about 12 percent greater in Pigeon than in Buck Lake.

The size of young on hatching was apparently not related to egg size, but depended on the length of incubation. This, however, may merely mean that at hatching the young from the larger eggs retained more yolk in the yolk sac, which in these fish has not been fully resorbed at hatching, and which is noticeably larger in fish that hatch early.

Egg size and number in freshwater calanoid copepods

The small planktonic crustacea of the sub-

The enormous disproportion that subsists
Between the mother and the numerous brood,
Which her small bulk must quicken into life.
Fifteen white spherules, small as moorland hare-bell,
And prettily bespecked like fox-glove flower,
Complete her number. Twice five days she sits,
Fed by her partner, never flitting off,
Save when the morning sun is high, to drink
A dewdrop from the nearest flowret cup.
But now behold the greatest of this train
Of miracles, stupendously minute;
The numerous progeny, clamant for food,
Supplied by two small bills, and feeble wings
Of narrow range; supplied, aye, duly fed,
Fed in the dark, and yet not one forgot!

25. R. D. Lord, Litter size and latitude in North American mammals. *Amer. Midl. Natural.*, 64:488–99, 1960. See also D. Lack, The significance of litter size. *J. Anim. Ecol.*, 17:45–50, 1948.

26. A. Määr, Fertility of char (*Salmo alpinus* L.) in the Faxälven water system. *Inst. Freshw. Res. Drottningholm Rep.*, 29:57–70, 1949.

G. Svärdson, Natural selection and egg number in fish. *Inst. Freshw. Res. Drottningholm Rep.*, 29:115–22, 1949.

27. B. F. Bidgood, Reproductive potentials of two lake whitefish (*Coregonus clupeaformis*) populations. *J. Fisher. Res. Board Can.*, 31:1631–39, 1974.

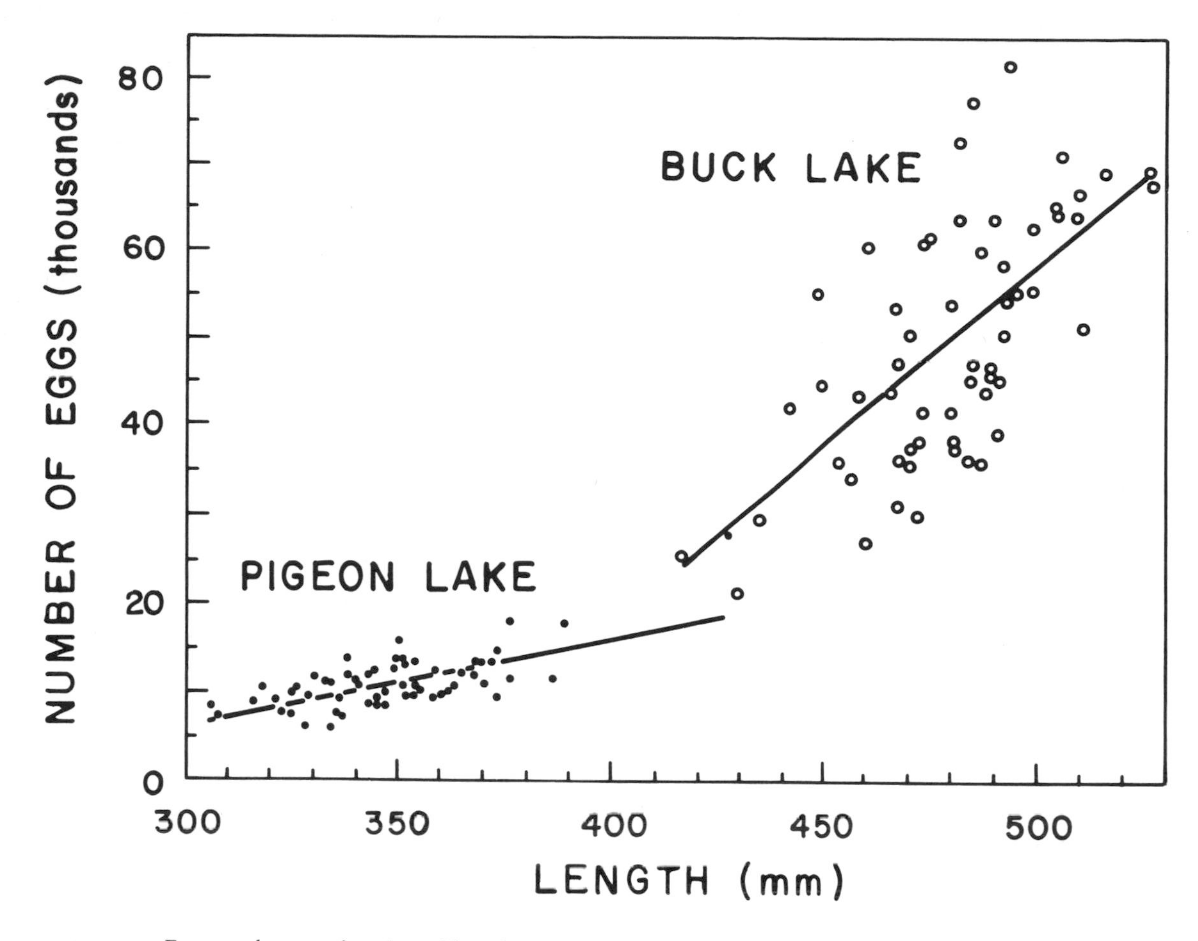

FIGURE 77. Egg number as a function of length in two populations of *Coregonus clupeaformis* in Alberta, Canada. The eggs in Pigeon Lake, where intraspecific competition is more severe, have a volume about 12 percent greater than those in Buck Lake (Bidgood, very slightly modified).

class Copepoda are very common in fresh waters. The females usually carry their eggs for a time in egg sacs, one in the Calanoida and Harpacticoida, two in the Cyclopoida. More than 70 years ago Wesenberg-Lund[28] pointed out that in several species living in lakes in Denmark, the rare specimens found with eggs in summer carried fewer eggs than would be found in spring, but each egg was larger. In *Eudiaptomus graciloides*, the only species for which he gave measurements, the summer eggs appeared to have had a volume at least twice as great as the spring egg. No further study of the matter was made for more than half a century when Czeczuga[29] published an account of egg

28. C. Wesenberg-Lund, *Plankton Investigations of the Danish Lakes*. Special part. Copenhagen, Gyldendalske Boghandel, 1904, 223 pp.

Wesenberg-Lund was perhaps the greatest of all students of the natural history of freshwater invertebrates. He retained an extremely conservative attitude to biological theory, in one case figuring two male specimens of *Daphnia* attempting to mate with the same female as an indication of the irrelevance of Mendelian genetics to what happens in nature. His observational work, however, was superb and of perennial value, even though he would probably have greatly disapproved of some of the uses to which his observations have been put by later, more theoretically minded biologists, including the present writer.

Wesenberg-Lund's observations were apparently forgotten until 1951, when I, strongly influenced by Lack and Svärdson, discussed them in a summary of what was then known about egg number in copepods. See G. E. Hutchinson, Copepodology for the ornithologist. *Ecology*, 32:571–77, 1951.

29. B. Czeczuga, Oviposition in *Eudiaptomus gracilis* G. O. sars and *E. graciloides* Lilljeborg (*Diaptomidae, Crustacea*) in relation to season and trophic level of

Lake	Transparency m	Mean number of eggs per ♀			Mean volume of each egg $mm^3\ 10^{-3}$			Mean volume of eggs per ♀		
		Spring	Summer	Summer/Spring	Spring	Summer	Summer/Spring	Spring	Summer	Summer/Spring
					Eudiaptomus gracilis					
Białe	4.5	15.5	6.4	0.41	0.66	0.83	1.26	10.2	5.3	0.52
Rajgrodskie	3.5	17.3	7.5	0.43	0.63	0.87	1.38	10.9	6.5	0.60
Dręstwo	3.0	19.8	9.7	0.49	0.68	0.90	1.32	13.1	8.8	0.67
					Eudiaptomus graciloides					
Białe	4.5	6.4	4.5	0.70	0.83	0.97	1.18	5.3	4.4	0.83
Rajgrodskie	3.5	9.5	6.8	0.72	0.75	0.90	1.19	7.1	6.1	0.86
Dręstwo	3.0	11.0	6.0	0.55	0.65	1.05	1.61	7.2	6.3	0.88

TABLE 2. Size and number of eggs in spring and summer borne by two species of the copepod (*Eudiaptomus*) in three lakes in Poland. The transparency is measured as the depth at which a white disc (Secchi disc) becomes invisible. It provides a rough inverse measure of productivity, being greatest when the plankton crop in the water is least.

number and size in *Eudiaptomus gracilis* and *E. graciloides* in Poland; some of his results are given in table 2.

The two species behave rather differently, the reduction in egg number and the increase in egg volume being much greater in *E. gracilis* than in *E. graciloides*. There must be marked differences in the biology of the species, which would be expected in any two species living together (see chap. 4), but the nature of these differences is not clear. The increase in egg volume in *E. gracilis* compensates much less completely for the large reduction in egg number than in *E. graciloides*, where the reduction is less great; in the latter species, a summer female may carry 83 to 88 percent of the egg load of a spring female.

The general pattern and the differences between the two species are maintained, even though the three lakes studied, as judged by their transparency, differ considerably in productivity. The adaptive significance of the variation becomes apparent when it is remembered that the food supply of the immature stages of the copepods is much greater in spring than in summer. In the spring the waters of the lake have fully circulated, are rich in plant nutrients, and support a dense growth of minute algae. The larvae pro-

lakes. *Bull. Acad. Pol. Sci.*, Cl. II, Sér. Sci. biol., 7:227–30, 1959.

B. Czeczuga, Zmiany plodności niektórych przedstawicieli zooplanktonu: I. *Crustacea* Jezior Rajgradzkich. *Polsk. Archwm. Hydrobiol.*, 7:61–89, 1960.

Czeczuga also gives data for the autumn; in both species the number of eggs carried is further reduced, though their volumes tend to be intermediate between those of spring and summer eggs.

The evidence for the differences in environment and fate of spring and summer copepod larvae is mostly in:

H. J. Elster, Über die Populationsdynamik von *Eudiaptomus gracilis* Sars und *Heterocope borealis* Fischer in Bodensee-Obersee. *Arch. Hydrobiol. Suppl.*, 20:546–614, 1954.

R. Eichhorn, Zur Populationsdynamik der calanoiden Copepoden in Titisee und Feldsee. *Arch. Hydrobiol. Suppl.*, 24:186–246, 1959.

In general, the copepods hatching from the eggs produced in spring by the small overwintering population show a much greater survival than do those produced from the large, but relatively infertile, summer population. No measurements of egg sizes were, however, made in these studies.

The phenomenon studied by Czeczuga is probably widespread; comparable variation in egg size occurs in the North American *Leptodiaptomus siciloides;* see G. Comita, The energy budget of *Diaptomus siciloides* Lilljeborg. *Verh. Int. Verein. theor. angew. Limnol.*, 15:646–53, 1964. It is, however, clearly not universal in freshwater calanoids, as Czeczuga did not observe it in *Eurytemora lacustris*.

duced from the spring eggs enter a world full of food; and the greater the number of larvae, the greater is the chance of a female leaving descendants. In the summer the warmer surface waters do not mix with the deeper water, which remains cool. Feces, dead plant cells, and animals fall into the cool deep water, decompose, and liberate their nutrient materials, which have thus been removed from the surface layer and are inaccessible to the current algal crop in the illuminated upper water. Consequently the crop is diminished. In the more productive lakes it is, moreover, apt to consist in summer largely of blue-green algae, which are hard to eat. The copepod larvae entering the lake at this season are therefore confronted by a less favorable environment, at a higher temperature at which their metabolic requirements are greater. The better prepared they may be, the greater is the chance of some of them surviving. Though the interspecific differences indicate that there are details of the story that are unanalyzed, its main drift is plain. The general situation is comparable to what happens geographically, rather than temporally, in char and whitefish. Whether there is any significance in both examples being derived from freshwater organisms is uncertain. It may perhaps be easier to make an aquatic egg that can easily be varied in diameter than a terrestrial one, where there must be some impermeable and often relatively elaborate egg shell or membrane, which may be less susceptible to intraspecific variation in size than is the simpler, less impermeable covering of aquatic eggs. In a very remote way the generally larger eggs of nidifugous, compared to those of nidicolous, birds provide an analogy to what has been described in the last two sections.

The cases in both the fishes and the copepods exemplify *r*- and *K*-selection occurring as alternatives in the same species of animal in different circumstances.

Botanical analogies

Studying the buttercup, *Ranunculus flammula*, Johnson and Cook[30] observed that carpel number declined with increasing altitude in the Cascade Mountains. Harper had pointed out that seed size is one of the least variable characters of any species of flowering plant. Both genetic and environmental factors were shown to influence the carpel number. It is reasonable to suppose that during the frost-free season a certain amount of material can be synthesized to make carpels and that fewer of a standard size are made when the period of synthesis is short than are made when it is long. This situation is compared with the regulation of clutch size to correspond to the maximum number of young to which food can be brought.

Solbrig,[31] in the dandelion, *Taraxacum officinale*, compared two strains, one of which, growing primarily in disturbed conditions, produced more but smaller seeds which started to ripen earlier in the season; the other strain, though less fecund, always displaced the more fecund one in competition, while the first strain was a better colonizer, neatly demonstrating the results of *K*- and *r*-selection respectively. The situation is comparable to one involving two species, one competitively successful, the other fugitive and always keeping ahead of its competitor, which may explain some cases of the apparent co-occurrence of two species in the same niche (see chap. 5).

Territoriality and natality

That some animals behave like landowners has been known for a very long time,[32] but

30. M. P. Johnson and S. A. Cook, Clutch size in buttercups. *Amer. Natural.*, 102:405–11, 1968.

J. L. Harper, The nature and consequence of interference amongst plants. *Genetics Today (Proc. XI Int. Congr. Genet.)*, 465–82, 1965.

31. O. T. Solbrig, The population biology of dandelions. *Amer. Scientist*, 59:686–94, 1972. See also W. M. Schaffer and M. D. Gadgil, Selection for optimal life histories in plants. In: *Ecology and Evolution of Communities*, ed. M. L. Cody and J. M. Diamond. Cambridge Mass., Belknap Press of Harvard University Press, 1975, pp. 142–57.

32. The historical information on knowledge of territoriality was first reviewed by M. M. Nice, The role of territoriality in bird life. *Amer. Midl. Natural.*, 26:441–87, 1941. Some further information has be-

come available since this article was written, and the interest of the question has not always been fully appreciated.

In antiquity Aristotle knew of territoriality in eagles and ravens, and there seems to have been a proverb that one bush does not hold two robins. These cases are discussed by D. Lack, *The Life of the Robin*. London, Witherly, 2nd ed., 1946, xvi, 224 pp.

There is a single known medieval reference to the phenomenon, in a poem variously called *Susann*, *Sussanne*, *The Pistil of Susan*, or *The Pistil Suet Susane*, by the south Scottish poet Huchon, of whom virtually nothing is known save that he wrote late in the fourteenth century. The poem has been published in a critical edition:

A. Miskimin, *Susannah: An Alliterative Poem of the Fourteenth Century*. New Haven and London, Yale University Press, 1969, xvii, 255 pp.

Describing the garden in which the events of the story of Susannah and the Elders begin, the poet writes of the birds: "On firres & fygges thei fangen her fees" (l. 86; stanza vii l. 8). Of the expression "fangen her fees" the editor writes that it is "semi-legal terminology, and its technical meaning is 'to take or have possession of . . .'; it is used here . . . as a mildly witty elevation of the birds to lordship over their domain." Mrs. Miskimin now agrees with me that a reference to territorial behavior is likely (see G. E. Hutchinson, Marginalia. *Amer. Scientist*, 58:528–35, 1970).

There is an early seventeenth-century reference in section 21 of *The orders, lawes, and ancient customes of swanns*, printed by order of John Witherings in 1632, which I have not seen but which is discussed by N. F. Ticehurst (Letters, *Brit. Birds*, 27:308, 1934). In this document it is stated that neither "the master of the Game, nor any Gamster may take away any swanne which is in broode with any other mans, or which is coupled, and hath a walke, without the others consent." Ticehurst indicates that this is based on regulations existing at least 50 years earlier. It would be most unlikely that the manifestation of territoriality implied by a swan having a walk would have been unknown in the middle ages.

The first essentially modern statement of the phenomenon, as far as it applies to a single species, namely, the nightingale (*Luscinia megarhyncha*), is from G. P. Olina, *Vccelliera overo discorso della natura e proprieta di diverse vccelli e in particulare di que chi cantano, con il modo di prendergli, conoscergli, allivargli, e mantenergli*. Roma, Appresso Andrea Fei, 1622, 81 pp. All the copies of the first Roman edition in the Beinecke Library differ; a fourth copy has the place of publication as Bracciano. The plates illustrating the preparation of food for captive nightingales and the musical training to which they were subjected are fantastically elaborate. A second edition appeared in 1684 in Rome. A French version (Paris, P. Fr. Didot, 1774) contains no explicit description of territoriality.

In Olina's work (on p. 1 recto) we read of the nightingale "nel suo arrivo ha per proprio il pigliarsi vn luogo, come sua franchigia, nel quale non ammette altri Russignuoli, che la propria femina, e in quello d'ordinario canta." John Ray, *The Ornithology of Francis Willughby* . . . (London, John Martin [12], 441 [6], 80 pl, 1678), translates this on p. 222: "It is proper to this Bird at his first coming (saith *Olina*) to occupy or seize upon one place as a Freehold, into which it will not admit any other *Nightingale* but its Mate." Ray gives no indication of the bird ordinarily singing in its territory.

An admirable history of ornithology has lately become available in translation: Erwin Stresemann, *Ornithology from Aristotle to the Present*, tr. H. J. and C. Epstein, ed. G. W. Cottrell, with a foreword and an epilogue on American ornithology by Ernst Mayr. Cambridge, Mass., Harvard University Press, 1975, xii, 432 pp. Stresemann gives a remarkable account of the researches of Freiherr Johann Ferdinand Adam von Pernau (1660–1731), who induced birds to regard cages, which could be left open, as the centers of their territories and to come back to them after flights in the surrounding country. He published his observations in 1702 in an anonymous book: *Unterricht, Was mit dem lieblichen Geschöpff, denen Vögeln, auch ausser dem Fang, nur durch Ergründung deren Eigenschafften und Zahmmachung oder anderer Abrichtung man sich vor Lust und Zeitvertreib machen können* (not seen, ref. as given in Stresemann). Later editions appeared in 1707 and 1716.

François Leguat, *Voyage et avantures de François Leguat & de ces compagnons en deux isles désertes des Indes Orientales* (Londres, David Mortier, marchand librairie, 1707, tome 1, 164 pp.; tome 2, 180 pp.; and index) gives one of the most extraordinary accounts of territoriality in his description of the natural history of the solitaire, *Pezophaps solitarius*, a now extinct flightless bird that was living on Rodriguez in the seventeenth century. See also:

M. Hachisuka, *The Dodo and Kindred Birds or the Extinct Birds of the Mascarene Islands*. London, Withery, 1953, 250 pp.

E. A. Armstrong, Territory and birds: A concept which originated from study of an extinct species. *Discovery*, 14:223–24, 1953.

G. E. Hutchinson, The dodo and the solitaire (essay review of Hachisuka). *Amer. Scientist*, 42:300–08, 1954.

The ethological aspects of Leguat's account are considered at the end of this book. They seem to have made little impression on later writers until the middle of this century.

There are two English eighteenth-century references to territoriality which are important since they appeared in very well-known works and refer to birds in general rather than to particular species. The first, in a letter from the Rev. Gilbert White to the Hon. Daines Barrington, is dated February 8, 1772. White is writing about winter flocks of birds and speculating as to its cause. He indicates that "the great motives that regulate the proceedings of the brute creation are love and hunger." However, as a cause of flocking, "love . . . is out of the question at a time of year when the soft passion is not indulged; besides during the amorous season, such a jealousy prevails between the male birds, that they can hardly bear to be together in the same hedge or field. Most of the singing and elation of spirits of that time seems to me to be the effect of rivalry and emulation; and it is to this spirit of jealousy that I chiefly attribute the equal dispersion of birds in the spring over the face of the country." Incidentally, he concluded that desire for food is not likely to be involved in flocking except in some special cases of mixed flocks. He wonders whether need for protection against predators may not be involved. These speculations appeared in [Gilbert White], *The natural history and antiquities of Selborne, in the county of Southampton:* with engravings, and an appendix. London, T. Bensley for B. White and Son, 1774, v, 468 [12], pp. In this, the first, edition, the letter to Barrington, numbered xi, appears on pp. 145–47; in later editions the letters are numbered consecutively and not separately for the two correspondents, so that this letter is XLVIII.

The second reference is to Oliver Goldsmith, *An history of the earth and animated nature* (London, J. Nourse, 1774, vol. 5, 400 pp.), who in Part IV, chapter 1 of this volume wrote: "All birds, even those of passage, seem content with a certain district to provide food and center in. The red-breast or the wren seldom leaves the field where it has been brought up, or where its young have been excluded; even though hunted it flies along the hedge, and seems fond of the place with an imprudent perseverance. The fact is, all these small birds mark out a territory to themselves, which they will permit none of their own species to remain in; they guard their dominions with the most watchful resentment; and we seldom find two male tenants in the same hedge together." This passage appears in the American edition (Philadelphia, Mathew Carey, 1795) in vol. III, p. 159, which thus contains the first account of avian territoriality printed in the New World. The passage is omitted from the brief general introduction to the passerine birds in the 2-volume edition, abridged for the use of schools by Mrs. Pilkington (Philadelphia, J. Johnson, 1804), even though its contents are applicable to American birds, while most of the material retained pertains to species that almost all American children would never see.

Dr. Johnson was doubtless unfair when he said to Boswell, "Goldsmith, sir, will give us a very fine book on the subject; but if he can distinguish a cow from a horse, that, I believe, may be the extent of his knowledge of natural history" (John Forster, *The Life of Oliver Goldsmith.* London, Hutchinson and Co., 3rd ed., 1855, reprinted 1905, 460 pp.; see pp. 369–72). A. H. Macpherson, Territory in bird life (*Brit. Birds*, 27:266–67, 1934), indicates that Goldsmith's main source, Buffon's *Histoire Naturelle*, contains no reference to territory. The fact that he makes no mention of song may suggest an influence from Ray. There is a remote possibility that White's letter to Barrington had been circulated in manuscript. I am, however, inclined to think that Goldsmith was reflecting some fairly general knowledge of the phenomenon widely held by people with much experience of the countryside.

Early in the nineteenth century the German ornithologist Johann Andreas Naumann, *Naturgeschichte der Vögel Deutschlands* (vol. I, Leipzig, E. Fleischer, 1820 [not seen]), noted that the male usually arrives first and seeks a breeding place, suffering no other pairs within a certain distance. There is no evidence that Naumann's remark influenced later German writers, though it is possible.

The first observations made on territory in the New World concern hummingbirds. W. Bullock, *Six months' residence and travel in Mexico* . . . (London, J. Murray, 2nd ed. 1825, vol. 2, vii, 264 pp. [1st ed., 1824, not seen]), gives on pp. 3–14 observations on several undesignated species in Jamaica and Mexico, showing marked territorial behavior. He compares this to the behavior of "the robin and other birds of Europe" and uses both "dominions" and "territory" to indicate the defended areas. There is probably some influence of Goldsmith here; Bullock was, however, clearly a keen and competent observer. He was originally a jeweller in Liverpool, later running a private museum in London. Though he has a column in the *Dictionary of National Biography*, neither his date of birth nor the place and date of his death appear to be known. He may have emigrated to the United States and settled in Cincinnati.

There are two modern nonscientific references to territorial concepts which seem to reflect traditional observations. The first is by Roy Campbell, writing of Lorca. "Andalusia is Lorca's *querencia*. The *querencia* is the exact spot which every Spanish fighting bull chooses to return to, between his charges in the arena. It is not marked by anything but the bull's preference for it and may be near the centre, near the barricade, or between the two, as the bull chooses. The nearer the bull is to his *querencia* or stamping-station the more formidable he is, the more full of confidence, and the more difficult to lure abroad into the territory of the bullfighters, for *their* territory is wherever the bull is most vulnerable, and least sure of himself" (Roy Campbell, *Lorca: An appreciation of His Poetry.* New Haven and London, Yale University Press, 1952, p. 8).

The second is William Lyon Phelps's explanation of why he believed that his friend Helen Wills Moody was a greater tennis player than Suzanne Lenglen, though on the only occasion that they met, Lenglen won. Phelps believed that this was because they played in France and that "one should be successful not only at home but abroad and under varying conditions" (W. L. Phelps, *Autobiography with Letters.* Oxford University Press, 1939, p. 897).

The uses of the concept before Altum's work, published in 1868, are of varying degree of explicitness and generality, but they indicate that the essential features of territorial behavior had often been observed, the scientific interest of this kind of behavior was generally unappreciated until 1868 when the first edition of B. Altum's *Der Vogel und sein Leben* appeared.[33] This book gave a clear and concise account of the role of defended territories in the lives of birds, indicating how the male bird proclaims his ownership by song and how an intruding male is chased until it has retreated beyond the boundary of the invaded territory. Altum's book, which reached its seventh edition in 1903, seems to have been widely read in Germany, but hardly at all outside that country. At least one Irish and four American writers used the concept independently, after Altum, but before the summary on territory in the fourth volume of H. Eliot Howard's *British Warblers*, published in 1914. It was that work, and Howard's subsequent *Territory in Bird Life*, that really made the concept well known throughout the English-speaking world. Moffat, the Irish writer, is, however, particularly important because his paper of 1906, reprinted in 1934, suggested the demographic importance of territoriality. Moffat envisioned a population of breeding pairs in territories accompanied by unmated defeated territoriless birds, on land "not adapted for nesting though providing plenty of food and shelter." Burkitt,[34] another Irish naturalist, who had banded a population of robins in his garden, concluded that about one-quarter of the males at any time were unmated. Later investigators concerned with the role of birds in controlling spruce-budworm infestations, in a rather murderous experiment,[35] shot all the breeding passerine birds in a tract of coniferous forest, hoping to see how the infestation developed in the absence of birds. Almost at once another population turned up to occupy all the desirable freehold properties that had suddenly become vacant. It would seem therefore that for many species of birds, Moffat was right in his fundamental contention that many individuals lead a skulking kind of life around and between the territories that provide possible nesting sites. This means that the whole system of taking and holding territories assures optimal reproductive possibilities, or in evolutionary terms, maximal fitness, to those birds that can take and hold territories while effectively sterilizing those that fail. A very strong selection in favor of any genetic basis for territoriality is therefore continually operating. An incidental result of this may be to lower the birthrate of the population as a whole, even though the breeding individuals may exhibit optimal fecundity. These phenomena are not confined to birds; there is evidence[36] of non-

---

apparently over many centuries, and had been noted in books that were widely known and read. It is, however, clear that in the absence of any adequate intellectual context, the behavior was not regarded as of real scientific importance. The great advance made by later investigators, primarily Altum, Moffat, and Howard (see note 33), was not to discover territoriality, though none of them seemed to have known of the work of their predecessors, but to appreciate the significance of what the bird was doing in taking territory, in Chaucer's words, to "knowe his menyng openly and pleyn." A comparable case is provided by the development of ideas about niche specificity, discussed in chapter 5. It is quite likely that as the history of ecology is better understood, it will appear that most ecological discoveries consist of the recognition of significance rather than in the discovery of facts.

33. B. Altum, *Der Vogel und sein Leben*. Münster, W. Niemann, 1868, 240 pp. See also:

E. Mayr, Bernard Altum and the territory theory. *Proc. Linn. Soc. N.Y.*, 45, 1935.

H. E. Howard, *The British Warblers, Part 2*. London, R. H. Porter, 1908, 31 pp. The account of the chiffchaff on p. 8 seems to be the first mention of territory in this book.

H. E. Howard, *Territory in Bird Life*. London, J. Murray, 1920, xiii, 308 pp.

C. B. Moffat, The spring rivalry of birds: Some views on the limit to multiplication. *Irish Natural.*, 12:152–66, 1906, (reprinted in *Irish Natur. J.*, 5:84–87, 115–20, 155–56, 1934).

34. J. P. Burkitt, A study of the robin by means of marked birds. *Brit. Birds*, I, 17:294–303; II, 18:97–103; III, 18:250–57; IV, 19:120–24; V, 20:91–101; 1924–26. A touching short account of Burkitt is given in D. Lack, Some British pioneers in ornithological research, 1859–1939. *Ibis*, 101:71–81, 1959.

35. R. E. Stewart and J. W. Aldrich, Removal and repopulation of breeding birds in a spruce-fir forest community. *Auk*, 68:471–82, 1951.

M. M. Hensley and J. B. Cope, Further data on removal and repopulation of the breeding birds in a spruce-fir community. *Auk*, 68:483–93, 1951.

E. O. Wilson, *Sociobiology* (Cambridge, Mass., Belknap Press of Harvard University Press, 1975, p. 276), gives an admirable review of the later work.

36. C. M. King, The home range of the weasel

territorial individuals wandering among the territories of residents in the European weasel, *Mustela nivalis*.

This demographic aspect of territoriality excited less interest than the other meanings of the phenomenon until 1962, when Wynne-Edwards published his large and controversial *Animal Dispersion in Relation to Social Behaviour*.[37] Wynne-Edwards claimed that not only does territoriality incidentally limit population growth but in many cases special kinds of behavior, termed by him *epideictic*, have been evolved solely to promote such limitation. Such behavior is believed to give information on population density to the various individuals forming the population. A typical example given by Wynne-Edwards would be the continual whirligig dance of gyrinid beetle swarms on the surface of water. The beetles are known to respond to movements of water in their immediate environment; the continual dance might therefore indicate that a population of a certain density had been achieved. No other explanation seems to have been given for this very conspicuous type of behavior familiar to every aquatic naturalist; this does not mean that Wynne-Edwards's interpretation is correct, but it must be taken more seriously than has generally been the case.

It is possible to see that a species, living in a region of variable climate, and feeding in its territory, which not all birds do, might evolve so as to claim more territory than it needs in an average year in order to have enough in a bad year. If in an average year the brood size depended on the time available to the parents to bring food to the nestlings, as we have seen is often the case, then territoriality, not strictly adjusted to immediate, but rather to potential demand, would have a powerful if undemocratic effect in conserving the natural resources of the species.

Moreover, it is clear from the theoretical investigations of Hamilton[38] that a gene promoting altruism in a parent, even at great cost, could be fixed by selection if the *inclusive fitness* or fitness of the individual and that of its descendants is increased by the act of parental self-sacrifice. This process is a special case of what is now called *kin selection*. What is hard to understand is how any genetically determined mechanism can produce an altruistic population, refraining from overreproduction, in a way that will ultimately benefit all descendants, when any individual not limiting its reproduction is bound to leave more descendants than one that is practicing the limitation. Actually a number of possibilities have been suggested,[39] involving the development of small related groups or demes, which become extinct if they exhaust their resources by reproducing too fast, and are later replaced by less fecund groups as the food supply increases again. In such cases, however, very special circumstances, involving partial barriers between the demes and very rapid extinction in some of them, have to be postulated. No clear cases that fit any of the theoretical schemes seem to have been discovered in nature, and it is unlikely that the conditions that they imply would be realized on the scale demanded by Wynne-Edwards. Moreover, these theoretical investigations tend to consider a single "altruistic" gene as all that is needed, whereas to Wynne-Edwards epideictic behavior is usually quite complicated, so that additional strains are put on the explanations.

There is a good deal of evidence, mainly derived from rodents, that overcrowding even in the presence of abundant food can have a marked effect in reducing fecundity, slowing population growth and at times causing a decrease in numbers. This effect may be formally compared with the reduced

(*Mustela nivalis*) in an English woodland (*J. Anim. Ecol.*, 44:639–65, 1975), seems to have identified directly underweight individuals not holding territories moving around in an area otherwise owned by territorial individuals.

37. V. C. Wynne-Edwards, see note 5.

38. W. D. Hamilton, The genetical theory of social behavior, I, II. *J. Theort. Biol.*, 7:1–52, 1964.

39. Two recent contributions containing references to earlier work, are:

D. S. Wilson, A theory of group selection. *Proc. Nat. Acad. Sci. U.S.A.*, 72:143–46, 1975.

M. E. Gilpin, *Group Selection in Predator-Prey Communities*. Princeton University Press, 1975, xii, 108 pp.

fecundity of female *Drosophila* when they are continually disturbed in an overcrowded population.

The simplest kinds of limitation due to increasing size of a population in a restricted space are demonstrated in a study by Southwick[40] on the house mouse. Here as the population grew, the tendency to fight also increased and when the aggressiveness reached a level at which there was one fight per hour per mouse, which happened as the population reached a level between 50 and 100 in different experimental pens, the survival of young began to suffer. This was due partly to mechanical disturbance, but may also have involved trauma due to psychophysiological effects. Further more, there was evidence that a tendency to huddle together, displayed by all populations, became greater as the population increased and had an unfavorable effect, apparently due to disturbance of nursing females by other mice intent on social contact. The presence of males in nesting boxes caused poorer survival of the young. Large populations, moreover, build less good nests than do sparse, and this is also a cause of poor litter survival. There is a tendency to form communal nests, which may contain up to 58 young; in such a case no young survived. All of these effects are reasonably, if not obviously, attributable to crowding; they certainly will produce a density-dependent brake on successful reproduction, but are hardly likely to have been evolved by a recondite process of natural selection to prevent a population explosion. More curious is the evidence that the presence of males that are strangers to a mated female may block pregnancy in the latter, apparently by an olfactory stimulus.[41]

Christian and his co-workers[42] have put forward the hypothesis that the inhibition of reproduction in mammalian populations as they become very dense is an expression of what has been called, rather euphemistically, the *general adaptation syndrome*. The social tensions resulting from excessive contact are believed to produce a hypertrophy of the adrenal cortex, which inhibits reproductive activity. Christian and Davis do give fairly clear data that in 35- to 40-gram *Microtus pennsylvanicus*, the individuals with largest adrenals are not sexually functional. Much of the other evidence seems rather fuzzy. The supposed relation is regarded as an adaptation to prevent environmental deterioration, and is regarded as an example of the kind of process that Wynne-Edwards postulates. The area is clearly not yet fully explored, and though most investigators are likely to regard the evolutionary mechanisms postulated as too improbable, the subject should not be regarded as closed.

Putting death and birth together

Having considered both mortality or survivorship and natality separately, we may now consider treating them together to construct some hypothetical populations which will prove to be quite like what is seen in nature. We do this by specifying a schedule of specific mortality and specific natality by intervals of time, usually most conveniently a year. In the first example we will suppose that during the time between the birth of young, restricted to a definite breeding season, and the time when that season comes around again, the mortality of the young is 0.8 (or 80%), while for every subsequent year it is 0.5 (or 50%). We also suppose initially that every individual produces 3 young; this is equivalent in birds to every pair producing fledged young from 6 eggs. We proceed as in table 3. The original population is entered as 1,000 in the top left-hand corner of the table; by the end of the 1st year, in which it does not breed, it is reduced to 200 individuals; we follow this population down obliquely as in subsequent years it is annually reduced by half. Meanwhile after entering its 2nd year, it starts breeding, producing 600 young, so now early in the 2nd year there

40. C. H. Southwick, Regulatory mechanisms of house mouse populations: Social behavior affecting litter survival. *Ecology*, 36:627–34, 1955.

41. H. M. Bruce, Pheromones. *Brit. Med. Bull.*, 20:10–13, 1970.

42. J. J. Christian and D. E. Davis, Endocrines, behavior and population. *Science*, 146:1550–60, 1964.

Year of observation	Age classes 0–1	1–2	2–3	3–4	4–5	5–6	6–7	7–8
0–1	1,000							
1–2	600	200						
2–3	660	120	100					
3–4	726	132	60	50				
4–5	798	145	66	30	25			
5–6	882	160	73	33	15	13		
6–7	975	176	80	37	17	8	7	
7–8	1,083	196	88	40	18	8	4	3
$\mathbf{l}_x$	752	136	61	28	12	6	3	2

TABLE 3. Growth of an ideal avian population with a specific mortality of 0.8 in the 1st year and thereafter 0.5; all individuals after 1st year reproducing, with 6 eggs producing fledged chicks per pair per year. Final age distribution given in bottom line as $\mathbf{l}_x$.

are 800 individuals. The 600 new young are reduced by the end of the 2nd year to 120, which now produce 360 young, to which are added the 300 produced by the 100 survivors of the original cohort. In this way we have built up a population that, with the passage of time, continues to increase.

It will be noted that since every reproducing individual produces 3 young, the total number of individuals in the first age class is always 3 times the number of all older individuals. Moreover, since the individuals in the 2nd age class have been produced from all the individuals in the 3rd and higher age classes as the result of a constant specific natality and have been reduced in number by constant specific mortalities, they will also form a constant proportion of the population, namely, 14 percent of the whole. The same argument applies to all higher age classes, as the right-hand side of the table is filled up. The age distribution is therefore stable. Soon after reaching the 8th year, as indicated in the table, the additions to still higher age classes become so small as to be negligible.

We may now consider what the effect would be of reducing the fecundity of the pairs in their 1st year of breeding; it is very usual to find fewer eggs laid by birds breeding for the first time (table 4). We therefore set the number of viable young in the 1st year of breeding as 4 per pair rather than 6, while retaining the specific mortalities of table 3. With this new schedule of reproduction we find that the population is in essential equilibrium, with a growth rate of only a fraction over 1 percent per year instead of a little over 10 percent per year when there are 6 young rather than 4 in the first brood. These situations are quite realistic for passerine birds in temperate regions. They indicate how what may seem to be a very small change in reproductive physiology may produce striking changes in demography.

It is important to note that, as before, the age distribution rapidly becomes stable. That this should be the case is not intuitively obvious by inspection, though it can easily be established from the table. A formal proof has been given by Lotka,[43] for any schedule of age-specific natalities.

In table 5 the same survivorship schedule is used with a constant production of 1 egg per adult per year, corresponding to 2 fledged birds, from 2 eggs per pair per season. Such a reproductive performance is seen to be

43. A. J. Lotka, The stability of the normal age distribution. *Proc. Nat. Acad. Sci. U.S.A.*, 8:339–45, 1922.

Year of observation	Age classes 0–1	1–2	2–3	3–4	4–5	5–6	6–7	7–8
0–1	1,000							
1–2	400	200						
2–3	460	80	100					
3–4	454	92	40	50				
4–5	455	91	46	20	25			
5–6	458	91	46	23	10	13		
6–7	463	92	46	23	12	5	7	
7–8	465	93	46	23	12	6	3	4
$\mathbf{l}_x$	750	142	71	35	18	9	5	6

TABLE 4. Growth of an ideal avian population, with a specific mortality of 0.8 in the 1st year and thereafter 0.5; all pairs produce 4 eggs in their 2nd year of life, which is their 1st year of breeding, but in later years the surviving pairs produce 6 eggs per year. Final age distribution given in bottom line as $\mathbf{l}_x$.

Years of observation	Age classes 0–1	1–2	2–3	3–4	4–5	5–6	6–7	7–8
0–1	1,000							
1–2	200	200						
2–3	140	40	100					
3–4	98	28	20	50				
4–5	69	20	14	10	25			
5–6	49	14	10	7	5	13		
6–7	36	10	7	5	4	3	7	
7–8	27	7	5	4	3	2	2	4
$\mathbf{l}_x$	500	130	93	74	56	37	37	74

TABLE 5. Growth of an ideal avian population, with a specific mortality of 0.8 in the 1st year and thereafter 0.5; all individuals after 1st year reproducing, with 2 eggs producing fledged chicks per pair per year. Final age distribution given in bottom line as $\mathbf{l}_x$.

quite inadequate to permit the survival of the species under consideration, which loses around a quarter of its numbers each year. In tropical passerine birds in which 2 eggs commonly form a clutch, the number of clutches laid throughout the year is usually greater than in temperate species. It must be borne in mind in these examples that many temperate species habitually rear several broods per year. The examples do emphasize that with the observed fecundity of temperate passerine birds, equilibrium populations imply very great and continuous mortality.

In contrast to this, table 6 gives the ideal development of a mammal population. The mortality in the year following birth is taken as 0.5, in the subsequent 7 years as 0.05, while in the 9th and 10th years it is 0.8. Breeding is taken to occur when the animals are 4 years old and to continue till they are in their

Year of observation	Age classes 0–1	1–2	2–3	3–4	4–5	5–6	6–7	7–8	8–9	9–10	10–11
0–1	1,000										
1–2	0	500									
2–3	0	0	475								
3–4	0	0	0	461							
4–5	219	0	0	0	*438*						
5–6	208	110	0	0	0	*416*					
6–7	198	104	104	0	0	0	*396*				
7–8	188	99	99	99	0	0	0	*376*			
8–9	256	99	94	94	*95*	0	0	0	*357*		
9–10	90	113	94	89	*90*	*90*	0	0	0	71	
10–11	128	46	107	89	*85*	*85*	*85*	0	0	0	14
11–12	134	64	44	102	*85*	*81*	*81*	*81*	0	0	0
12–13	204	67	61	41	*97*	*81*	*77*	*77*	*77*	0	0
13–14	177	102	64	58	*39*	*92*	*77*	*73*	*73*	15	0
14–15	161	88	97	60	*55*	*37*	*87*	*73*	*69*	15	3
15–16	148	80	84	92	*57*	*52*	*35*	*83*	*69*	14	3
16–17	152	74	76	80	*88*	*55*	*49*	*34*	*79*	14	3
17–18	145	76	71	72	*76*	*83*	*52*	*47*	*32*	16	1
18–19	157	72	72	67	*69*	*72*	*79*	*49*	*45*	6	2
19–20	160	78	69	69	*64*	*65*	*68*	*75*	*47*	9	2
20–21	162	80	75	65	*65*	*61*	*62*	*65*	*71*	9	3
21–22	151	81	76	71	*62*	*62*	*58*	*59*	*62*	14	2
22–23	148	76	77	72	*67*	*59*	*59*	*55*	*56*	12	2
23–24	148	74	72	73	*68*	*64*	*56*	*56*	*52*	11	2
24–25	151	74	70	68	*70*	*65*	*61*	*53*	*53*	10	2
25–26	151	75	70	67	*65*	*66*	*62*	*58*	*51*	11	2
$\mathbf{l}_x$	221	111	103	98	103	97	91	85	74	15	3

TABLE 6. Growth of an ideal mammalian population, with a specific mortality of 0.5 for the 1st year, 0.05 for the 2nd to 9th years and thereafter 0.8. Reproduction starts in the 5th year and continues to the 9th, 1 young being produced annually by each pair. Reproductive part of population italicized. Age distribution of 26th year given as $\mathbf{l}_x$ in the bottom row.

9th year, the rapidly dying animals in the 10th and subsequent years being postreproductive. The initiation of the population with 1,000 young is artificial, but is the easiest way of presenting the development. The numbers of the different age groups vary greatly at the beginning, and although after 25 years the population is practically stabilized, there are slight irregularities which produce a transitory maximum in the survivorship curves which would persist till the 31st year. This hypothetical population gives a good idea of what happens in animals such as the Dall sheep when they have achieved a stationary population.

SUMMARY

The forces that regulate the size of a population may be *density-independent*, operating without regard to the number of individuals occupying a space or volume, or *density-dependent*, in which case the action on an individual depends on the size of its population. Density-independent mortality, due to storms, droughts, frost, and the like is obviously important, but its importance is probably much greater in organisms which have annual cycles than those which have several generations in a year or live for a number of years. The probability that any organism completely regulated by density-independent factors could exist indefinitely is small. If density-dependent mortality was never possible, it would imply that the density-independent factors always prevent overcrowding, though the only regulatory mechanism would be the probability distribution of variously sized catastrophes. Ultimately, a catastrophe exterminating the species would occur. The existence of density-dependent mortality or density-dependent limitation of reproductive activity implies that the species exhibiting it have survived the past, random, unfavorable periods in their environment, against which they have become at least somewhat adapted by natural selection.

Variation in fecundity, though often largely controlled environmentally, is also found to be genetically determined to some extent in the cases that have been studied. This means that unless the offspring produced by more fecund parents are inherently less fit, such offspring will tend to displace those of less fecund parents. Since there will always be such selection in favor of those parents having most viable offspring, regulation of numbers is more likely to be due to density-dependent death rates rather than birthrates. The inevitable reduction of fecundity through starvation does not imply any modification of this; in such a case the reduction is kept as low as possible by natural selection. From a human point of view, density-dependent regulation by removal of the offspring in excess of the carrying capacity of an area seems immoral and wasteful, but without a compensating regulation of natality such wasteful immorality is ultimately inevitable. Foresight and contraception provide the compensating regulations where they are adequately practiced, but as yet this is true only of a limited part of the human population of the earth.

There are two extreme patterns of reproduction in multicellular organisms, which may be called *prodigal* and *prudent*. In prodigal reproduction, an enormous number of small and easily dispersed seeds, eggs, or larvae are produced and scattered widely, so that the probability of a few finding proper habitats is very great, even though the probability of any one surviving is very small. In prudent reproduction, few seeds, eggs, or larvae are produced, but each is provided with food reserves in the seed or egg and by parental care of the juveniles, so that the probability of survival of any one is much increased. Though there is a complete set of transitions between the two types, the dichotomy is useful and may be recognized in various contexts throughout this chapter. It is obviously closely related to the dichotomy between *r*- and *K*-selection.

In multicellular organisms capable of living for more than a few months, there are two main patterns of reproduction. Some, like annual plants and many insects, produce all their seeds or eggs at one time and then die. Such organisms are said to be *semelparous*. Others live some years, producing young every year, and are said to be *iteroparous*. If the young have the same specific mortality as the adults, which as we have seen is rare, the only advantage of iteroparity is that it adds an extra pair of adults to the brood that will carry on the population in the succeeding year. In a more general case, the advantage depends on the ratio of survivorship of the adults to that of the young. In pelagic fishes like the mackerel, in which some hundred thousand eggs are laid by each female, but with no survival in the not-infrequent poor years, iteroparity is obviously necessary to maintain the species. Iteroparity is also of great advantage in species in which learning and experience count in reproduction.

The fecundity of animals is often density-dependent. The simplest situation is when a large population is restricted by a limited supply of food. In such cases partial starvation limits reproduction. This happens in experiments on *Daphnia* no less than in lions in nature. Sometimes closely related species differ considerably in their reaction to changes in food. In the American warblers or Parulidae, the bay-breasted warbler, *Dendroica castanea*, and probably a couple of other species, lay more eggs when insects, such as spruce budworms, produce an abundant supply of food. At such times clutches of up to 7 eggs may be laid. Most species of the genus lay 4 or quite exceptionally 5, as may the bay-breasted in years when the budworm is scarce.

In species that rear their young, the amount of work that the parents can do in a day may limit the size of the brood. This is very clearly seen in birds. In the alpine swift in Switzerland, 4-egg clutches produce no more fledged young than do 3-egg clutches. In the allied European swift in England, 3-egg clutches produce no more, and in fact very slightly fewer, fledged young than do 2-egg clutches in poor years, though in years with much sunshine they are a little superior. Adding extra eggs is not merely pointless, but is likely to be deleterious if attempts are made to share an inadequate food supply among all mouths.

There is evidence that clutch size, though often sensitive to food supply, can also be determined genetically. In areas where the weather differs greatly from year to year, there is probably a genetic polymorphism in egg number, so that small clutches are favored in some years, large in others.

Though in some cases the larger broods produce no more fledged young than the small, there are numerous cases where this is not so. At least in some of these cases it has been shown that a juvenile from a large brood has a lower chance of survival during the first few months of independent life than one from a small brood. This is presumably because it starts as a light and slightly undernourished bird. Complications are raised by the fact that in a cool environment a large brood huddled together may lose, per nestling, less heat than a small brood. The balance of the evidence, however, strongly suggests that an optimal clutch size exists and that going beyond it is pointless, wasteful, and so deleterious.

There is a strong tendency to geographical variation in avian clutch sizes within a species. Usually the number increases, in either hemisphere, on moving poleward from the equator. Two factors are probably involved here. One is the greater day length during which the parents can work in the temperate summer as compared to the short equatorial day; the other is the greater number of predators in the tropics, which puts a premium on rapid development and inconspicuous behavior.

In general, clutch size is smaller in areas with oceanic climates such as the western areas of Europe or the coastal part of western North America than in areas at the same latitude but with more continental climates. It is suggested that this is due to more intense competition under more stable conditions; or, in other words, that *r*-selection is less and *K*-selection more important in such circumstances.

A few cases are known among freshwater animals, notably diaptomid copepods and salmonid fishes, in which numerous small eggs are produced when there is abundant food and little competition, while the eggs are fewer and larger when resources are scarce and competition great. In such cases *r*-selection will be operating when the food supply is abundant, *K*-selection when competition is greatest. There are also a few cases known where the number of seed set by different populations of the same species of plant is apparently regulated in a comparable way.

Territoriality in the higher animals may have the effect of limiting the mean reproductive ability of a population as a whole, though increasing the capacity of those pairs that have taken territory to produce young. It is very doubtful if any satisfactory mechanism can evolve by natural selection, unaided by conscious foresight, to limit growth of a whole population in such a way that resources are conserved for future generations. It has, however, been pointed out that in any organism, selection will tend to operate most strongly on those age classes that have highest reproductive value. This, in general, means that young adults will tend to become less and less vulnerable as evolution proceeds, while predation and other sources of death fall most heavily on the individuals of least reproductive value, notably numerous very young and a limited number of old individuals. This implies that predators are forced by natural selection into being prudent, for what they can get most easily ends up by being the least valuable part of the population from the standpoint of its stability.

Given constant schedules of deaths and of births with increasing age of the individuals dying or giving birth, the age structure becomes constant. This means that the intrinsic rate of increase of a population, whether it is positive, zero or negative, also becomes constant so long as the age-specific mortality and natality do not change.

Chapter Four

Living Together in Theory and Practice

WE will now return to the logistic, which we have already found to be a good point of departure, and ask ourselves what would happen to a population of two species living together and using partly or wholly the same resources. We evidently should now have to include a term for the amount of vacant space used by both species, and not merely one, when estimating how much of K was still available. Moreover, we do not know a priori how a specimen of the second species alters the rate of increase of the first. It may merely provide an extra hungry mouth, but it also might secrete some material inhibitory to the first species. We strictly should write then

$$\frac{dN_1}{dt} = r_1 N_1 \left[\frac{K_1 - N_1 - f_2(N_2)}{K_1}\right] \qquad (4.1\text{ a, b})$$

$$\frac{dN_2}{dt} = r_2 N_2 \left[\frac{K_2 - N_2 - f_1(N_1)}{K_2}\right]$$

as before $f_1(N_1)$ and $f_2(N_2)$ could be expanded as power series, without absolute terms, giving

$$f_1(N_1) = \alpha_1 N_1 + \beta_1 N_1^2 + \gamma_1 N_1^3 + \ldots$$

$$f_2(N_2) = \alpha_2 N_2 + \beta_2 N_2^2 + \gamma_2 N_2^3 + \ldots$$

Although there are probably cases where the first two terms are needed, we may start the investigation with only the first term of each series and write

$$\frac{dN_1}{dt} = r_1 N_1 \left(\frac{K_1 - N_1 - \alpha_2 N_2}{K_1}\right) \qquad (4.2\text{a, b})$$

$$\frac{dN_2}{dt} = r_2 N_2 \left(\frac{K_2 - N_2 - \alpha_1 N_1}{K_2}\right)$$

The easiest and most interesting procedure is to enquire what will happen as equilibrium is approached. At equilibrium we suppose that some constant number, which may be zero, of each species is present. When that is the case, it is evident that neither N_1 will change relative to N_2 nor the reverse. We may therefore write, dividing (2a) by (2b) and the reverse,

$$\frac{dN_1}{dN_2} = \frac{N_1 r_1}{N_2 r_2}\frac{K_2}{K_1}\left(\frac{K_1 - N_1 - \alpha_2 N_2}{K_2 - N_2 - \alpha_1 N_1}\right) = 0 \qquad (4.3\text{a, b})$$

$$\frac{dN_2}{dN_1} = \frac{N_2 r_2}{N_1 r_1}\frac{K_1}{K_2}\left(\frac{K_2 - N_2 - \alpha_1 N_1}{K_1 - N_1 - \alpha_2 N_2}\right) = 0.$$

We are obviously not interested in those cases in which the rates of increase, the saturation populations, or both the actual populations are zero. The conditions for equilibria of interest are therefore defined by

$$K_1 - N_1 - \alpha_2 N_2 = 0 \qquad (4.4\text{a, b})$$

$$K_2 - N_2 - \alpha_1 N_1 = 0$$

If we have two rectangular Cartesian axes for

N_1 and N_2, we can plot the lines representing these equations. Since the first line defines the points where $dN_1/dN_2 = 0$, the second where $dN_2/dN_1 = 0$, any trajectory that describes the mutual variation of N_1 and N_2 will cross the first line vertically, N_1 not changing whatever happens to N_2, while it will cross the second line horizontally since N_2 does not change whatever happens to N_1. Such lines of constant slope are called isoclines.

Notice for the first isocline, if

$$N_1 = 0,\ N_2 = K_1/\alpha_2$$
$$N_2 = 0,\ N_1 = K_1.$$

while for the second, if

$$N_1 = 0,\ N_2 = K_2$$
$$N_2 = 0,\ N_1 = K_2/\alpha_1.$$

If we examine figure 78, it is evident that there are four possible qualitative situations:

(a) $\alpha_1 > K_2/K_1,\ \alpha_2 < K_1/K_2$

(b) $\alpha_1 < K_2/K_1,\ \alpha_2 > K_1/K_2$

(c) $\alpha_1 > K_2/K_1,\ \alpha_2 > K_1/K_2$

(d) $\alpha_1 < K_2/K_1,\ \alpha_2 < K_1/K_2$.

If we have any two populations N_1 and N_2, where these numbers are small so that (N_1, N_2) lies near the origin, both populations may be expected to grow, describing a trajectory which in fact corresponds to an integral solution of the simultaneous differential equations (4.2a, b). Examples corresponding to the four possibilities are give in figure 78. If we measure populations in terms of fractions of K values, we may write $K_1/K_2 = K_2/K_1 = 1$. Thus in figure 78a, $\alpha_1 > 1$ and $\alpha_2 < 1$. In such a case it is evident that at equilibrium only species 1 remains, with a population of 1 K, just as if the second species had not been present. In figure 78b, the reverse holds, only species 2 surviving.

In the other two cases the isoclines cross in the all-positive quadrant, which alone concerns us in biology since we cannot have negative populations. In figure 78c, $\alpha_1 > 1$ and $\alpha_2 > 1$; equilibrium either involves the survival of one species or the other, depending on the initial ratio of the numbers of the two species. The diagonal divides the diagram into two fields; starting from any point in the upper left-hand field, only species 2 survives, while from the lower right-hand field, only species 1. Though at the point where the isoclines cross there is no change of either species with respect to the other, this equilibrium is unstable; the slightest displacement to one side or other of the diagonal causes a drift to a unispecific population.

In the fourth case, in which $\alpha_1 < 1$, $\alpha_2 < 1$, the equilibrium is stable, all trajectories leading toward it. The biological interpretation of these four cases is formally quite simple. In the first case, a member of the first species depresses one of the second species more than it does one of its own species, while a member of the second species depresses one of the first less than it does one of its own. In such a case the first species excludes the second. The reverse is of course true of the second case. In the third case, each member of either species depresses one of the other species more than does a conspecific individual. In such a case the species present initially in excess of a critical ratio, depending on α_1 and α_2, excludes the other. In the fourth case, an individual of neither species depresses individuals of its competitor as strongly as one of its own species; with this situation both persist.

The first two cases might be expected to exemplify situations where two species are introduced into a vessel of some homogeneous nutrient medium and are allowed to compete. The more efficient species, with the higher value of α, will take over the vacant spaces more rapidly and so outcompete the less efficient, which will disappear.

The case in which each species has more effect on the other than on itself could correspond to that of two microorganisms producing antibiotics effective against each other.

The fourth case, in which the interspecific effects are smaller than are the intraspecific, is usually thought of as involving either some geometrical separation of the species, the more vulnerable one having some sort of refuge to which it can go but its competitor cannot, or having some resource available that the otherwise better adapted competitor cannot use.

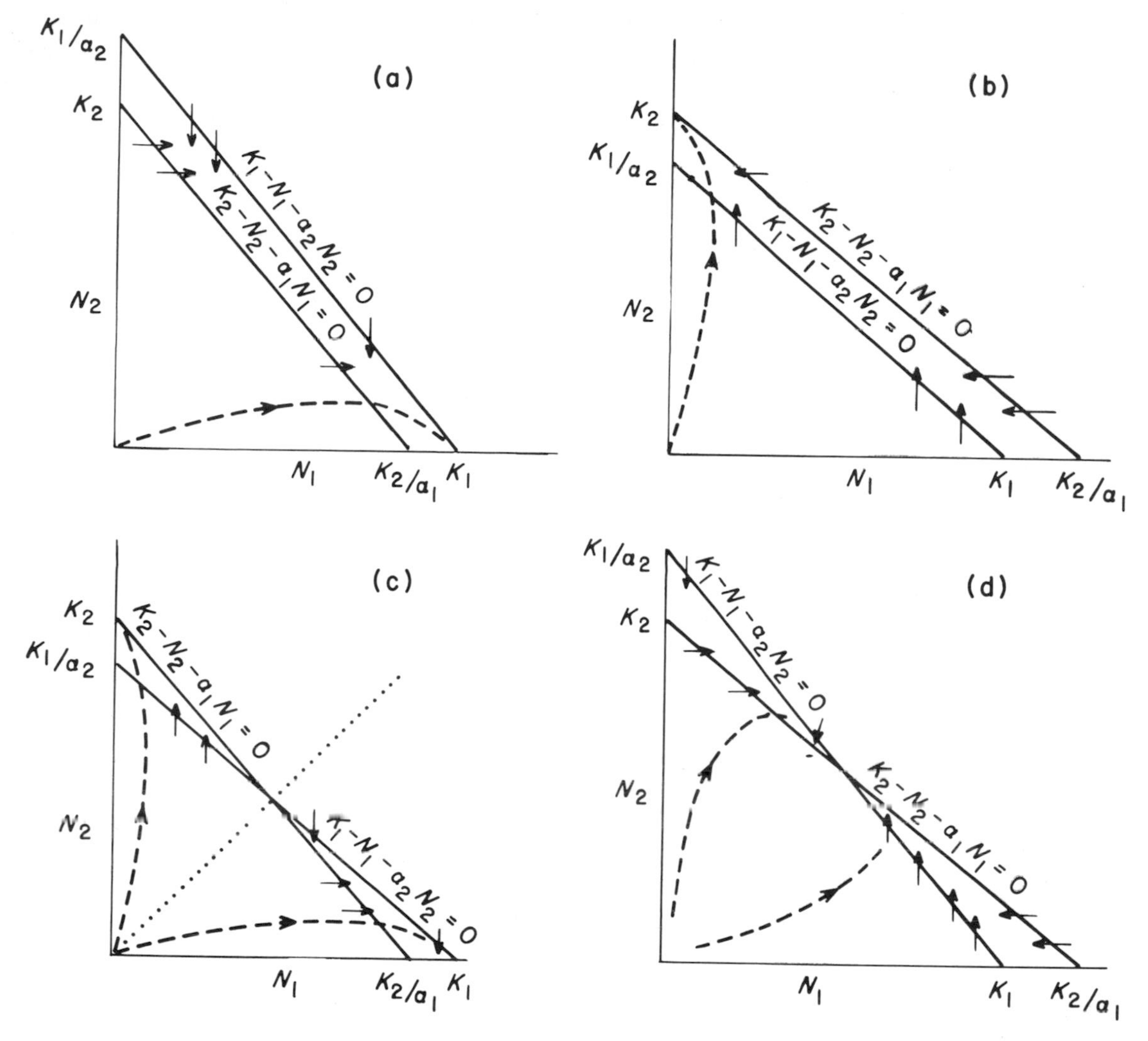

FIGURE 78. The four possible outcomes of competition: (a) $\alpha_1 > K_2/K_1$, $\alpha_2 < K_1/K_2$, and N_1 displaces N_2; (b) $\alpha_1 < K_2/K_1$, $\alpha_2 > K_1/K_2$, and N_2 displaces N_1; (c) $\alpha_1 > K_2/K_1$, $\alpha_2 > K_1/K_2$, and outcome depends on initial conditions; (d) $\alpha_1 < K_2/K_1$, $\alpha_2 < K_1/K_2$, and the species coexist. The broken lines indicate possible trajectories for the relative numbers of the two species; in (c) the dotted line divides the field, so that all trajectories on its upper left side lead to N_2 displacing N_1, all on its lower right side to N_1 displacing N_2.

In either case, there is a limitation on interspecific competition which does not apply in the other three cases. In its baldest form the results reduce to Hardin's[1] statement, *complete competitors cannot coexist.* This he calls the principle of *competitive exclusion*.

An algebraic treatment of competition was put forward by Nägeli in 1874,[2] but this work was almost entirely overlooked until it was rediscovered by Harper in 1974. The principle of competitive exclusion was apparent from the mathematical treatment of selection by

1. G. Hardin, The competitive exclusion principle. *Science*, 131:1292–97, 1960.

2. C. Nägeli Verdrängung der Pflanzenformen durch ihre Mitbewerber. *Sitzb. Akad. Wiss.*, München, 11:109–64, 1874. For a more detailed discussion of Nägeli's ideas of what we now call density dependence and of other aspects of his study see:

J. L. Harper, A centenary in population biology. *Nature*, 252:526–27, 1974. Nägeli's paper was completely neglected by nineteenth-century biologists. It was, however, cited in Gause's *Struggle for Existence* (see note 7). This neglect, as is clear from Harper's

J. B. S. Haldane[3] in 1924, a basic document in population genetics, but one not appreciated by ecologists at the time. The theoretical contribution that really started investigation of the subject was that of Vito Volterra[4] in 1926. Volterra, described somewhere by Robert May as the greatest mathematician[5] to make a contribution to biology, was stimulated to develop his ideas by the changes in proportion of fishes in the Mediterranean that had resulted from suspension of commercial fisheries during the war years of 1915–18, to which phenomenon his attention was drawn by his son-in-law, Umberto D'Ancona,[6] a very eminent hydrobiologist.

Volterra's work stimulated the detailed and incisive experimental studies of G. F. Gause and it was from his book, *The Struggle for Existence*,[7] that the concept of competitive exclusion really entered the minds of biologists. In its less analytic and often weaker form, that closely allied species living together occupy different niches, the principle had been held by a number of naturalists working at least as early as 1857, but until the mathematical and experimental work of Volterra and of Gause respectively, such a generalization evoked little sustained interest.

Experimental study of competition

The laboratory study of competing populations that might be expected to exhibit the relationships deduced by Volterra was begun by Gause in 1932 and was summarized by him in two short books, one in English in 1934, the other in French in 1935. The first of these, *The Struggle for Existence*, has rightly become a foundation stone of ecology; the French book, *Vérifications expérimentales de la théorie mathématique de la lutte pour la vie*,[8] however,

account, was primarily due to the inability of most nineteenth-century naturalists, in spite of Darwin, to think in terms of population dynamics. The same inability undoubtedly was responsible for the related persistent failure to understand the significance of the ideas about related species not co-occurring in the same habitat that were put forward over and over again at least from 1857 (see chap. 5). Other cases of the same disease in increasing order of seriousness are the failure to understand the significance of territoriality, the neglect of Verhulst's theoretical treatment of population growth, and, of course, above all the total disregard of Mendel's work.

3. J. B. S. Haldane, A mathematical theory of natural and artificial selection: Part 1. *Trans. Cambridge Phil. Soc.*, 23:19–41, 1924.

4. V. Volterra, Variazioni e fluttuazioni del numero d'individui in specie animali conviventi. *Mem. R. Acad. Naz. dei Lincei* (ser. 6), 2:31–113, 1926. A translation of much of this paper, by M. E. Wells, appeared as an appendix to R. Chapman, *Animal Ecology*. New York and London, McGraw-Hill, 1931, pp. 409–48.

5. G. H. Hardy, whose name is familiar to biologists as one of the two original proponents of the Hardy-Weinberg law, is at least a good runner-up. Two of Hardy's colleagues at Cambridge, his collaborator, Professor J. E. Littlewood, F.R.S., and Professor W. A. H. Rushton, F.R.S., have told me that Hardy was somewhat embarrassed at being immortalized in so simple an application of the binomial theorem.

6. Volterra indicates this in his first paper. See also:
U. D'Ancona, Intorno alle associazioni biologiche e a un saggio di teoria mathematica sulle stesse con particolare riguardo all'idrobiologia. *Int. Rev. ges. Hydrobiol. Hydrogr.*, 17:189–225, 1927.
V. Volterra and U. D'Ancona, *Les associations biologiques au point de vue mathématique*. Actualités scientifiques et industrielles, 243. Paris, Hermann, 1935, 96 pp.

Six years after Volterra's first publication, A. J. Lotka examined the problem in: The growth of mixed populations: Two species competing for a common food supply. *J. Wash. Acad. Sci.*, 22:461–69, 1932. Lotka seems to have been the first to use isoclines in the exposition of biologically significant equations. Lotka's paper was published in full knowledge of Volterra's work. It made certain advances in the theory of competition, but certainly not more than are found in Gause and Witt's paper (note 46).

In view of the gradual development of the subject, the term Lotka-Volterra equations for the competition equations is inappropriate; this phrase should be restricted to the prey-predator equations with oscillatory solutions studied by Lotka in 1920 and independently by Volterra in 1926 (see chap. 6).

7. G. F. Gause, *The Struggle for Existence*. Baltimore, Williams and Wilkins, 1934, ix, 163 pp.

8. G. F. Gause, *Vérifications expérimentales de la théorie mathématique de la lutte pour la vie*. Actualités scientifiques et industrielles, 277. Paris, Hermann, 1935, 62 pp., Gause states that he had discussed the mathematical developments with A. A. Witt, with whom he wrote, almost or quite simultaneously, an important theoretical paper (note 46). It is reasonably clear that Witt played a significant part in the development of the modern theory. The method of isoclines is ultimately due to Poincaré. Gause indicates that the competition equations are identical to equations used earlier in the study of oscillating currents.

has been unjustifiably neglected in America, so I have taken my examples of Gause's experiments largely from it, rather than from the better known and earlier English work. The French book also added greatly to the theory of the matter as it contained the inequalities of p. 118.

Gause's work was done mainly, though not exclusively, with liquid media in which food is suspended and which usually constitute homogeneous or isotropic environments. His first study was of two species of yeast, but in these organisms there is a complication that when the ethyl alcohol that they produce increases in the medium above a critical concentration, division and so increase of the population cease, but the cells do not die. The species with the higher ethanol tolerance goes on increasing for a time after its competitor has stopped dividing, but does not eliminate it and soon itself stops reproducing.

The most critical work was done with ciliates, and in this work a small fraction of the entire population was removed daily, so introducing an artificial death rate, which in nature would no doubt be represented by various forms of natural death. This procedure complicates the mathematics and the empirical results a little.[9] It permits approach to an equilibrium in less time than were it not employed.

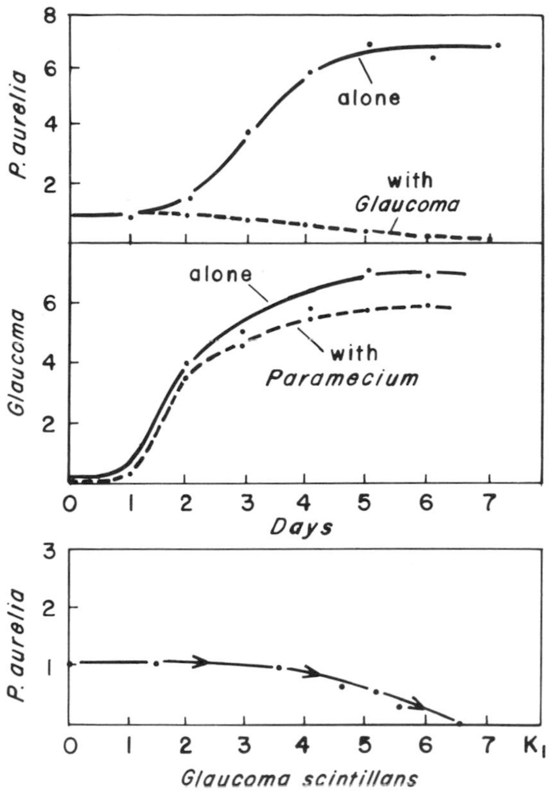

FIGURE 79. Upper panel, *Paramecium aurelia* cultivated without (solid line) and with (broken line) *Glaucoma scintillans*; middle panel, *G. scintillans* cultivated without (solid line) and with (broken line) *P. aurelia*; bottom panel, relative trajectory, as in figure 78 a (Gause, modified).

The first paper published by Gause on mixed populations was Experimental studies on the struggle for existence: I. Mixed populations of two species of yeast. *J. Exp. Biol.*, 9:389–402, 1932.

Most of the other work of this period is summarized in the two books cited; of individual papers one deserves particular mention: G. F. Gause, O. K. Nastukova, and W. W. Alpatov, The influence of biologically conditioned media on the growth of a mixed population of *Paramecium caudatum* and *Paramecium aurelia. J. Anim. Ecol.*, 3:222–30, 1934. This paper is historically significant since Alpatov, with whom Gause studied, had worked with Raymond Pearl. It also provided very clear evidence in experiments involving reproducing cultures of the effect of environmental factors on the direction of competition. That this was to be expected was known to the authors from two earlier pieces of work:

R. S. A. Beauchamp and P. Ullyott, Competitive relationships between certain species of freshwater triclads. *J. Ecol.*, 20:200–08, 1932. These two authors concluded from work in the field that the zonation of species of triclads observed along a stream depends not on the direct effect of temperature, but on the control of the directions of competition by temperature and current speed.

N. W. Timoféeff Ressovsky, Ueber die relative Vitalität von *Drosophila melanogaster* und *Drosophila funebris* unter verschiedenen Zuchtbedingungen, in Zusammenhang mit den Verbreitungsareal dieser Arten. *Arch. Naturgesch.*, Abt. B, 2:285–90, 1933. This work reported single-generation experiments which showed clearly that *D. funebris* did better than *D. melanogaster* at low temperatures and with damper larval food, the reverse at higher temperatures with drier larval food.

The concept of environmental control of competition had been investigated much earlier by Nägeli (note 15) and Tansley (note 16) in plants, but had made little impression on biologists.

9. This will be discussed in greater detail in chapter 5.

In figure 79 are plotted the growth of populations of two ciliates, one the very small *Glaucoma scintillans*, the other the much larger *Paramecium aurelia*. Populations are measured in volumes, the unit being the volume of a single *P. aurelia*. One hundred *Glaucoma* have a volume of 2.6 paramecium volume units. The solid curves of the top two panels of figure 79 represent respectively the growth of *P. aurelia* and *Glaucoma* populations. The saturation values are about the same when measured in volumes, but *Glaucoma* approaches saturation much more rapidly than does *P. aurelia*. When the two species are cultivated together, *Glaucoma* rapidly increases as if the *P. aurelia* was not there, while the latter species is inhibited right from the start of the experiment.

Gause also did experiments with three species of *Paramecium*. Using *P. aurelia* and *P. caudatum*, he found that under most circumstances *P. caudatum* was replaced by *P. aurelia*; but if special care was taken to remove metabolic products by completely and repeatedly replacing the medium after centrifugation, this did not happen, and *P. caudatum* started to inhibit *P. aurelia*. Evidently, *P. aurelia* is considerably more resistant to pollution by metabolic products of other organisms than is *P. caudatum*; yet, when such products are removed, *caudatum*, though larger than *aurelia*, can multiply faster than can that species. This experiment is of great importance in showing that the direction in which competition proceeds may be altered by changes in the environment. The phenomenon of the replacement of one species by another in a uniform microcosm can be predicted from theory; that the direction in which competition proceeds may depend on external factors required an experimental investigation. Further aspects of this phenomenon appear in the work on flour beetles to be discussed later.

When *P. bursaria* was used with *P. aurelia*, it was possible to get apparently stable mixed cultures. The integral paths and one actual pair of population growth curves are shown in the middle and upper panels of figure 80. It is evident that the trajectories converge on a stable point, as in the fourth case considered

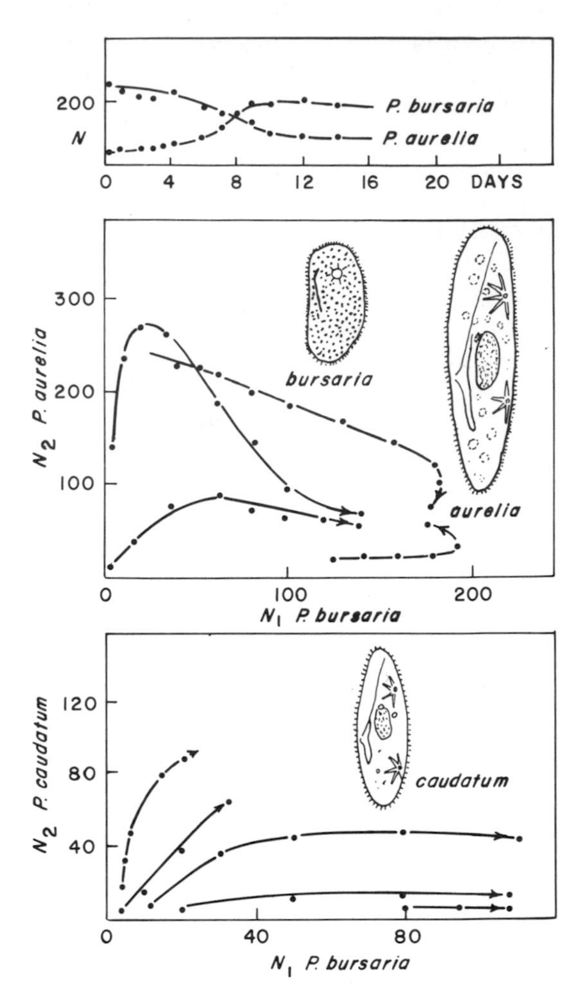

FIGURE 80. Upper panel, population of *Paramecium aurelia* and of *P. bursaria* introduced into the *P. aurelia* culture. Middle panel, trajectories converging on a stable point when *P. aurelia* and *P. bursaria* are cultivated together. Bottom panel, trajectories when *P. caudatum*, much more sensitive to metabolic products, is cultivated with *P. bursaria* (Gause, modified).

by Volterra. This is due to a marked difference in the behavior of *P. aurelia* and *P. bursaria*. The latter species is provided with symbiotic algae or *Zoochlorella*. In any culture in which the vessels are not continually agitated, some of the food bacteria fall to the bottom. *P. bursaria* tends to feed on this sedimented material, and is able to do so partly because the presence of the algae permit the ciliate to enter a thin, somewhat deoxygenated zone full of food at the bottom of the vessel. The sedimented bacteria deplete the oxygen; the algae in the protozoa, if illuminated, compensate for this depletion. We have in fact here essentially imperfect competition in that there is a zone to which *P. bursaria* is adapted but into which *P. aurelia* cannot enter.

When the same experiment is performed with *P. caudatum* and *P. bursaria*, a rather different result is obtained. Though mixed cultures are possible, there seems to be no unique proportion of the two species that is stable. Above a certain concentration of *P. bursaria*, *P. caudatum*, which is (as we have noted) very sensitive to pollution by the metabolic products of other organisms, is inhibited by *P. bursaria*. This can be demonstrated by using medium in which the latter species had previously lived, for the cultivation of *P. caudatum*. No reciprocal repressive effect of *P. caudatum* on *P. bursaria* occurs. In cultures very rich in *P. bursaria*, the latter can continue increasing in spite of the persistence of a small *P. caudatum* population. Presumably the latter ultimately disappears. This type of case requires more study; one might well enquire whether a symmetrically opposite inhibition, with another special zone against the N_2 axis, might not sometimes be present.

This work of Gause and his associates provided cases where the Volterra equations clearly suggested the right answers, showing that when competition is complete only one species survives, but when it is easy to see that it is incomplete, two species can coexist. It also indicated the very important fact that the direction of competition may be reversed by imposition of environmental changes. All subsequent work has shown the existence of these phenomena, even though no universally true propositions can be established as the result of such experiments. This is of course the usual situation in biology; once an existential proposition has been confirmed for a number of instances, further insight is most naturally obtained from the cases where the generalization does not work. If the generalization is usually true, it should not be abandoned merely because the exceptions are interesting.

Much later experimental work has been done with various metazoa. Flour or grain insects, mainly beetles, have proved to be particularly suitable animals for such investigations. Though found primarily in artificial situations, they must by now be well adapted to life with man, as they have been found in grain offered for the dead in ancient Egyptian tombs.

In populations of these animals eating their way through flour or grain, the main sources of mortality are accidental cannibalism. The feeding stages, both larval and adult, will eat eggs that they encounter and also eat into pupae. The density of any population will depend on a balance between oviposition rate and rate of destruction of eggs and pupae. Since the latter process takes place at random, it will increase as the density of the nonmotile stages increases. In mature populations such cannibalism may prevent 99 percent of the eggs that are laid from reaching the adult stage. In the simple environment of flour, which provides no spatial diversification, the primary determinant of survival between competitors is voracity, leading to a higher destruction of eggs and pupae of both species, but a higher production of eggs of the more voracious species. While the growth of flour beetles living in this way is best expressed by equations considerably more complicated than the logistic, the behavior of the equivalent competition equations leads to the same conclusions as the Volterra equations, as Crombie has shown.[10]

10. A. C. Crombie, Further experiments on insect competition. *Proc. Roy. Soc. Lond.*, 133B:76–109, 1946. Crombie writes (see p. 37) for the rate of change of a population of a graminivorous insect

$$\frac{dN}{dt} = N[rf(N)(1 - qn)^h(1 - pn)^k - m]$$

where $f(N)$ is an undetermined function, n is the part of N that is actually feeding, namely, larvae and adults but not pupae, q is the egg consumption per unit feeding individual per day, and h the duration of the egg stage, p and k are the same parameters for the pupal stage, and m is density-independent mortality. The equivalent competition equations are

$$\frac{dN_i}{dt} = N_i[r_i(N_i + N_j)(1 - q_{ii}n_i)^{h_i}(1 - q_{ij}n_j)^{h_j}$$

$$(1 - p_{ii}n_i)^{k_i}(1 - p_{ij}n_j)^{k_j} - m_i].$$

Assuming m_i and m_j are very small compared to the

In a very extensive study of competition between two species of *Tribolium*, *T. confusum* and *T. castaneum*, Park[11] has studied the effect of simultaneous variation of temperature and humidity on the outcome of competitive interaction.

Three temperatures, cool at 24°C, temperate at 29°C, and hot at 34°C, and two humidities, dry at 30 percent relative humidity and wet at 70 percent relative humidity, were used; the six possible combinations were indicated as HW, HD, TW, TD, CW, CD. Twenty replicate cultures starting with 4 ♂♂ and 4 ♀♀ individuals were set up as unispecific controls, and 30 mixed cultures of 2 ♂♂ and 2 ♀♀ of either species were used for the study of competition.

In the unispecific cultures of *T. confusum*, the mean size of the total population decreased HW > TW, TD, CW, CD, HD; only the first pair, as indicated by the greater than sign, differed significantly.

The corresponding order for *T. castaneum* was TW TC > HW > TC > HD > CD, all but the first pair being significantly different. *T. castaneum* is clearly more sensitive to microclimatic differences than is *T. confuseum*. When the species were cultivated together, *T. castaneum* was the sole survivor

		H(34°C)	T(29°C)	C(24°C)
T. confusum	Wet (70%)	0.00	0.14	0.71
	Dry (30%)	0.90	0.87	1.00
T. castaneum	Wet (70%)	1.00	0.86	0.29
	Dry (30%)	0.10	0.13	0.00

TABLE 7.

in all hot-wet cultures and *T. confusum* in all cool-dry ones. The latter result might have been expected as *T. castaneum* did very poorly in cool-dry conditions. The displacement of *T. confusum* from hot-wet cultures is more surprising, because HW was the optimum condition for this species but not for *T. castaneum* in unispecific cultures. In all the other categories, competition proceeded till only one species was present, but the identity of the species varied among the replicates. The resulting frequencies, which may be regarded as estimates of the probabilities of outcome, are shown in table 7.

The origin of the variation in the less extreme circumstances has been attributed by a number of investigators[12] to chance genetic differences in the very small original popula-

other terms, that the action of individuals of the two species affect the egg-production rate equally, and that larvae and adults eat eggs and pupae in the same amounts, we may write

$$\alpha_{ij} = \frac{1 - (1 - q_{ij})^{h_i}(1 - p_{ij})^{k_i}}{1 - (1 - q_{ii})^{h_i}(1 - p_{ii})^{k_i}}.$$

The assumptions are not quite accurate, but the treatment indicates the robustness of the concept of the competition coefficients α_{ij}.

11. T. Park, Experimental studies of interspecific competition: II. Temperature, humidity, and competition in two species of *Tribolium*. *Physiol. Zool.*, 27:177–238, 1954.

T. Park, Beetles, competition and populations. *Science*, 138:1369–75, 1962.

12. In general, genetic variation in *T. castaneum* is more important in determining the outcome of competition than is genetic variation in *T. confusum*; see T. Park, P. H. Leslie, and D. B. Mertz, Genetic strains and competition in populations of *Tribolium*. *Physiol. Zool.*, 37:97–162, 1964.

The principal proponents of the view that indeterminacy is caused by random genetic differences among the members of small founding stocks are:

I. M. Lerner and E. R. Dempster, Indeterminism in interspecific competition. *Proc. Nat. Acad. Sci. U.S.A.*, 48:821–26, 1962.

P. S. Dawson and I. M. Lerner, The founder principle and competitive ability in *Tribolium*. *Proc. Nat. Acad. Sci. U.S.A.*, 55:1114–17, 1966.

P. S. Dawson, A further assessment of the role of founder effects in the outcome of *Tribolium* competition experiments. *Proc. Nat. Acad. Sci. U.S.A.*, 66: 1112–18, 1970.

C. E. King and P. S. Dawson, Population biology and the *Tribolium* model. *Evol. Biol.*, 5:133–227, 1972.

The more important contributions emphasizing largely nongenetic stochastic aspects are:

J. Neyman, T. Park, and E. L. Scott, Struggle for existence. The *Tribolium* model: Biological and statistical aspects. *Proc. 3rd Berkeley Symposium on Mathematical Statistics and Probability*. Berkeley, University of California Press, 1956, pp. 41–79.

P. H. Leslie, A stochastic model for studying the properties of certain biological systems by numerical methods. *Biometrika*, 45:16–31, 1958.

P. H. Leslie and J. C. Gower, The properties of a stochastic model for two competing species. *Biometrika*, 45:316–30, 1958.

tions used. Very recently, Mertz, Cawthorn, and Park have looked into this kind of indeterminacy and concluded that it is due largely to chance events or, as they put it, the ecological stochasticity of the system. Cultures were set up in which both the initial ratios of the two species and the coefficients of inbreeding of the *castaneum* used to found the cultures were systematically varied. The initial ratios had a great effect on outcome. When the experiment started with 40 *confusum* and 10 *castaneum*, *confusum* won in 94 percent of the contests, while with a reciprocal arrangement, it won in only 9 percent. There is a possibility that this is due in part to the presence in the pupae of substances that make attack by a larva on a pupa of the opposite species more probable than on one of its own species. This would produce the situation of (c) in figure 78.

When the founding *castaneum* had a coefficient of inbreeding of 1/2 per pair, *confusum* won in about two-thirds of the cultures; the reverse was true when the *castaneum* came from a stock culture that was practically panmictic. Intermediate degrees of inbreeding gave essentially complete indeterminacy.

These results suggest to Mertz, Cawthorn, and Park that random variations in the populations due to ecological accidents are of major importance in determining the outcome of competition in *Tribolium* when the environmental determinants are weak. There does also seem to be some genetic effect, but it is not what would be expected. Park's work therefore not merely confirms on an extended scale Gause's basic concept that the direction of competition is under environmental control, but introduces a new concept: that where the environmental control is weak, the outcome, though always involving elimination of one species, depends to some extent on chance.

Park[13] also found that under conditions in which *T. confusum* usually eliminates *T. castaneum* if the beetles are infected with the sporozoan parasite *Adelina tribolii*, *T. castaneum* usually eliminates *T. confusum* in the absence of the parasite.

Another elaborate set of experiments using flour beetles was performed by Crombie. Some of these experiments provide a parallel to Gause's work on *Paramecium bursaria* and *P. aurelia*. Crombie[14] found that if the two beetles, *Tribolium confusum* and *Oryzaephilus surinamensis*, were cultivated together in flour, *Tribolium* always displaced *Oryzaephilus*, though the elimination of the last individuals of the latter might be a very slow process. *Tribolium* is more voracious than *Oryzae-*

P. H. Leslie, A stochastic model for the competing species of *Tribolium* and its application to some experimental data. *Biometrika*, 49:1–25, 1962.

P. H. Leslie, T. Park, and D. B. Mertz, The effect of varying the initial numbers on the outcome of competition between two *Tribolium* species. *J. Anim. Ecol.*, 37:9–23, 1968.

D. B. Mertz, D. A. Cawthon, and T. Park, An experimental analysis of competitive indeterminacy in *Tribolium*. *Proc. Nat. Acad. Sci. U.S.A.*, 73:1368–72, 1976.

Leslie put forward as competition equations

$$E[N_{t+1}|N_t, M_t] = \lambda_1 N_t/(1 + \alpha_1 N_t + \beta_1 M_t)$$

$$E[M_{t+1}|N_1, M_t] = \lambda_2 M_t/(1 + \alpha_1 N_t + \beta_1 M_t),$$

where the terms on the left are the conditional expectations of populations N and M of two species at time t + 1, given the values of N_t and M_t. The terms λ_1, λ_2 are essentially equivalent to r_1 and r_2, while the α and β terms are dependent on intra- and interspecific competition.

If the main part of the indeterminacy is genetic, one would expect it to be greatest when at least one species is started with a very small population; replication of this condition would produce more genetic inhomogeneity in the cultures with small than with medium-sized founding populations. If purely environmental change events were involved, one would expect indeterminacy to be greatest when the numbers of individuals in the two populations were about even; see:

D. B. Mertz, The *Tribolium* model and the mathematics of population growth. *Ann. Rev. Ecol. Syst.*, 3:51–78, 1972.

The effect of pupal extracts is discussed in M. F. Ryan, T. Park, and D. B. Mertz, Flour beetles: Responses to extracts of their own pupae. *Science*, 170:178–79, 1970.

13. T. Park, Experimental studies of interspecies competition: I. Competition between populations of the flour beetles *Tribolium confusum* Duval and *Tribolium castaneum* Herbst. *Ecol. Monogr.*, 18:265–307, 1948.

14. A. C. Crombie, see note 10. See also A. C. Crombie, The competition between different species of graminivorous insects. *Proc. Roy. Soc. Lond.*, 132B: 362–95, 1945.

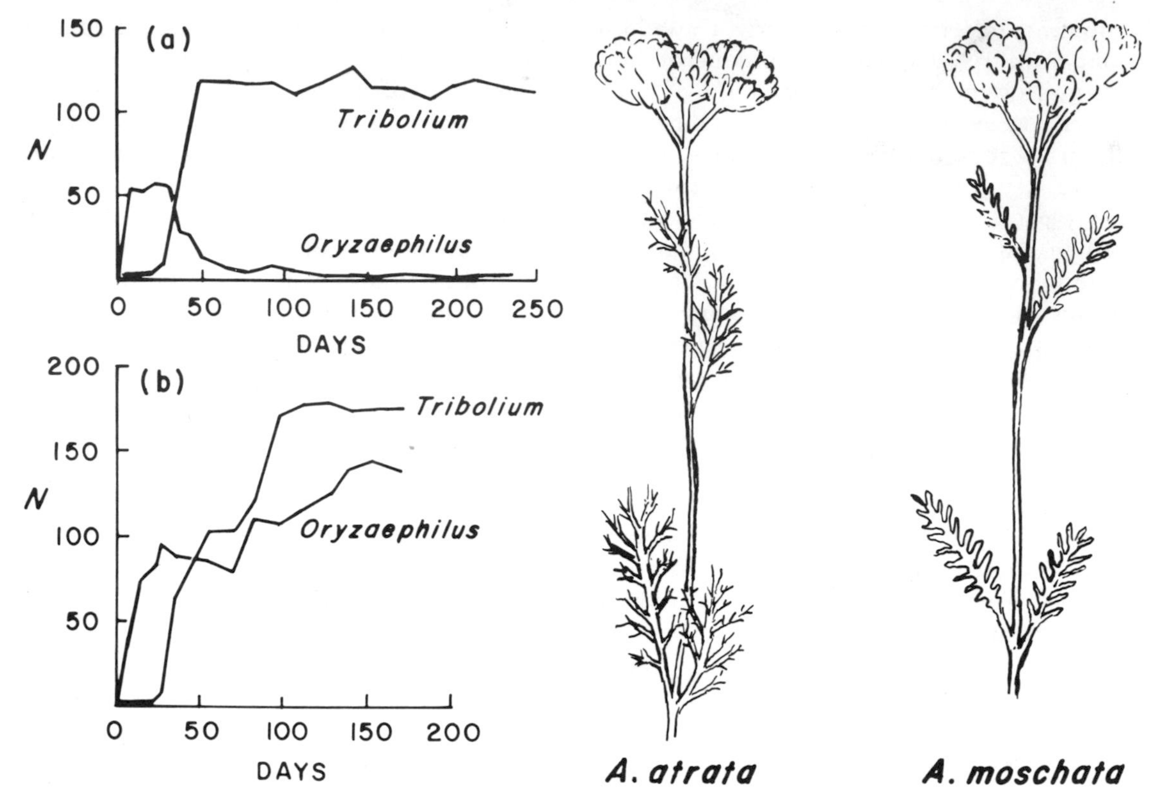

FIGURE 81. (a) Populations of the two grain beetles, *Tribolium confusum* and *Oryzaephilus surinamensis*, cultivated together in flour; (b) the same when short lengths of capillary tubing are mixed with the flour (Crombie).

FIGURE 82. *Achillea moschata* and *A. atrata*.

philus, so that more of the eggs of the latter are eaten by the former than vice versa. Moreover, *Tribolium* eats *Oryzaephilus* pupae, but *Oryzaephilus* cannot eat the pupae of *Tribolium*. These relationships seem adequate to explain the observations on mixed culture in flour. When the insects are raised together in cracked wheat, they form a persistent mixed population. Detailed study of the cultures showed that *Oryzaephilus* pupated in the cracked wheat grains, which afforded protection from *Tribolium*. Short lengths of 1-mm.-bore capillary tubing mixed with flour provided a comparable refuge and permitted a mixed population to persist (figure 81).

Some examples of environmentally determined competition in plants

The determination of the local distribution of flowering plants by environmental regulation of competition was apparent to Nägeli[15] over a century ago. He observed that in the mountain regions of Central Europe, the milfoil (figure 82), *Achillea moschata*, was ordinarily calcifuge, while *A. atrata* was calcicole; however, if *A. atrata* was absent from a particular small area, *A. moschata* could grow on calcareous soil, such as that lying over a limestone erratic. A third species, *A. millefolium*, was apparently indifferent to the calcareous or noncalcareous nature of the soil. Similar phenomena were noted by Nägeli in pairs of species of *Rhododendron*, *Saussurea*, *Gentiana*, *Veronica*, *Erigeron*, and *Hieracium*. The matter was discussed by some later German writers on plant ecology without their adding much new knowledge.

15. C. Nägeli, Ueber die Bedingungen des Vorkommens von Arten und Varietäten innerhalb ihres Verbreitungsbezirkes. *Sitz. K. Akad. Wiss. München*, 1865 (2): 228–84.

In 1917, Tansley[16] reported the results of an experiment initiated at his suggestion by Hume and continued by Marsh and by Tansley himself (figure 83). In this two species of bedstraw (figure 84), *Galium saxatile* and *G. pumilum*, were studied growing separately and together on various kinds of soil. *G. saxatile* is ordinarily calcifuge, *G. pumilum* calcicole. If the two species are grown from seed on a calcareous soil, *G. pumilum* develops normally, while *G. saxatile* is initially chlorotic, though such seedlings as survive for 3 to 5 months become normal plants. In competition, however, the calcicole *G. pumilum* overgrows such plants of *S. saxatile*. When the seeds are sown in peat, both species grow slowly, though individuals of either may flower in the second year. After that *G. saxatile* overtops *G. pumilum*, though odd healthy individuals of the latter may grow up out of the mat of *G. saxatile*. If the experiments had been conducted longer, *G. pumilum* would undoubtedly have disappeared. In sandy loam a comparable, but less extreme dominance of *G. saxatile* was observed. Though these experiments, and the earlier field work on which they were founded, were widely known, they did not initially appear to mark a major advance in ecology.

Experimental competition in neustonic plants

Clatworthy and Harper[17] did a number of experiments on the growth of *Salvinia natans* and several species of *Lemna* (figure 85). In table 8, the specific initial rate of increase, corresponding to *r*, and the maximum yield of *K* are given.

	r	K(8 weeks)
L. minor	$0.243\ g \cdot g^{-1} \cdot day^{-1}$	0.480 gr.
L. gibba	0.218	0.394
L. polyrhiza	0.188	0.457

TABLE 8.

Clatworthy and Harper indicate that *L. polyrhiza* would have a higher saturation population than the others after 12 weeks, though by this time *L. gibba* was probably declining. *L. polyrhiza*, moreover, seems to increase in the later stages of growth faster than the others. In spite of this, in competition *L. gibba* is definitely superior to *L. polyrhiza*. In competition between *L. polyrhiza* and *L. minor* the two species are rather evenly balanced and a stochastic outcome may be expected. Difficulties in certain identification of all plants precluded using *L. gibba* with *L. minor*. *Salvinia natans* behaves rather like *L. gibba*, never being the most effective species by itself but outcompeting *L. polyrhiza*. The direction of competition appears to depend on morphological factors determining which competitor can most easily occupy the surface layer of a mat of plants, rather than on those physiological differences that determine growth rates in unispecific cultures. As in *Tribolium* under warm-moist conditions, prediction of what may happen in competition cannot be made simply by looking at unispecific cultures. Clatworthy and Harper compare their findings with similar results that have been obtained intraspecifically in one-generation competition

16. A. G. Tansley, On competition between *Galium saxatile* L. (*G. hercynium* Weig.) and *Galium sylvestre* Pall. (*G. asperum* Schreb.) on different types of soil. *J. Ecol.*, 5:173–79, 1917. The second of these species appears to be correctly named *G. pumilum* Murray.

The caricature of Tansley and Blackman was evidently one of a series, but no others appear to have survived. Denis Gascoigne Lillie was a student at Cambridge and later became a marine biologist on the Terra Nova Expedition under the command of Captain Scott, who was lost on his return from the South Pole. Lillie suffered a mental collapse in 1919 and never recovered; he died in 1962. About fifteen of his caricatures are believed to survive; of these, four are in the National Portrait Gallery, London, and four in the Scott Institute for Polar Research, Cambridge, England, including a most beautiful one of the heroic Dr. E. A. Wilson, the medical officer of Scott's Expedition and a distinguished ornithologist, depicted as a penguin. See also G. E. Hutchinson, A Cambridge caricaturist. *Country Life*, 144:644–45, 1968.

17. J. N. Clatworthy and J. L. Harper, The comparative biology of closely related species living in the same area: V. Inter- and intraspecific interference within cultures of *Lemna* spp. and *Salvinia natans*. *J. Exp. Bot.*, 13:307–24, 1962.

FIGURE 83. "Cambridge Illustrated Lecture Guide No. 2. Mr. Blackman and Mr. Tansley. General Botany (Intermediate). Times to be arranged. £1.1s." Sir Arthur G. Tansley, F.R.S. (1871–1955), who introduced the word "ecosystem," founded and long edited the *New Phytologist* (the title comes from Sir Thomas Browne) and was important for his part in establishing the *Journal of Ecology*, is the tall thin figure on the right; his works on British vegetation are classical. Frederick Frost Blackman, F.R.S. (1866–1947), who became Tansley's brother-in-law, is best known for his work on photosynthesis. Both men had immense influence as teachers (Caricature by D.G. Lillie, National Portrait Gallery, London.)

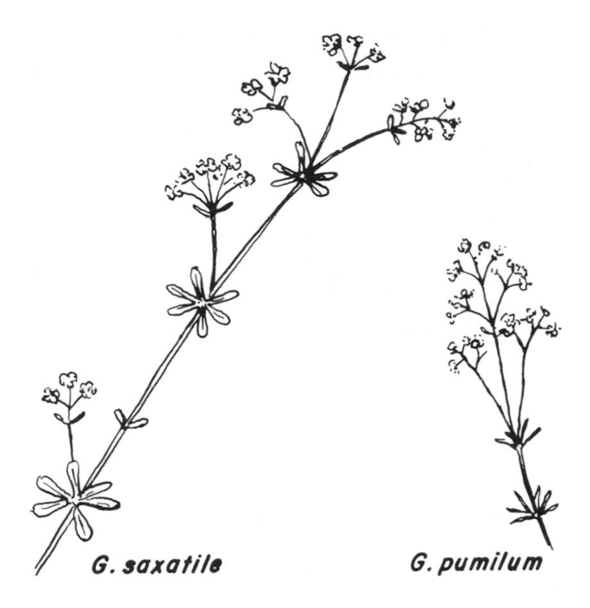

FIGURE 84. *Galium saxatile* appearing in nature as calcifuge and *G. pumilum* as calcicole.

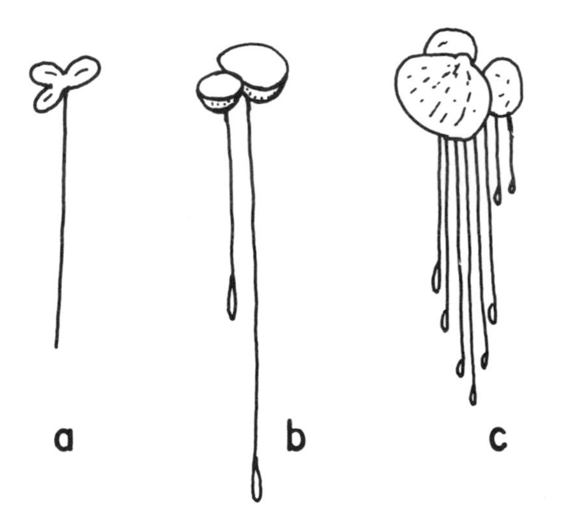

FIGURE 85. The three species of *Lemna* used in Clatworthy and Harper's experiments. (a) *Lemna minor*, (b) *L. gibba*, (c) *L. polyrhiza*.

experiments with barley, in which the variety doing best by itself was by no means the most successful competitor. All such cases emphasize the enormous potential difficulty of constructing a predictive theory of the evolutionary process.

Interspecific competition in the Hydridae
Slobodkin[18] has done some fascinating and important experiments with a brown hydra, *Hydra littoralis*, in competition with the green *Chlorohydra viridissima*, which contains symbiotic algae. In competition experiments the animals were fed on a roughly constant ration of *Artemia* nauplii, made up to contain 0.05 ml of live food, added from above. The cultures were left for half an hour, after addition of food, under a towel; this reduction of illumination prevented aggregation of the nauplii. After feeding, the water containing excess food was poured off and replaced.

When cultures were kept permanently in darkness except for brief periods of manipulation, a stratification, comparable to that in a forest on a minute scale, developed. The brown *H. littoralis* grew to be taller than *C. viridissima*, and so could expand its tentacles nearer the surface, from which food was added. Enough food, however, descended to the understory, composed of *C. viridissima*, to permit the latter's continued existence (figure 86, a). Artificial predation in this case favored *C. viridissima*, apparently by creating more gaps in the *H. littoralis* canopy through which food could fall.

When the mixed culture was illuminated, *C. viridissima*, now having an additional energy source as the result of the photosynthetic activity of its algal symbionts, grew better than did *H. littoralis* and displaced the latter species in about 50 days (figure 86, b). If artificial predation was applied to the system by removing the same proportion of both species, *H. littoralis* persisted longer, and with intense enough predation, by which 90 percent of the hydrids produced were removed, both species persisted together, *H. littoralis* forming a small and fairly stationary population established in company with a very variable population of *C. viridissima* (figure 86, c). Slobodkin[19] had already developed a theory, based on Gause's modification of the Volterra equations, by the

18. L. B. Slobodkin, Experimental populations of Hydrida. *Brit. Ecol. Soc. Jubilee Symposium, Suppl. J. Ecol.*, 55 or *J. Anim. Ecol.*, 33 (issued separately): 131–48, 1964.

19. L. B. Slobodkin, *Growth and Regulation of Animal Populations.* New York, Holt, Rinehart, and Winston, 1961, vi, 184 pp.; see pp. 75–80.

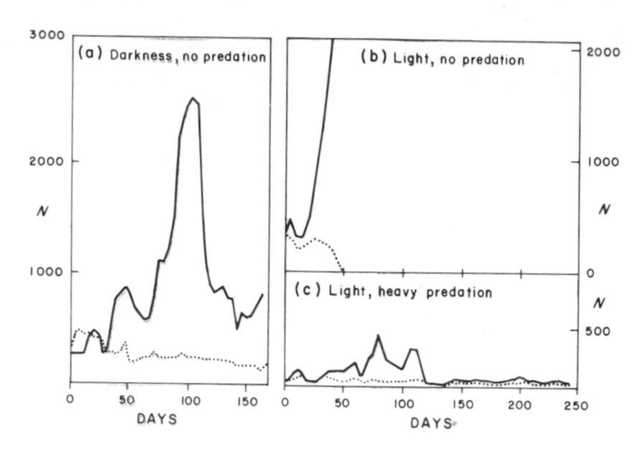

FIGURE 86. Competition between populations of *Hydra littoralis* (broken line) and *Chlorohydra viridissima* (solid line); (a) in darkness, (b) in light without predation, (c) in light with the equivalent of 90 percent of the number of new mouths produced in the previous 4 days removed, and regarded as predation.

introduction of a rarification or death rate imposed unselectively on both species, which showed that at the right level of unselective predation, the two species could coexist. Subsequent studies in nature have confirmed on a large scale the role of predation in maintaining the diversity of ecosystems.

Interspecific competition in Cladocera

Some very interesting work, which gives hints as to what may happen in nature, has been done by Frank.[20] In the simpler of his two sets of experiments he studied two species of *Daphnia*, *D. magna* and *D. pulicaria*. The former species is mainly found in small, very shallow, and often temporary bodies of water, not inhabited by fish; it is a large species which is vulnerable to predation and has been widely used and distributed by aquarists as a food for fish. *D. pulicaria* is smaller and occurs in a variety of ponds and lakes; in general, the two species are not likely to co-occur, though they perhaps may do so occasionally.

In the first experiments with these two species, the animals were reared in small vessels containing 50 ml of water, and were fed on the motile green alga, *Chlamydomonas*. Though initially *D. pulicaria* grew slightly less rapidly than in the unispecific control, at all times it increased more rapidly than the larger *D. magna* and soon displaced it completely (figure 87).

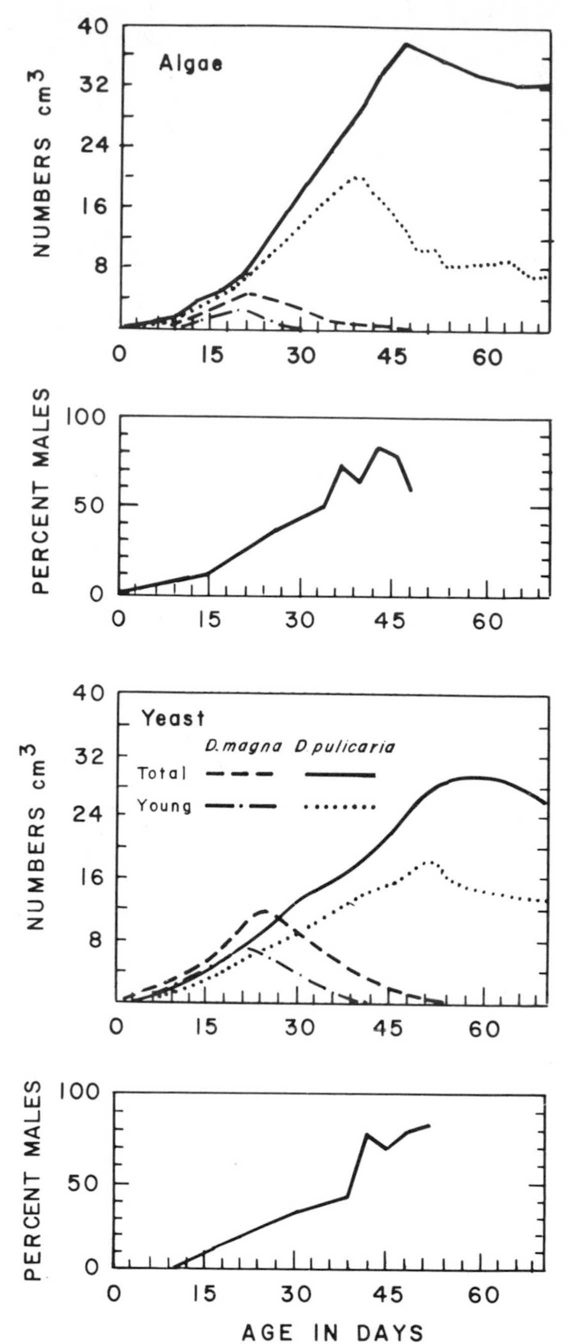

FIGURE 87. Population changes in a mixed culture of *Daphnia magna* and *D. pulicaria* fed on *Chlamydomonas*; the same fed on yeast. The percentage of males produced by *D. magna* is shown in the second panel of each part of the figure.

20. P. W. Frank, Coactions in laboratory populations of two species of *Daphnia*. *Ecology*, 38:510–19, 1957.

When the same kind of experiment was done using yeast as a food, *D. magna* initially grew faster than did *D. pulicaria*, but after about 24 days started to decline. Meanwhile the population of *D. pulicaria* continued to grow and finally, as in the previous experiment, displaced *D. magna* completely (figure 87). The explanation of the initial rapid growth of *D. magna* lies in the fact that this species is more tolerant of low oxygen concentrations than is *D. pulicaria.* When the *Daphnia* concentrations are small, food added at regular intervals tends to persist in the culture medium. When the food is yeast, its respiration reduces the oxygen concentration in the medium and this favors *D. magna.* When the food is the photosynthetic *Chlamydomonas*, this effect is much more transitory, and *D. magna* is thus initially much less favored. As the population of *D, magna* grows in the yeast-fed culture, it assists in removing the respiring food, so that the oxygen rises faster than it would in a unispecific yeast-fed culture of *D. pulicaria* in which a large population takes a long time to develop. *D. magna* by its initial strong growth at rather low oxygen concentrations thus prepares the medium for the growth of *D. pulicaria.*

The actual elimination of *D. magna* is due much less to its failure to reproduce as the culture becomes dense, than to its greater proneness to produce males, which may form 80–90 percent of the *D. magna* population in mixed cultures, only about 20–30 percent in old unispecific cultures, and is of negligible importance in any of the *D. pulicaria* cultures.

These results may be compared with what happens in mixed-mass cultures in vessels containing 8 liters of water. In such cultures *D. pulicaria* may persist for 11 to 18 months, though it finally disappears. The very long period of coexistence suggests some environmental heterogeneity in the mass cultures: the final disappearance of *D. pulicaria* may well be due to critically low oxygen concentrations occurring with a low but finite probability.

After the *D. pulicaria* had disappeared from the water, the cultures were allowed to dry; the vessels were then filled up with water into which many resting or ephippial eggs hatched. In the resulting population of *Daphnia, D. pulicaria* was several times as common as *D. magna.* In a natural environment in which the water level was low during a dry season, it might be possible to have a regular succession of the two species, *D. pulicaria* appearing as the water rose after rains, *D. magna* replacing it as the water evaporated and became deoxygenated by decay of vegetation.

Frank[21] also worked with *Daphnia pulicaria* in competition with *Simocephalus vetulus*; the two species belong to the same family but differ somewhat in structure and behavior, *Daphnia* being a continuous obligate swimmer, whereas *Simocephalus* can attach to the surfaces of solids by the secretion of its cervical gland, in which position it produces a feeding current. In consequence of these differences *Simocephalus* tends to be littoral. In Frank's study both species came from the same body of water near Chicago.

The natality, as is usual in the Cladocera, is strongly density-dependent; if comparison of this effect is made between the two species in unispecific cultures, the general inhibition of birth as sparse populations become moderately dense is about the same in the two species (figure 88), but when densities in excess of 6 or more individuals per ml are considered, the natality of *Simocephalus* is reduced far more than that of *Daphnia.* In mixed cultures at low densities *D. pulicaria* inhibits its own natality more than that of *Simocephalus*, but as the cultures become denser, increasing inhibition of *Simocephalus* occurs. The effect of *Daphnia* on *Simocephalus* is frequency-dependent, being greatest when the population contains 10 percent *S. vetulus* and 90 percent *D. pulicaria.* If the proportion of *Simocephalus* decreases further the birthrate of the latter rises. Moreover, there is a clear maximum death rate of *Simocephalus* in populations containing equal numbers of both *S. vetulus* and *D. pulicaria*, while in

21. P. W. Frank, A laboratory study of intraspecies and interspecies competition in *Daphnia pulicaria* (Forbes) and *Simocephalus vetulus* O. F. Muller. *Physiol. Zool.*, 25:178–204, 1952.

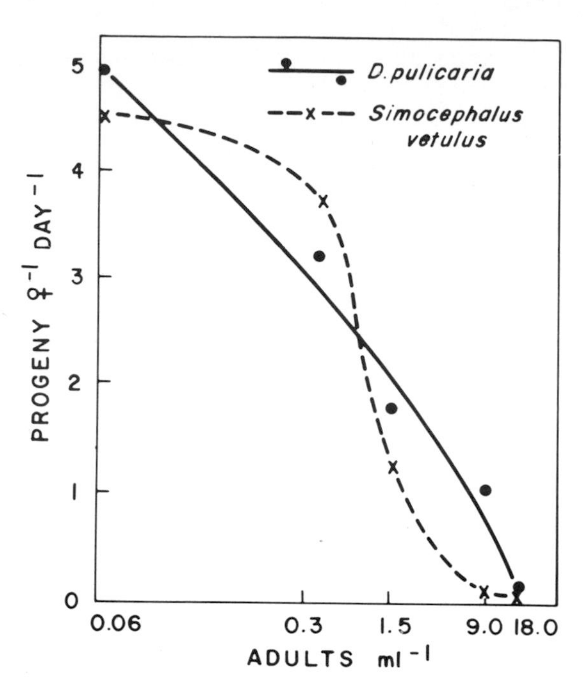

FIGURE 88. Dependence of natality on density in *Daphnia pulicaria* (solid line) and *Simocephalus vetulus* (broken line) (Frank, modified).

Daphnia, longevity, though greatest when the animal is common, is not clearly minimal in a population of equal proportions of the two species. These observations suggest that in crowded cultures of *Daphnia*, a few *Simocephalus* may persist for a very long time and do a little breeding, as indeed is often observed to be the case.

Competition in experimental populations of Drosophila

A great deal of work has been done on competition in *Drosophila*. With this insect, in which the adults spend most of their time on the interface between the air space and the solid nutrient medium, while the larvae burrow near the surface of the latter, it is impossible to avoid the introduction of heterogeneity into the system. This heterogeneity may be of a very subtle kind; the edge of a nutrient surface may provide a different habitat from its center. Moreover, the experiments intended to reach an equilibrium population must be run for many months, so that it is extremely difficult to provide temporal as well as spatial homogeneity. These limitations must be born in mind in interpreting the results.

The first work was done by L'Héritier and Teissier[22] in 1935, who cultivated *D. melanogaster* and *D. funebris* together in population cages and found that using wild-type strains the population after about 50 days settled down to containing between 1 percent and 6 per cent *D. funebris*, the proportions oscillating irregularly over a period of about 14 months. When the same experiment was done with *D. melanogaster* w, the white-eyed mutant, which is less fecund and has a lower life expectancy than the wild type, the equilibrium proportion of *funebris* was higher, around 17 percent. L'Héritier and Teissier pointed out that their observations did not accord with theory, though they mentioned that Gause had obtained comparable results with protozoans; presumably they were referring to the experiments with *Paramecium aurelia* and *P. bursaria*.

The same two species were used again by Merrell,[23] who found the same general result, *funebris* persisting in most cultures after 200 days as 3–30 percent of the population. Merrell, however, found that this was duc to the regular addition of food. If food was not added, *melanogaster* succumbed more rapidly than *funebris*, so that such starved cultures consisted entirely of the latter species. Neither species depresses the fecundity of the other, but in well-fed cultures the larvae of *melanogaster* are inimical to those of *funebris*, so that only 62 percent of the eggs laid produce pupae in the competitive situation, while 95 percent of the *funebris* controls pupate. There is no effect of *funebris* on the larval survival of *D. melanogaster*. In experiments in which *funebris* persisted, the *melanogaster* were shown to result from rapid

22. P. L'Héritier and G. Teissier, Recherches sur la concurrence vitale: Étude de populations mixtes de *Drosophila melanogaster* et de *Drosophila funebris*. *C. R. Mem. Soc. Biol.*, 118:1396–98, 1935.

23. D. J. Merrell, Interspecific competition between *Drosophila funebris* and *Drosophila melanogaster*. *Amer. Natural.*, 85:159–69, 1951.

development on fresh food, while the *funebris* grew up on food that was too old adequately to support *D. melanogaster* larvae. The direction of competition was thus reversed about once a month when new food was added. Merrell concluded that his results in no way contradicted the principle of competitive exclusion. Actually his experiment forms a very real model of one explanation of the diversity of the phytoplankton to be discussed on a later page.

A significant advance in the analysis of competition in *Drosophila* cultures was made by Miller[24] who studied the details of the process in the sibling species, *D. melanogaster* and *D. simulans*. If the effect of crowding in unispecific cultures is studied by rearing larvae at a variety of larval densities, it is found that initially the number of flies emerging increases with the number of larvae almost linearly and with only a very small advantage in larval survival in favor of *D. melanogaster*. When more than 120 larvae per vial were used, a reduction of the flies emerging took place, but this was strikingly greater in *D. simulans*. Miller concluded that so long as the larval concentration was below about 120 larvae per vial, the competition coefficients were both essentially equal to unity and that both species could probably survive. This might occur if the reproductive activities of the adults were inhibited by the deterioration of the surface of fresh medium, a process that reduces the rate of egg laying and prevents the full expression of the reproductive potential of the adults. When a very dense population develops, *D. melanogaster* would presumably have a marked advantage over *D. simulans*. Experiments indicate, however, that the interaction between the two species is more complex than might be assumed from the finding that at low or moderate densities the population biology of the two species is about the same. Actually the survival of individuals of both species is frequency-dependent, so that both species do rather less well when they are present in equal numbers than when the proportion is very disparate. This suggests a situation not unlike that observed in *Simocephalus*.

Ayala[25] and his associates have further studied situations comparable to that initially recognized by Miller. Such situations are evidently common in *Drosophila* and may lead to coexistence under conditions in which competitive exclusion would be expected. Ayala has pointed out that in the classical case of imperfect or, as he terms it, conditional competition (as illustrated in figure 78, d), the point corresponding to equilibrium lies outside the line K_1K_2. Moreover, as we have seen, in such a case $\alpha_1\alpha_2 < 1$. Knowing the final outcome of an experiment that has reached equilibrium, we can obtain from the equations of the isoclines, knowing K_1 and K_2 from unispecific populations, and N_{1_e} and N_{2_e} the equilibrium populations in a mixed culture,

$$\alpha_1 = (K_2 - N_{2_e}) / N_{1_e}, \qquad \alpha_2 = (K_1 - N_{1_e}) / N_{2_e}. \tag{4.5a, b}$$

In eight cases involving competition between various strains of *D. pseudoobscura* and *D. serrata*, *D. melanogaster* and *D. pseudoobscura*, *D. nebulosa* and *D. serrata*, and *D. pseudoobscura* and *D. willistoni*, in only one case was $\alpha_1\alpha_2 < 1$. It is evident that in these cases the simple-minded formulation of figure 78 does not apply.

In a study of coexistence of *D. willistoni* strain M11 and *D. pseudoobscura*, Ayala found that the competitive fitness of each species is related to its relative frequency. Cultures were maintained in such a way that each population consisted of the inhabitants of five bottles, one containing adult reproducing

24. R. S. Miller, Larval competition in *Drosophila melanogaster* and *D. simulans*. *Ecology*, 45:132–48, 1964.
R. S. Miller, Interspecific competition in laboratory populations of *Drosophila melanogaster* and *Drosophila simulans*. *Amer. Natural.*, 98:221–38, 1964.

25. F. J. Ayala, Experimental invalidation of the principle of competitive exclusion. *Nature*, 224:1076–79, 1969.
F. J. Ayala, Competition between species: Frequency dependence. *Science*, 171:820–24, 1971.
F. J. Ayala, Competition between species. *Amer. Scientist*, 60:348–57, 1972.
See also M. E. Gilpin and K. E. Justice, Reinterpretation of the invalidation of the principle of competitive exclusion. *Nature*, 36:273–74, 299–301, 1972.

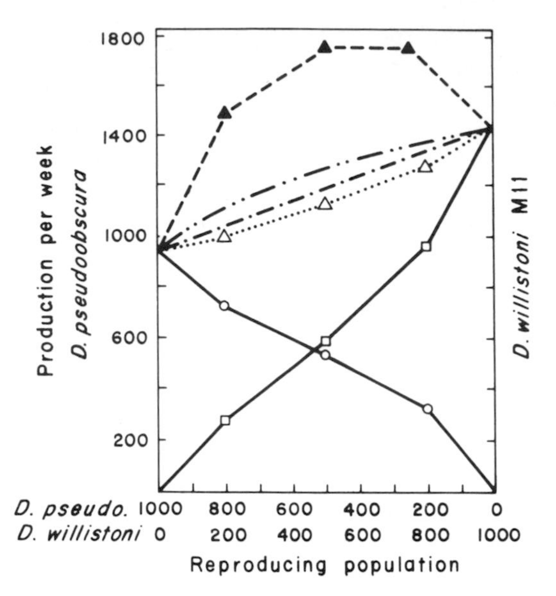

FIGURE 89. DeWit diagram showing the relation of the number of individuals produced per week to the reproducing populations in mixtures of *Drosophila pseudoobscura* and *D. willistoni*. The solid lines give the number of flies of each species produced, while the dotted line gives the sum. Note that this line lies below the dashed and dotted line joining the points corresponding to production of each species by itself from 1,000 flies. The uppermost broken line gives the sum of individuals produced in two single-species cultures with the same numbers of reproducing individuals as in the mixed cultures; here there is no interspecific competition and intraspecific competition is reduced, so that 500 flies by themselves produce more than half the progeny of 1,000 flies. The dashed and double-dotted line gives the sort of result to be expected if the two species had separate niches, not completely eliminating competition, but allowing coexistence (Ayala, slightly modified).

flies, the other four with eggs, larvae, pupae, and newly emerged adults.

If the production of flies per week is plotted against the number used in establishing the culture, a curve is obtained for each species (solid lines in figure 89) and for the two together (dotted line of figure 89). If neither species had any effect on the other, the dotted line should fall on a line giving the sum (dashed line of figure 89) of the populations of controls of each species reared under identical conditions as the experimentals. If competition were complete so that the two species shared the resources in proportion to their initial numbers, the resulting population would fall on the broken and dotted line joining K_1 and K_2. Incomplete competition of the kind ordinarily permitting coexistence would correspond to some such curve as the broken double-dotted curve of figure 89. Actually the dotted line for the sum of the observed coexisting populations falls significantly below the line K_1K_2 rather than above it. This must mean that when the two species are about equally common, they exert a greater unfavorable effect on each other than when one or the other is abundant; rareness thus favors the competitive process. In such a mixed population, the rarer species will start with an advantage that it loses as it becomes commoner. *D. pseudoobscura* has at all densities a slightly better adult survivorship than *D. willistoni*. The frequency dependence is shown primarily by larval survival and the equilibrium is determined by the point at which the frequency-dependent larval advantage of *D. willistoni* balances the adult advantage of *D. pseudoobscura*. The appropriate mathematical models for this process are discussed later.

Competition in poppies

A very important case, which bears a formal relationship to the examples of coexistence in *Drosophila*, later analyzed by Ayala, had been discovered by McNaughton and Harper[26] in the genus *Papaver*, the small field poppies of the temperate Old World.

In Britain there are five species (figure 90) of the genus, arranged in two sections. The commonest species is *P. rhoeas* which has the largest and most showy flowers. Within the same section Orthorhoeades, there are two other species with rather more divided leaves

26. J. H. McNaughton and J. L. Harper, The comparative biology of closely related species living in the same area: I. External breeding barriers between *Papaver* species. *New Phytol.*, 59:15–26, 1960.

J. L. Harper and J. H. McNaughton, The comparative biology of closely related species living in the same area: II. Interference between individuals in pure and mixed populations of *Papaver*. *New Phytol.*, 61: 175–88, 1962.

J. H. McNaughton and J. L. Harper, Biological flora of the British Isles: *Papaver* L. *J. Ecol.*, 52:767–93, 1964.

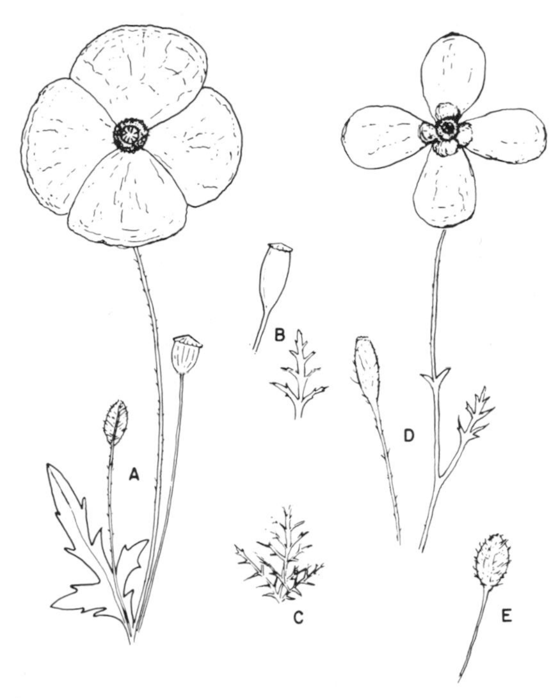

FIGURE 90. Species of *Papaver* occurring in Britain. Section Orthorhoeades: wide petals, capsule not bristly; A, the common red poppy, *P. rhoeas*; B, the long smooth-headed poppy, *P. dubium*, capsule and leaf; C, Babington's poppy, *P. lecoqii*, leaf grown under same conditions as that of *P. dubium* in B. Section Argemonorhoeades: narrower petals, capsule with spinous bristles; D, long prickly-headed poppy, *P. argemone*; E, round prickly-headed poppy, *P. hybridum*, capsule (from material in the herbarium of the Peabody Museum, Yale University, except leaves of B and C redrawn from McNaughton and Harper).

and more elongate seed capsules. These two species are primarily distinguished by the commoner of them, *P. dubium*, having latex that remains white when exposed to the air, while that of the rarer *P. lecoqii* turns dark yellow; minor differences are recorded in the other structures of the plant and the two species are cytologically distinct (*rhoeas* 2n = 14, *lecoqii* 2n = 28, *dubium* 2n = 42). These two species alone can produce at least partly fertile F_1 hybrids, but in nature this rarely if ever happens.

The other two species, in the section Argemonorhoeades, are smaller and less conspicuous, *P. argemone* (2n = 12) has a long capsule, *P. hybridum* (2n = 14) an ovoid one. *P. dubium* and *P. lecoqii* ordinarily flower earlier than *P. rhoeas*; *P. argemone* may be still earlier. *P. rhoeas*, *P. dubium*, and *P. argemone* are widely distributed in England, though less common in Wales, Scotland, and Ireland. *P. lecoqii* and *P. hybridum* are much more local, mainly in southeast England; their range appears to be contracting, though that of *P. dubium* may be increasing. Studying the poppies of 75 habitats in southeast England, McNaughton and Harper found the distribution given in table 9.

It is evident that only *P. rhoeas* occurs commonly by itself, though at least *dubium*, *lecoqii*, and *argemone* as well as the Mediterranean *apulum* can grow in pure stands in cultivation. When populations of two species are set up, there is a considerable variety of response, but at least in the populations

rhoeas + *dubium*
rhoeas + *lecoqii*
lecoqii + *argemone*
dubium + *argemone*
lecoqii + *argemone*

the final dry biomass (figure 91) is well below the value expected by weighting the *K* values, and in only one case, namely, the very closely allied *lecoqii* and *dubium*, does a mixture clearly yield more than the *K* value of the more productive species. When the number of individuals (figure 92) rather than mass is considered, in every case except *lecoqii* and *argemone*, there is a mixture which gives a clear minimum number of individuals. Pure stands thus suffer excessive mortality from the substitution of a small number of a related species. It is therefore quite clear that striking frequency dependence is involved in determining the outcome of competition in *Papaver*. If we suppose that *P. rhoeas* is more eurytopic and better dispersed than the other species, there is probably nothing mysterious about it occurring in many places by itself. That the other species always co-occur with it may be merely due to them always finding *P. rhoeas* wherever they go, but also always being able to coexist with it.

Genetic changes in competition experiments

Haldane[27] pointed out over 40 years ago that when two species are competing, one being

27. J. B. S. Haldane, *The Causes of Evolution*. London, New York, and Toronto, Longmans, Green, 1932, 235 pp.

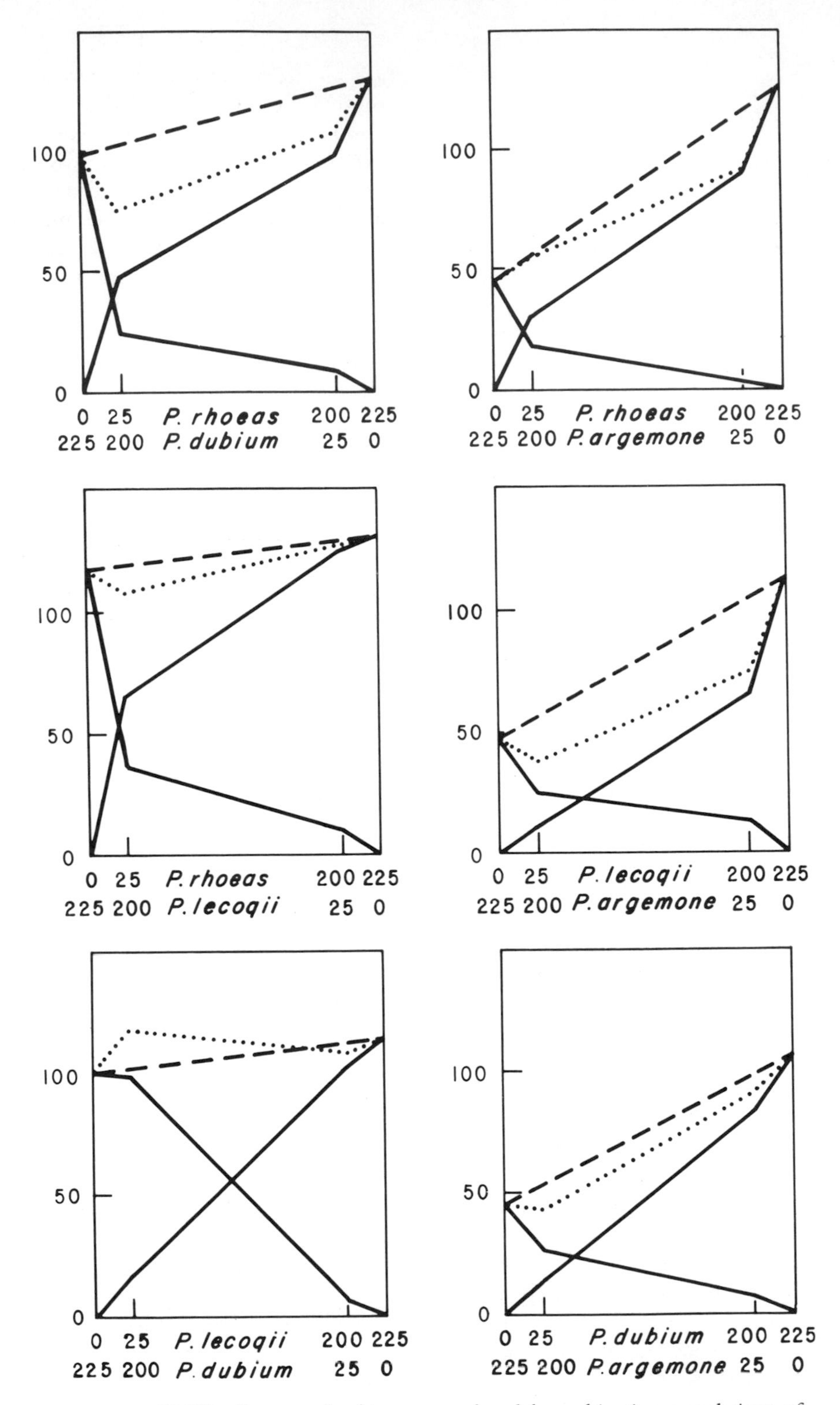

FIGURE 91. DeWit diagrams for biomass produced by cultivating populations of pairs of species of poppy, initiated by sowing seed of the two species in different proportions. The symbolism of the lines as in figure 89. Note that only in the case of the pair of sibling species, *P. dubium* and *P. lecoqii*, does the line giving the sum lie above the line joining the points for the pure species (Harper and McNaughton, modified).

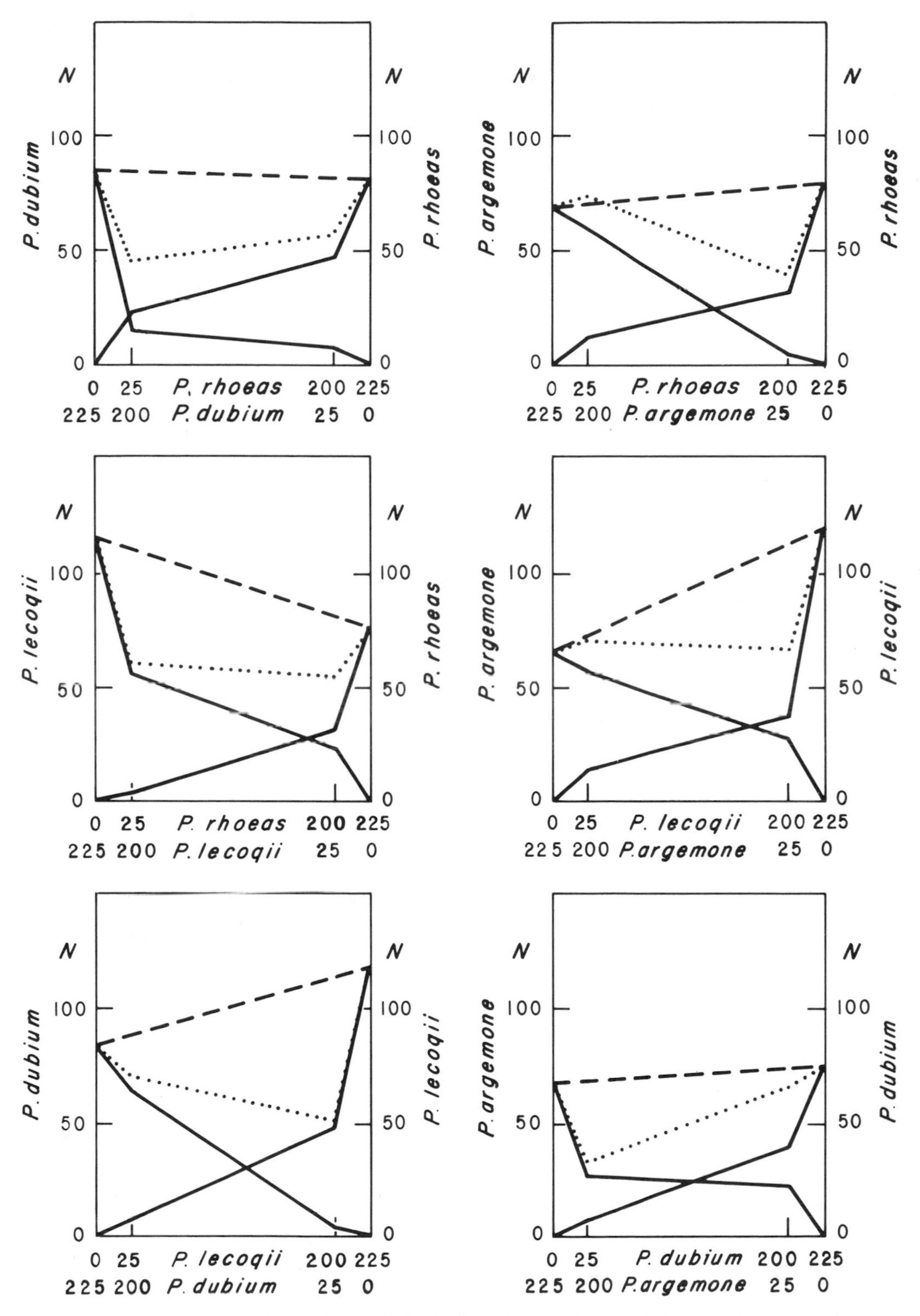

FIGURE 92. DeWit diagrams for number of individuals in the experiments of figure 91. In no case does the line giving the sum of the two species significantly lie above that joining the points for the pure species (Harper and McNaughton, modified).

	P. rhoeas	*P. dubium*	*P. lecoqii*	*P. argemone*	*P. hybridum*
P. rhoeas	30				
P. dubium	33	0			
P. lecoqii	11	6	0		
P. argemone	22	17	3	0	
P. hybridum	9	2	0	5	0

TABLE 9. Co-occurrence of the British species of *Papaver* in 75 habitats in southern England.

rare and the other common, any given individual of the rare species will have predominantly interspecific contacts, while any given individual of the common species will have mainly intraspecific contacts. Selection will therefore improve the interspecific competitive ability of the rare species a great deal more rapidly than it will improve that of the common species. The rare species will therefore become commoner, though its rate of improvement as a competitor will decrease.

Pimentel, Feinberg, Wood, and Hayes[28] did an elaborate experiment in which house flies (*Musca domestica*) and blowflies (*Phaenicia sericata*) were released in opposite corners of a 16-cell population cage formed of plastic boxes connected by plastic tubes. Control populations reared together in ordinary cages showed a strong tendency for *M. domestica* to displace *P. sericata*, though under the conditions used there is some indeterminacy.

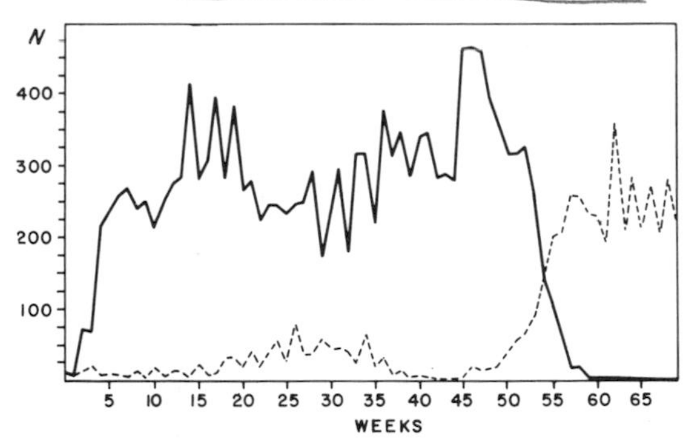

FIGURE 93. Population of the housefly, *Musca domestica* (solid line), competing in a 16-cell population cage with one of the blowfly, *Phaenicia sericata* (broken line). The former species almost displaced the latter after 40 weeks of coexistence, but then a dramatic increase in the blowfly population eliminated the housefly. There is evidence that the process involved natural selection (Pimentel, Feinberg, Wood, and Hayes).

In the many-chambered cages, the structure of which greatly slowed the processes of interspecific interaction, *M. domestica* initially tended to increase, but was later overtaken by *P. sericata* (figure 93). After some indication of oscillation in the relative sizes of the populations, *M. domestica* was displaced by *P. sericata*. Cultures from the surviving *P. sericata* were shown to have a greater capacity for competition with *M. domestica* in ordinary single-chambered cages than did the original *P. sericata* stocks. This type of competition, if it took place in a more elaborate environment, in which the free interaction of the two species is reduced even more than in a 16-cell cage, might lead to indefinite oscillation, though presumably some limit to the adaptation possible in each species would ultimately lead to the one that first failed to keep up in this process becoming extinct.

It is to be noted that although rarity ultimately confers a competitive advantage in such a case, the mechanism must be different from that observed in single-generation experiments such as Harper and McNaughton's study of *Papaver*. Selection of the kind found in *Phaenicia* competing with *Musca* may underlie a little of the frequency dependence observed in some of the experiments with *Drosophila* or with *Simocephalus*, but is unlikely to be a major determinant of what is observed in any experiments in which there is free access over a short distance to all parts of the habitat.

28. D. Pimentel, E. H. Feinberg, P. W. Wood, and J. T. Hayes, Selection, spacial distribution, and the coexistence of competing fly species. *Amer. Natural.*, 99:97–109, 1965.

Single-generation studies on competition in vertebrates

A good many experiments have been done in which populations of two or more species of vertebrates have been established in isolated areas or volumes of natural habitat, previously cleared of the competing species. In such experiments it has been usual to follow competition over relatively short times, seldom more than one year. In such cases at least most of the competition does not involve modification of breeding success. Though elimination of one competitor, or its marked reduction, would indicate that in the course of time only one species would remain, it is by no means certain that all cases where rather feeble competitive interactions or possibly none at all occurred, indicate clearly the possibility of coexistence over many generations.

A number of cases have been reported in mammals and have been reviewed by Grant.[29] As in all fairly large-scale field experiments, complete uniformity of all conditions other than those being manipulated, so permitting the establishment of adequate controls, is very difficult.

The best experiments are those of Grant himself, who used the common North American meadow vole, *Microtus pennsylvanicus*, which is ordinarily found in grasslands, and the red-backed vole, *Clethrionomys gapperi*, and the deer mouse, *Peromyscus maniculatus*, which ordinarily occur in more wooded areas. The three enclosures used each had an area of an acre and were half grassland and half woodland.

When only one species is present, it is found in both types of habitat. If *P. maniculatus* is first established, addition of a small population of *M. pennsylvanicus* restricts the deer mice to the woods and adjacent grass; if a large addition of meadow voles is made, the *P. maniculatus* is completely restricted to the areas with trees. Removal of *M. pennsylvanicus* permits a spread of the deer mice into the grassy area. The same sort of change occurs when *Clethrionomys gapperi* is used as the woodland species. Observations of the behavior of animals of these three genera, when brought together in pairs in the laboratory, are consistent with the hypothesis that *Microtus* can establish dominance and drive off members of the other two genera in grass, while in a more wooded situation the reverse occurs. The outcome of interspecific competition in such a case is settled before there is any important competition for food, let alone any depression of fecundity.

A most peculiar case has been reported by Davis[30] in a comparable experiment done with birds. He found the golden-crowned sparrow, *Zonotrichia atricapilla*, feeding on seeds and sprouting plants along a limited stretch, with continuous cover, of the margin of a field on the Hastings Reserve, Carmel Valley, California. The junco, *Junco hyemalis*, fed solely on seeds along the rest of the margin. Davis found a partial exclusion of the junco from the area occupied by the golden-crowned sparrow. When the latter birds were removed by trapping, juncos moved into their area. Release of the sparrows caused a retreat of the juncos.

The golden-crowned sparrows occupied an area providing them with succulent shoots as well as seeds; in consequence they had no need to drink. Juncos eating only seeds, and so obtaining a much drier diet, needed a supply of liquid water, but found that the best source lay in a trough in the *Zonotrichia* area from which they were largely excluded. The outcome competition can thus be due to very specific circumstances, unlikely to be repeated when the next case of close association of the two species is encountered. Bearing in mind this lack of generality, antimonarchists and Marxists can draw any moral that seems appropriate.

Exploitation and interference

In passing from the experiments with protozoa, through those on a variety of multicellular plants and animals till we reach the birds and mammals, the general pattern of

29. P. R. Grant, Interspecific competition among rodents. *Ann. Rev. Ecol. Syst.*, 3:79–106, 1972.

30. J. Davis, Habitat preferences and competition of wintering juncos and golden-crowned sparrows. *Ecology*, 54:174–80, 1973.

competition changes somewhat. In the ciliates and Cladocera in which feeding involves the collection of small suspended particles, the winning competitor is either the one that collects most food in a given time, or uses that food, under the conditions of the experiment, most efficiently in making new food-gatherers in its own image. In the competition between woodland and grassland rodents, quite elaborate patterns of behavior seem to insure that the inappropriate species is driven out by the appropriate one long before any critical competition for food develops. These two types of competition have been termed[31] *exploitation* and *interference* respectively.

The distinction is not absolute. In Slobodkin's experiments in which *Chlorohydra viridissima* formed an understory below *Hydra littoralis* when cultivated in darkness, the latter was presumably interfering and exploiting at the same time. It will, however, sometimes be useful to make the distinction between the two processes in the theoretical discussion of the next section.

A similar, but by no means identical, distinction has been made by Nicholson who recognized two possible kinds of competition, which he called *scrambles* and *contests*. The human model for a scramble would be a group of boys to whom a number of pieces of candy are thrown. The model for a contest would be any athletic event conferring on the winner a valuable prize.

In a scramble in its ecological sense, a certain number of individuals will gain enough resources to reproduce as well as to survive; others will only be able to survive without reproducing; still others will fail altogether. The resources which are used in maintaining the nonreproductive survivors are wasted from the point of view of the species. In a contest, however, the outcome is either acquisition of nothing or of all of what is required, a host for a parasite, a defended territory for a bird or mammal. Many individuals may die, but those that survive utilize the resources fully, in a way that permits a maximum number of descendants.

While it is obvious that exploitation may often involve a scramble, and interference superficially appears to imply a contest, this is not invariably true. In particular, it would seem that acquiring a feeding ground of a sort, which clearly involves interference by

31. A. C. Crombie (Interspecific competition. *J. Anim. Ecol.*, 16:44–73, 1947) wrote: "Individuals of one species may then be superior to those of another in two different ways: (a) in the combined *rate of reproduction and survival* independent of interspecific interference: this will depend on adaptation to the environment, or (b) in *interference* (through direct attacks, conditioning the medium, consumption of food, etc.) which reduces the rate of reproduction and survival of the competitor." Crombie indicates that these two processes were distinguished by Darwin and other nineteenth-century naturalists. Although it is possible that the use of interference in a limited technical sense is derived from this passage, the meaning Crombie gives to the term includes exploitation, the competitor interfering with the reproduction of the other species by eating its food.

Crombie's distinction would be equivalent to the difference between *Lemna gibba* and *L. polyrhiza* grown separately when the superiority of the latter is indeed shown in "the combined rate of reproduction and survival," and grown together when *L. gibba* interferes with *L. polyrhiza* by growing over it. What Crombie noted was the difference between species that, appearing only in competition, often makes it impossible a priori to predict which way the competition will go.

The distinction between exploitation and interference was formally made by T. Park in his 1954 paper on *Tribolium* (see note 11) and was sharpened by M. W. Brian, Exploitation and interference in interspecies competition. *J. Anim. Ecol.*, 25:339–47, 1956. Brian's belief that Winsor's competition equations (see note 32) refer only to exploitation is correct, but there is no reason to suppose that the Volterra equations refer only to interference (see Schoener, note 40).

C. Elton and R. S. Miller (The ecological survey of animal communities: With a practical system of classifying habitats by structural characters. *J. Ecol.*, 42:460–96, 1954) use only the term interference.

An excellent account of these aspects of competition is given in R. S. Miller, Pattern and process in competition. *Advances in Ecological Research*, 4:1–74, 1967.

The comparable terms *scramble* and *contest* were introduced by A. J. Nicholson, An outline of the dynamics of animal populations. *Austr. J. Zool.*, 2: 9–65, 1955. See also G. de Jong, A model of competition for food: I. Frequency dependent viabilities. *Amer. Natural.*, 110:1013–27, 1976.

the preemption of an area and its resources, is by no means a pure case of a contest. The occurrence of territoriless, nonreproductive individuals, which we have seen can occur in both birds and mammals, is formally equivalent to what Nicholson regards as characterizing a scramble. Since, however, the presence of such individuals supplies replacements, should any of the territorial and reproductive individuals die or otherwise fail, it is probable that a formalized kind of scramble, producing a category of territoriless unmated adults, could evolve as part of the inclusive fitness of the individuals concerned as soon as the number of pairs surviving to the first breeding season had reached the maximum number of possible territories in the species and area under consideration.

Further theoretical formulations

The experimental material that has just been presented, which is but a sample of what has been published and could have been used, shows that when an isotropic habitat is set up experimentally, the survival of one and the extinction of the other is the usual fate of a pair of competitors. It is obvious, however, that there are exceptions to this even when the habitat has a quite simple structure. Before proceeding to a study of the situation in nature, where the simplicity possible in the laboratory is quite exceptional, it will be well to examine certain other theoretical formulations that well might be useful in the interpretation of field data. In general, these will be stated and the significance of their predictions explained without formal mathematical proofs; the sources of all such proofs are, however, given in the notes. In all cases the symbolism has been made conformable with that employed elsewhere in this book.

Volterra equation for n species

The general equation given by Volterra is

$$\frac{dN_i}{dt} = N_i r_i \left(\frac{K_i - N_i - \sum_j \alpha_{ij} N_j}{K_i} \right) \quad (4.6)$$

where N_i is the population of the *i*-th species, N_j that of any other species and α_{ij} the coefficient of competition indicating the effect of a single individual of N_j on the population N_i.[32]

If there is always a transitive relation between any three species S_1, S_2, and S_3, so that S_1 displacing S_2 and S_2 displacing S_3 implies that S_1 displaces S_3, it can be shown that in a uniform habitat only one species of the many that initially compose the system will ultimately survive. In competition by exploitation, this is almost certain to be the case.

In a multispecies system competing transitively in a complex habitat, it can also be shown[33] that the existence of as many different modes of limitation as there are species is a necessary, though obviously not a sufficient, condition for stable coexistence. In competition by interference, involving specific kinds of behavior elicited in one competitor by specific sign-stimuli given by the other, it would be quite possible to have S_1 displacing S_2 and S_2 displacing S_3, but S_3 displacing S_1. Gilpin[34] has provided a very interesting mathematical analysis of this possibility. Under certain circumstances unstable types of coexistence, including limit cycles, may be set up. The overall effect of a number of species oscillating with different periods might give an appearance of randomness.

The possibility of this kind of coexistence in *Drosophila* has been studied in an elaborate investigation by Richmond, Gilpin, Perez Salas, and Ayala,[35] but no cases were identified. It is probably more likely to occur in vertebrates in which nontransitive competi-

32. C. P. Winsor, Mathematical analysis of growth of mixed populations. *Cold Spring Harbor Symp. Quant. Biol.*, 2:181–87, 1934.

33. S. A. Levin, Community equilibria and stability, and an extension of the competitive exclusion principle. *Amer Natural.*, 104:413–24, 1970.

V. G. Haussmann, On the principle of competitive exclusion. *Theor. Popul. Biol.*, 4:31–41, 1973.

34. M. E. Gilpin, Limit cycles in competition communities. *Amer. Natural.*, 109:51–60, 1975.

35. R. C. Richmond, M.E. Gilpin, S. Perez Salas, and F. J. Ayala, A search for emergent competitive phenomena: The dynamics of multispecies *Drosophila* systems. *Ecology*, 56:709–14, 1975.

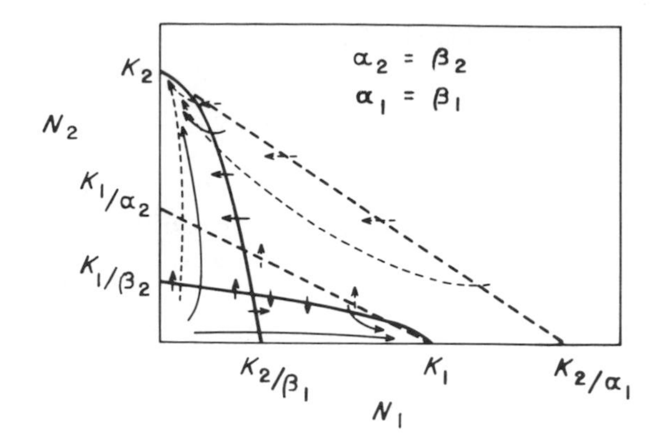

FIGURE 94. Parabolic isoclines (solid lines) for the competition equations (4.7a, b) compared with the linear isoclines (broken line) obtained if second-order terms are omitted (Hutchinson).

tion based on elaborate, and in part arbitrarily symbolic, interspecific communication is possible.

Competition equations involving second-order terms

If we go back to equation (4.1a, b), we may note that the retention of second-order terms in expansion of the competition function

$$\frac{dN_1}{dt} = N_1 r_1 \left(\frac{K_1 - N_1 - \alpha_2 N_2 - \beta_2 N_2^2}{K_1} \right)$$

(4.7a, b)

$$\frac{dN_2}{dt} = N_2 r_2 \left(\frac{K_2 - N_2 - \alpha_1 N_1 - \beta_1 N_1^2}{K_2} \right)$$

gives a pair of equations that could express a social effect.[36] The effect of the competitor depends not only on single individuals, but on the augmentation of the action of such individuals by the support given by the whole community. Such competition may be described as positively density-dependent.

The isoclines for equations (4.7a, b) are parabolas intersecting at an unstable saddle, and the outcome of competition depends on the initial proportions of the two species (figure 94). At least qualitatively, populations of the sooty tern, *Sterna fuscata*, and the noddy *Anous stolidus*, appear to behave in this way on the low islands of the Indian Ocean. New unispecific colonies can be established by the species initially present in greater proportion.

A second-order interaction involving the product of the two competing populations N_1 and N_2 is believed by Wilbur[37] to be involved in accounting for the competitive relations of his experimental populations of *Ambystoma*. The species used include the very curious triploid, *A. tremblayi*, eggs of which must be activated by the sperm of *A. laterale*, though the latter contribute no genetic material to the resulting gynogenetic embryo. The data to be presented later suggest that there is some symbiosis between the species, ensuring the stability of this arrangement.

A comparable second-order equation in which the intraspecific competition increases with density has been put forward by Ayala, Gilpin, and Ehrenfeld along with a number of other alternatives (see the next section).

Wilbur, in the interpretation of his very interesting work on competition between salamander larvae belonging to different species of *Ambystoma*, concludes that a second-order term in N_iN_j is implied by an analysis of variance applied to his data. Since this case involves a peculiar commensal or symbiotic relationship, it is deferred till later in this chapter.

A further set of equations

In an important paper, Ayala, Gilpin, and Ehrenfeld[38] have examined 10 types of competition equations in addition to the original Volterra model. In the latter, coexistence, when competition is partial or conditional, involves an equilibrium point that lies above the line joining K_1 and K_2 (figure 89) on the coordinates of the two-dimensional space on which the trajectories of the two populations are plotted. Moreover, the product of the two competition coefficients α_{ij} and α_{ji} is less

36. G. E. Hutchinson, A note on the theory of competition between two social species. *Ecology*, 28: 319–21, 1947.

37. H. M. Wilbur, Competition, predation, and the structure of the *Ambystoma-Rana sylvatica* community. *Ecology*, 53:3–21, 1972.

38. F. J. Ayala, M. E. Gilpin, and J. G. Ehrenfeld, Competition between species: Theoretical models and experimental tests. *Theor. Popul. Biol.*, 4:331–56, 1973.

The full set of equations proposed is:

3 parameter equations

1. $dN_i/dt = r_iN_i(K_i - N_i - \alpha_{ij}N_j)/K_i$

than unity. As has already been noted, in cases of coexistence in *Drosophila*, these conditions are usually not fulfilled. When populations were studied in detail, so that actual trajectories could be drawn and the configuration of the isoclines estimated, it is clear that the latter are not rectilinear but are curves that are concave upward (figure 95).

The 10 equations were put forward as possible interpretations of these observations. The model preferred is one based upon the asymmetrical logistic (eq. 1.13), giving

$$\frac{dN_i}{dt} = r_i N_i \left(\frac{K_i^{\theta_i} - N_i^{\theta_i} - \alpha_{ij} N_j K_i^{\theta_i - 1}}{K_i^{\theta}} \right). \qquad (4.8)$$

The index θ turns out to be not far from 1/2 in some actual experiments. In this model, when $\theta < 1$, the growth of each population is faster before $K/2$, the half-saturation den-

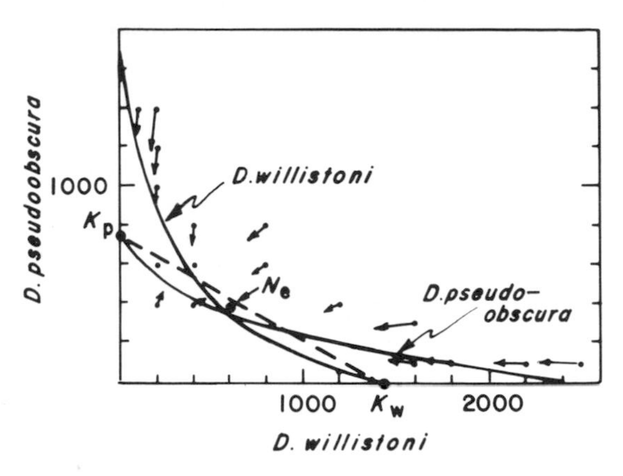

FIGURE 95. Trajectories of relative changes in populations of *Drosophila pseudoobscura* and *D. willistoni*, corresponding to curved isoclines, concave upward, determined empirically. The point N_e indicates an equilibrium mixture achieved in another experiment, very close to the equilibrium implied by the intersection of the isoclines. Note that such equilibrium points lie well below the broken line joining the K values (K_p and K_w) for the two species (Ayala, Gilpin, and Ehrenfeld).

---

2. $dN_i/dt = r_i N_i(\log K_i \quad \log N_i - \alpha_{ij} \log N_j)/\log K_i$
3. $dN_i/dt = r_i N_i(K_i^{1/2} - N_i^{1/2} - \alpha_{ij} N_i/K_i^{1/2})/K_i^{1/2}$

4 parameter equations

4. $dN_i/dt = r_i N_i(K_i - N_i - \alpha_{ij} N_j - \beta_{ij} N_i N_j)/K_i$
5. $dN_i/dt = r_i N_i(K_i - N_i - \alpha_{ij} - \beta_i N_i^2)/K_i$
6. $dN_i/dt = r_i N_i(K_i - N_i - \alpha_{ij} N_j - \beta_j N_j^2)/K_i$
7. $dN_i/dt = r_i N_i(K_i^{\theta_i} - N_i^{\theta_i} - \alpha_{ij} N_j/K_i^{1-\theta_i})/K_i^{\theta_i}$

5 parameter equations

8. $dN_i/dt = r_i N_i(K_i - N_i - \alpha_{ij} N_j - \beta_i[1 - e^{-\gamma_i N_i}])/K_i$
9. $dN_i/dt = r_i N_i(K_i - N_i - \alpha_{ij} N_j - \beta_j[1 - e^{-\gamma_j N_j}])/K_i$
10. $dN_i/dt = r_i N_i(K_i - N_i - \alpha_{ij} N_j - \beta_{ij} N_i N_j - \delta_i N_i^2)/K_i$

6 parameter equation

11. $dN_i/dt = r_i N_i(K_i - N_i - \alpha_{ij} N_j - \beta_{ij} N_i N_j - \delta N_i^2 - \gamma_j N_j^2)/K_i$

The first of these is the original Volterra competition model; it is a special case, with the appropriate coefficients being zero, of 4, 5, 6, 8, 9, 10, and 11, and of 7 when $\theta = 1$. Equation 2 is improbable and does not fit any of the *Drosophila* data; it may be abandoned. Equations 8 and 9, which are based on the law of diminishing returns in economics, are approximated by 5 and 6, since

$$1 - e^{\gamma_i N_i} = (1 - 1 + \gamma_i N_i - \gamma_i N^2/2 + \ldots).$$

Equation 6 is the one that formally may explain social aspects of competition (see note 36), while 4, 10, or 11 with second-order terms in $N_i N_j$ are the kind of equation evoked in general terms by Wilbur (note 37).

Ayala, Gilpin, and Ehrenberg dislike equation 5 on the ground that it is a parabola and reenters the all-positive quadrant. Two such equations could give parabolic isoclines intersecting twice, to give a node nearer the origin and a saddle further out, from which both populations might grow to infinity. Actually the isocline for dN_i/dt, N_i being plotted along the abscissa, is a parabola convex side up with its maximum in the N_i negative, N_j positive quadrant. It may, moreover, be questioned whether an isocline with a branch reentering the all-positive quadrant to give a biologically unacceptable result is any worse than isoclines entering a quadrant where the populations are unbiologically negative and perhaps intersecting there to give a mathematically stable, but biologically nonexistent node.

A further study of the preferred equation 7, derived from the asymmetrical logistic, is given by M. E. Gilpin and F. J. Ayala, Global models of growth and competition. *Proc. Nat. Acad. Sci. U.S.A.*, 70:3590–93, 1974. I regret that I cannot follow their argument about the constancy of dN/dt at its maximum.

sity, is reached and slows later, implying an increase in intraspecific competition per individual as the density increases. Though in a formal sense θ has this biological meaning, only further work will indicate how the change in intraspecific competition actually occurs. In the experiments reported, the best fit of the data to this model explains 95.3 percent of the observed change in the population, but some of the other equations are about as good, though they may be less satisfactory in requiring more than the four parameters r_i, K_i, α_i, and θ_i. It is evident that empirical tests of the models using contemporary techniques are likely to be difficult. What is of enormous importance is the demonstration, in a detailed and coherently intellectual manner, that the simple-minded statement of competitive exclusion is often false and that this is due to the very widespread occurrence of density-dependent or frequency-dependent regulation of competitive processes.

Equations involving energetics and the nature of the competitive process

Early in the history of competition theory Winsor[39] put forward a very simple pair of equations, in the notation here used,

$$\frac{dN_1}{dt} = r'_1 N_1 F - m_1 N_1$$

$$\frac{dN_2}{dt} = r'_2 N_2 F - m_2 N_2 \qquad (4.9a, b)$$

in which F is a common source of food. Of the amount eaten, which depends entirely on the properties of the two species, a certain amount $m_i N_i$ is used for maintainance metabolism of the populations, the rest is responsible for reproduction, r'_i giving the efficiency of the conversion.

Eliminating F from the two equations and integrating

$$N_1^{r'_2} / N_2^{r'_1} = Ce^{(r'_1 m_2 - r'_2 m_1)t} \qquad (4.10)$$

Thus if $(r'_1 m_2 - r'_2 m_1)$ is positive, the ratio is positive and increasing, implying elimination of N_2, while, if this expression is negative, N_1 is being eliminated. This appears to be the simplest expression of competitive exclusion. It depends on competition being for a single resource and purely by exploitation, but is independent of any variation of the resource itself.

This mode of treatment introduces the concept of the partitioning of energy within the species; it involves a rather rigorous dependence on exploitation. Such a point of view, which Winsor felt was "at least amusing," remained unappreciated for a long time, but has finally been developed on a grand scale by Schoener.[40]

Schoener first considers the simplest possible situation. For any species S_i, the time spent in feeding over a period of observation by an individual is T_i, and the energy harvested per unit time is E_i. The product $E_i T_i$ is the amount of energy available to the individual during the period considered. If the food is constantly renewed and is present in considerable excess of the amount required by any population, E_i is essentially constant. This situation may be approximated in some flour-beetle cultures.

If, however, there is a constant input into the system which must be shared by all individuals, as is usual in experiments with *Daphnia* or *Hydra*, and two species are using this input, the share of the energy going to each individual of S_i will be $I_{oi} / (N_i + \beta N_j)$, and the share going to each individual of S_j will be $\beta I_{oi} / (N_i + \beta N_j)$, where I_o is the amount of resource available in unit time and β is the ratio of feeding efficiency of S_2 to that of S_1.

We may write therefore for the first species,

$$\frac{dN_1}{dt} = R_1 N_1 \left(\frac{I_{01}}{N_1 + \beta N_2} - C_1 \right) \qquad (4.11)$$

where C_1 is a term for metabolic maintenance and loss by death per individual, and R_1 is an energetic equivalent of r_1. Since at equilibrium such loss must balance the input of energy as food $K_1 C_1 = I_{01}$. For species 1 therefore we may write

$$\frac{dN_1}{dt} = R_1 N_1 \left(\frac{I_{01}}{N_1 + \beta N_2} - \frac{I_{01}}{K_1} \right)$$

39. C. P. Winsor, see note 32.

40. T. W. Schoener, Competition and the form of habitat shift. *Theor. Popul. Biol.*, 6:265–307, 1974.

$$= \frac{R_1 N_1 I_{01}}{(N_1 + \beta N_2) K_1} (K_1 - N_1 - \beta N_2) \quad (4.12a)$$

which has the same form as the Volterra equation.

For species 2 we obtain

$$\frac{dN_2}{dt} = \frac{R_2 N_2 \beta I_{02}}{(N_1 + \beta N_2) K_2} \left(K_2 - \frac{N_1}{\beta} - \beta N_2 \right) \quad (4.12b)$$

Since the two competition coefficients are reciprocals, the isoclines are parallel and no coexistence is possible.

Alternatively we may consider that competition is purely by interference. If we assume that this interference occurs as the result of random collisions between the two species involving some sort of agonistic or at least nontrophic behavior, the rate of food intake per individual will be

$$E_i (1 - \lambda_{ii} N_i - \lambda_{ij} N_j)$$

where λ_{ii} is a coefficient giving the effect of intraspecific encounters and λ_{ij} of interspecific encounters.

For pure interference between two species with an unrestricted food supply we have therefore

$$\frac{dN_1}{dt} = R_1 N_1 [E_1(1 - \lambda_{11} N_1 - \lambda_{12} N_2) - \gamma_{11} N_1 - \gamma_{12} N_2 - C_1]. \quad (4.13)$$

The terms $\lambda_{11} N_1$ and $\lambda_{12} N_2$ indicate the proportion of feeding time wasted by an individual of S_1 in interactions with its own species and with its competitor, while the terms $\gamma_{11} N_1$ and $\gamma_{12} N_2$ represent any loss of energy, including injury or death, resulting from such interference. C_1 is density-independent mortality and metabolism, which at equilibrium will be constant. In the vicinity of any equilibrium, we may therefore write

$$\frac{dN_1}{dt} = R_1 N_1 [(E_1 - C_1) - N_1 (E_1 \lambda_{11} + \gamma_{11}) - N_2 (E_1 \lambda_{12} + \gamma_{12})]$$

$$= R_1 N_1 \left[\frac{E_1 - C_1}{(E_1 \lambda_{11} + \gamma_{11})} - N_1 - N_2 \left(\frac{E_1 \lambda_{12} + \gamma_{12}}{E_1 \lambda_{11} + \gamma_{11}} \right) \right]. \quad (4.14)$$

This has the form of the Volterra equation with $\alpha_{12} = (E_1 \lambda_{12} + \gamma_{12})/(E_1 \lambda_{11} + \gamma_{11})$ and $\alpha_{21} = (E_2 \lambda_{21} + \gamma_{21})/(E_2 \lambda_{22} + \gamma_{22})$ which are not in general reciprocals. The isoclines therefore can, but need not, intersect in the all-positive quadrant to give a stable node or a saddle point; there can just as well be cases of exclusion of one and survival of the other species, as in the classical treatment.

When the food supply and so the energy input is limited and all the competition is by interference, we obtain

$$\frac{dN_1}{dt} = R_1 N_1 (IE_1/N_1 - \gamma_{11} N_1 - \gamma_{12} N_2 - C_1). \quad (4.15)$$

The first term in the parentheses is the share of the limited food available to N_1, and the two competitive terms are due solely to the loss of energy produced by competitive behavior.

The isoclines are of the form

$$I_{E_1}/N_1 - \gamma_{11} N_1 - \gamma_{12} N_2 - C_1 = 0$$

or

$$I_{E_1} - \gamma_{11} N_1^2 - \gamma_{12} N_1 N_2 - C_1 N_1 = 0.$$

On differentiating by N_1 we obtain

$$\frac{dN_2}{dN_1} = \frac{-\gamma_{11} 2 N_1 - \gamma_{11} N_2 - C_1}{\gamma_{12} N_1} \quad (4.16)$$

which is always negative; the second derivative is always positive. This means that the isocline decreases continuously, but this decrease is continually slowed (figure 96, a). Moreover, as N_1 approaches zero, N_2 approaches infinity. The isocline is therefore asymptotic to the N_2 axis but crosses the N_1 axis. This arrangement ensures that the isoclines cross, the point of intersection being a stable node. This very curious result, which depends on the food supply always supporting less biomass than would be possible in the absence of competition, is reminiscent of the effect of nonselective predation on a pair of competing species.

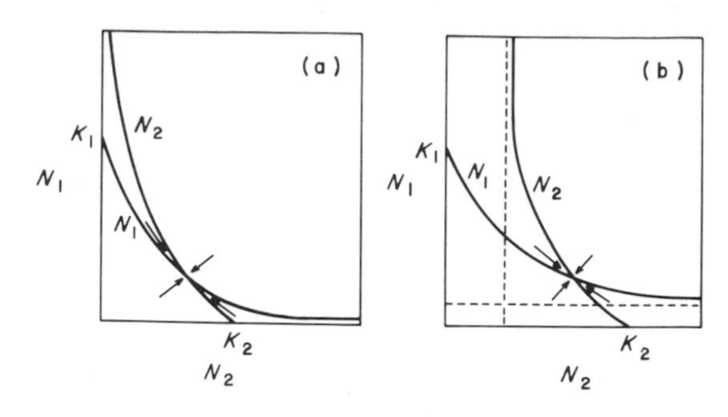

FIGURE 96. Isocline diagrams for some of Schoener's competition equations; (a) each isocline asymptotic to one axis but intersecting the other (4.15); (b) each isocline asymptotic to a line parallel to the axis (4.17).

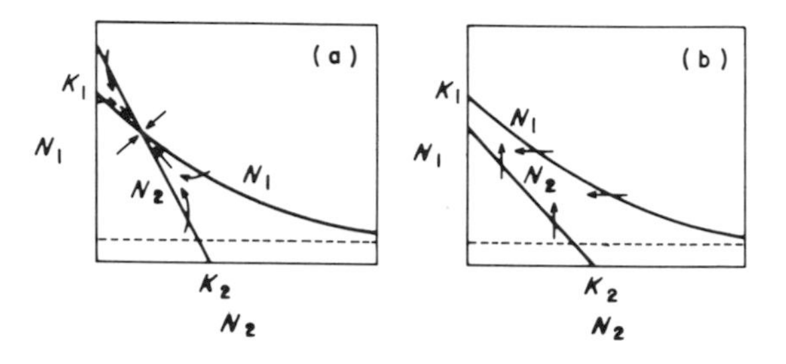

FIGURE 97. (a) One isocline asymptotic to a line parallel to an axis (4.17), the other a straight line cutting both axes (4.12a); $K_1 < I_2\beta/C_1$, isoclines intersecting at a stable node, species 2 in an included niche; (b) the same but $K_1 > I_2\beta/C_1$, isoclines not intersecting and species 1 displacing species 2.

The case of two species competing for a common resource, while each of them has also a second resource available, involves a modification of equation (4.11)

$$\frac{dN_1}{dt} = R_1N_1[(I_{E1}/N_1) + I_{01}(N_1 + \beta N_2) - m_1]. \qquad (4.17)$$

The isoclines (figure 96, b) are now not asymptotic to the axes, but approach the values of N_1 and N_2 supportable on the exclusive resources. The intersection corresponds to coexistence and is stable.

An equation of greater generality can be obtained by adding interference terms to equation (4.17), giving

$$\frac{dN_1}{dt} = R_1N_1[I_{01}/(N_1 + \beta N_2)] + (I_{E1}/N_1) - \gamma_{11}N_1 - \gamma_{12}N_2 - m_1. \qquad (4.18)$$

If interference tends to replace competition by exploitation,[41] an increase in γ_{12} will be accompanied by a decrease in β and the equation approximates equation (4.16).

Schoener points out that in his models one competitor may behave in one way, the other in another (figure 97). This happens when one species (S_1) has as its set of resources a subset of the resources of a second species (S_2). The first species will behave according to equation (4.12a) with a linear isocline, the second according to equation (4.17) with a concave isocline. If $K_1 < I_2\beta/C_1$, the isoclines cross at a stable node and both species survive; otherwise only the species with the wider range of resources survives. This case is of great interest as it corresponds to the "included niche" of Miller[42] (see chap. 5).

In a consideration of Schoener's approach it is important to remember that he has been mainly occupied with the study of the population ecology of lizards. The various terms for interference introduce a distinctly vertebrate cast to the theory.

Gause's rarefaction equation and Slobodkin's study of predation

In order to avoid changes in his populations of protozoa due to excessive crowding, Gause[43] often used a method of culture involving the frequent periodic removal of a small part of his medium with its contained organisms, replacing what was removed with fresh medium. The equations for this situation are of the form

$$\frac{dN_i}{dt} = r_iN_i\frac{K_i - N_i - \alpha_{ij}N_j}{K_i} - mN_i.$$

For two species the zero isoclines are

$$N_1 = K_1(1 - m/r_1) - \alpha_{12}N_2$$
$$N_2 = K_2(1 - m/r_2) - \alpha_{21}N_1.$$

41. R. H. MacArthur, *Geographical Ecology: Patterno in the Distribution of Species*. Philadelphia, Harper and Row, 1972, xviii, 269 pp.

42. R. S. Miller, Ecology and distribution of pocket gophers (Geomyidae) in Colorado. *Ecology*, 45:256–72, 1964.

43. G. F. Gause, *Vérifications expérimentales . . .*, see note 8.

Gause was aware that the species with the smaller intrinsic rate of increase would be depressed relative to the other species by rarefaction. Slobodkin[44] examined the situation further and pointed out that under certain conditions increasing m from zero to r_2, where $r_1 > r_2$, reversed the direction of competition.

If $r_1 > r_2$ but $\alpha_{12} > \alpha_{21}$, when m is small

$$\alpha_{12}\frac{1 - \frac{m}{r_2}}{1 - \frac{m}{r_1}} > \frac{K_1}{K_2},\ \alpha_{21}\frac{1 - \frac{m}{r_1}}{1 - \frac{m}{r_2}} < \frac{K_2}{K_1}.$$

In this situation S_2 wins since α_{12} is greater than α_{21}, while $(1 - m/r_2)(1 - m/r_1)$ and its reciprocal are nearly unity.

If m approaches r_2, the coefficient of α_{21} approaches infinity and that of α_{12} zero. As this condition is approached, the inequalities are reversed and S_1 wins in the competition. Slobodkin, using a geometrical argument which some may prefer to the algebra here used, likewise shows that provided the two competition coefficients are not reciprocals of each other, there is a range of values of m between zero and r_2, in which

$$\alpha_{21}\frac{1 - \frac{m}{r_2}}{1 - \frac{m}{r_1}} < \frac{K_1}{K_2},\ \alpha_{12}\frac{1 - \frac{m}{r_1}}{1 - \frac{m}{r_2}} < \frac{K_2}{K_1}.$$

and coexistence is possible. He believed that his experiments with *Hydra littoralis* and *Chlorohydra viridissima* competing in the light under conditions of extreme rarefaction or artificial predation provided a laboratory confirmation of this theoretical approach. It should be noted, as Slobodkin emphasizes, that the conditions for this to happen are that $r_1 > r_2$ but $\alpha_{12} > \alpha_{21}$, the intrinsic rate of increase being in favor of S_1, the competition coefficients in favor of S_2. This suggests that competition in such cases is primarily by interference.

Equation (4.13), expressing Schoener's concept of pure interference with a constant food input to the entire population, which also leads to coexistence, may provide an alternative model: in this case m is included in C.

When predation is selective and primarily on the more successful competitor, it is easy to see how coexistence is possible. The general role of predation in permitting coexistence and so increased diversity is now very well established in the field, as will appear in chapter 5. It is quite possible, as MacArthur[45] suggested, that this role is greater in more or less two-dimensional than in three-dimension habitats. This may perhaps be due to interference being of greater importance in the former than in the latter type of habitat.

Commensalism and symbiosis

If one or both of the competition coefficients is negative, we have situations that formally correspond to commensalism or symbiosis.[46] In the case of symbiosis with two negative coefficients, if $\alpha_{12} \cdot \alpha_{21} > 1$, the intersection is in the all-negative quadrant and has no biological meaning. When the product of the negative competition coefficients is between zero and 1, stable equilibrium is possible (figure 98).

Little experimental work has been done with symbiosis between physically discrete, closely allied, species as contrasted with endosymbiosis of one organism physically within another, or parasitism, which is the equivalent form of commensalism.

44. L. B. Slobodkin, *Growth and Regulation of Animal Populations*, see note 19.

45. R. H. MacArthur, Coexistence of species, in: *Challenging Biological Problems*, ed. J. A. Behnke. New York, Oxford University Press for A.I.B.S., 1972, pp. 253–59.

46. G. F. Gause and A. A. Witt, Behavior of mixed populations and the problem of natural selection. *Amer. Natural.*, 69:596–604, 1935.

An interesting, slightly earlier, precursor is V. A. Kostitzin, *Symbiose, parasitisme et evolution* (*Étude mathématique*). Actualités scientifiques et industrielles, 96. Paris, Hermann, 1934, 47 pp.

Gause and Witt's paper has the same sort of presentation as in the former's *Vérifications expérimentales* (see note 8) of the same year. This has become the standard approach to the mathematics of competition. It was doubtless worked out by them together on the basis of Volterra's and Lotka's investigations.

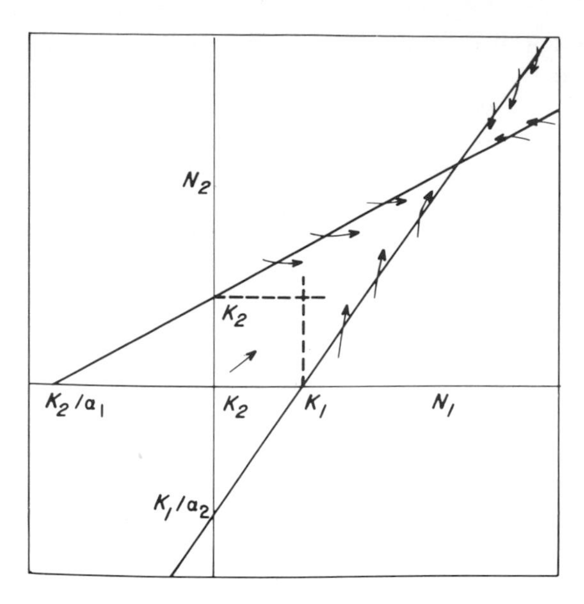

FIGURE 98. Isocline diagram with both competition coefficients negative, in the simple Volterra equations (4a, b), corresponding to symbiosis.

We shall see later that very elaborate kinds of symbiosis are known in nature, but are often hard to detect. One case of a set of results of field experiments, perhaps dependent on peculiar symbiotic relations, does, however, exist.

Wilbur[47] has studied competition among the larvae of salamanders of the genus *Ambystoma* in Michigan. He supposes that competition is only important between larval stages, and accordingly reared larvae of three species either separately or in pairs. The three species, *A. laterale*, *A. tremblayi*, and *A. maculatum*, were kept either singly or in combination in pens 240 × 60 × 60 cms, set in a pond at the Edwin S. George Reserve of the University of Michigan, in Livingstone County, Michigan. The walls and lids of the pens were made of fiberglass of 7 meshes per cm, which excluded predators but not small food organisms. In the unispecific pens, 32 or 64 viable eggs were introduced. When interspecific competition was studied, 32 larvae of one species were reared with 32 and 64 larvae of the other. It is important to bear in mind that *A. tremblayi* is a peculiar triploid of hybrid origin. It is always female and gynogenetic, but its eggs require activation by *A. laterale* sperm. It is obvious that *tremblayi* is dependent on the syntopic existence of *laterale*, though the two species appear to have slightly different requirements. If *tremblayi* is to persist, it must not completely out compete *laterale*. Uzzell[48] and Wilbur both indicate that it is less fecund than either of the other two species; so if it is to survive in competition, it must have some advantage over *laterale*. Wilbur interprets all his results with these three species in terms of analysis of variance and concludes that the effects of interactions between two species may be as significant as the direct effect of one or other species. This he expresses theoretically by suggesting that a term in N_1N_2 must be used in the Volterra equation. When we examine the actual data from Wilbur's 1972 paper, which is set out in table 10 in a rather different form from that of his own presentation, it becomes evident that for no species is the percentage survival greatest when it is cultivated by itself. This is also seen in table 11.

In the eight possible combinations of three species at densities of 32 or 64 per pen, Wilbur found the entire range of survival to be from 1 percent for *maculatum* with 64 *laterale*, 32 *tremblayi*, and 32 *maculatum* to 51 percent for the same species with 64 *laterale*, 32 *tremblayi*, and 64 *maculatum*. These quite different survivorships look odd and may suggest some caution in accepting the survivorship data as significant. Moreover, in his 1971 paper dealing only with *tremblayi* and *laterale*, the survivorships of larvae reared alone are different, averaging 84 percent for *laterale* and 64 percent for *tremblayi* in populations initially of 64 larvae and 76 percent for both species when the initial population was 32 larvae per pen. It is therefore difficult to judge how far the apparent mutualism, indicated by the

47. H. M. Wilbur, The ecological relationship of the Salamander *Ambystoma laterale* to its all-female gynogenetic associate. *Evolution*, 125:168–79, 1971. See also note 37.

48. T. M. Uzzell, Relations of the diploid and triploid species of the *Ambystoma jeffersonianum* complex (Amphibia. Caudata). *Copeia* 1964, 257–300.

	A. laterale		*A. tremblayi*		*A. laterale*		*A. maculatum*		*A. tremblayi*		*A. maculatum*	
unispecific	32	64	32	64	32	64	32	64	32	64	32	64
% survival	33	42	62	**63**	*21*	55	40	29	62	63	40	*29*
larval period (days)	**87**	94	**81**	92	**87**	94	95	**91**	**81**	92	95	**91**
wt. at metamorphosis (g.)	**0.94**	0.64	**0.79**	0.77	**0.94**	0.64	0.99	1.05	**0.79**	0.77	0.99	**1.05**
bispecific with 32 of other species initial no.	32	64	32	64	32	64	32	64	32	64	32	64
% survival	58	28	48	48	**62**	52	*24*	31	58	*22*	**47**	45
larval period (days)	98	88	91	84	91	102	*125*	105	**81**	97	94	112
wt. at metamorphosis (g.)	0.61	*0.52*	0.72	0.61	0.59	*0.48*	*0.61*	**1.12**	0.68	0.64	1.00	*0.71*
bispecific with 64 other species initial no.	32	64	32	64	32	64	32	64	32	64	32	64
% survival	**97**	*24*	34	*24*	39	26	**53**	29	**81**	24	39	40
larval period (days)	98	*100*	82	*99*	88	*117*	**91**	112	94	*99*	97	*122*
wt. at metamorphosis (g.)	0.56	0.57	0.73	*0.63*	0.77	0.47	0.73	0.68	0.68	*0.59*	0.77	0.87

TABLE 10. Survivorship, larval period and weight at metamorphosis of *Ambystoma* larvae of three species, reared separately or together; data of Wilbur. Boldface indicates **optimal** values, italics *pessimal* values for each species in each series.

later data for two-species interactions, is real. The time taken to reach metamorphosis is usually least and the weight at metamorphosis usually greatest in the unispecific and least in the most crowded bispecific cultures, and seem to give no suggestion of symbiotic interaction.

Experiments with *A. tigrinum* showed that it could be a competitor with the other species if it did not have a source of frog tadpoles as food, but the growth subsequent to feeding on such tadpoles enabled it to become rapidly large enough to be a predator on its congeners. It must be born in mind that Wilbur with his great experience of the animals did not even consider the possibility of a symbiotic interaction and believed the sympatric existence of the various species to be due to relatively small ecological differences between them and to the continual shifting nature of the environment, physical and biotic, provided by the temporary ponds of the region where the work was done. If there are really symbiotic interactions in *Ambystoma*, they may have facilitated the formation, in this genus, of the curious gynogenetic triploid hybrids *tremblayi* and *platineum*.

	unispecific		maximal
	32 per pen	64 per pen	
laterale	33%	42%	97% (32 *laterale*, 64 *tremblayi*)
tremblayi	62	63	81 (32 *tremblayi*, 64 *maculatum*)
maculatum	40	29	53 (32 *maculatum*, 64 *tremblayi*)

TABLE 11. Survivorship in larval life of different species of *Ambystoma* cultured separately and together.

SUMMARY

Testable hypotheses about competition and coexistence are most simply constructed by including in the logistic a term for the use, by a second species, of the vacant spaces, remaining at any time t. For two species living together we may write

$$\frac{dN_1}{dt} = r_1 N_1 \left(\frac{K_1 - N_1 - \alpha_{12} N_2}{K_1} \right)$$

$$\frac{dN_2}{dt} = r_2 N_2 \left(\frac{K_2 - N_2 - \alpha_{21} N_1}{K_2} \right).$$

These equations are usually termed Volterra equations after the Italian mathematician Vito Volterra, whose work first introduced them to biologists.

The coefficients α_{12} and α_{21} are called *competition coefficients* and indicate the strength of the effect of the second species on the first and vice versa.

In general, if the effect of an individual species on one of the second species is greater than the effect that it has on one of its own species, while an individual of the second species has a less effect on one of the first than it has on one of its own species ($\alpha_{12} > K_1/K_2$, $\alpha_{21} < K_2/K_1$ or the reciprocal of this), the species having the greater effect on its competitor than on itself will displace the other species.

If individuals of both species have a greater effect on those of their competitors ($\alpha_{12} > K_1/K_2$, $\alpha_{21} > K_2/K_1$) than on their own conspecifics, the winning species depends on the initial proportion of the two species in the growing mixed population.

If individuals of each species have a greater effect on their own conspecifics than on the individuals of the competing species ($\alpha_{12} < K_1/K_2$, $\alpha_{21} < K_2/K_1$), coexistence is possible. In these circumstances competition is said to be partial or conditional.

The two lines whose equations are

$$K_1 - N_1 - \alpha_{12} N_2 = 0$$

$$K_2 - N_2 - \alpha_{21} N_1 = 0,$$

which give respectively the values of N_1 and N_2 for which $dN_1/dN_2 = 0$ and $dN_2/dN_1 = 0$, are called *isoclines*. In the two cases in which either one or the other species alone survives, they do not intersect in the quadrant corresponding to both N_1 and N_2 being positive. For the case where either one or the other species wins, depending on the initial conditions, the intersection represents an unstable equilibrium or *saddle* point, while when the two species can coexist the intersection is a stable *node*.

Numerous experiments have been done to test the validity of these conclusions, which are often subsumed under the *principle of competitive exclusion*.

When two species are competing in a quite homogeneous medium, as is often the case with protozoa or flour beetles, favorite organisms for such study, it is usually found that only one species persists.

The identity of the surviving species depends greatly on the environmental conditions under which the experiment is done.

If one of two species always excludes the other under one set of environmental conditions, and the opposite exclusion occurs in another set, it is not unusual to find that in some intermediate environment the outcome is indeterminate. Only one species survives, but the identity of this species appears to depend to some extent on random factors. There is evidence that the probability of success can depend both on the relative numbers of the two species initially present and on random differences in the genetic constitution of the founding individuals.

The introduction of heterogeneity into the habitat of an experimental culture is apt to allow survival of two species, of which only one would survive if the heterogeneity had not been introduced.

Predation or its artificial equivalent, rarefaction, even if applied quite unselectively, can determine which of two species will survive and may permit both to survive as reduced populations.

There is an increasing amount of evidence that the outcome of competition can be *frequency-dependent*, individuals having a greater competitive ability when rare than they will have when common. This may lead to the prolonged persistence of a species in the late stages of its elimination as in the water fleas *Simocephalus* competing with *Daphnia*, or to the permanent coexistence of two competitors as in many species pairs in *Drosophila* or in the European poppies of the genus *Papaver*. This effect can be observed in experiments involving a single generation and so is not due to natural selection operating during the experiment. However, in long-term experiments, the fact that a rare species is more subject to interspecific than intra-specific contacts may stimulate it to evolve a mechanism of competition that results in its displacing its previously dominant competitor.

The elementary linear theory of Volterra has been reconsidered by many investigators. Second-order terms have been suggested to explain certain special situations, such as a competitor depending for its effectiveness on the presence of other members of its species. This may happen in the establishment of colonies of seabirds; the isoclines are parabolic, intersecting at a saddle.

A number of alternative equations have been suggested for frequency-dependent competition. The one most favored appears to be the simple Volterra equation derived from the asymmetrical logistic, namely,

$$\frac{dN_i}{dt} = r_i N_i (K_i^{\theta_i} - N_i^{\theta_i} - \alpha_{ij} N_j K^{\theta_i - 1}) / K_i^{\theta_i}$$

where the index θ is less than unity. This equation produces curved isoclines, concave side up, which may intersect in a stable node.

A distinction can be conveniently made between those kinds of competition in which two kinds of organisms, using a common food, take up all that they can efficiently use for maintenance growth and reproduction, but at different rates, and those kinds where one organism prevents the other from gaining access to the food. The first kind is usually termed *exploitation*, the second *interference*. Competition equations explicitly constructed on the basis of exploitation appear to be of the classical Volterra type with rectilinear isoclines, but when interference occurs, the isoclines may be curvilinear and may intersect at a stable node.

Negative competition coefficients may be used to express commensalism (or parasitism) and symbiosis. It is, however, easy to produce unbiological models in this way. There seems to be some experimental evidence of symbiosis in *Ambystoma* larvae. Though less universal than competition, symbiosis of an unexpected kind may provide explanations of curious situations observed in nature.

Chapter Five

What Is a Niche?

GAUSE,[1] in reviewing the information about competitive exclusion available before the publication of *The Struggle for Existence*, mentioned the work of Tansley on *Galium*, Timoféef Ressovsky on *Drosophila*, and Beauchamp and Ullyot on triclads in streams. He also considered the extensive evidence then available about plant succession. From this body of facts he went on to discuss the idea of the *niche*, taken directly from Elton's *Animal Ecology*;[2] writing, "A niche indicates what place the given species occupies in a community, i.e., what are its habits, food and mode of life. It is admitted that as a result of competition two similar species scarcely ever occupy similar niches, but displace each other in such a manner that each takes possession of certain peculiar kinds of food and modes of life in which it has an advantage over its competitors" (p. 19). This statement embodied what came to be known as Gause's Principle; Gause, however, clearly indicates by his phrase "it is admitted" that the principle was already known, from Volterra's mathematics. He goes on to discuss, from a then-unpublished manuscript by Formosov,[3] a striking case of niche specificity among coexisting members of the avian family Sternidae, the terns. This example made it possible to go from mathematical theory, through laboratory experimentation to the interpretation of nature. Though the concept of the niche was Elton's, it may be noted that competitive exclusion was not a prime concern of Elton's in framing this concept. Gause's great achievement was to give a clear exposition of the way that the competitive exclusion, so often previously noted, actually worked. By the wedding of theory and experiment he made the matter not only credible but also interesting and significant. Actually the full extent of his contribution did not become apparent until the *Vérifications expérimentales*[4] of 1935 and the paper by Gause and Witt in the same year; in these the theory was developed in

1. G. F. Gause, *The Struggle for Existence*. Baltimore, Williams and Wilkins, 1934. See chap. 4, notes 8 and 46 for other works cited.

2. C. S. Elton, *Animal Ecology*. London, Sidgwick and Jackson, 1927, xvii, 207 pp. This book was described by F. N. Egerton (Studies of animal populations from Lamarck to Darwin. *J. Hist. Biol.*, 1:225–59, 1968) as establishing the paradigm for ecology for much of the twentieth century. I obtained it first in South Africa shortly after it was published; it proved a stimulus by showing me that what I wanted to do in biology was indeed a significant part of the science. Elton refers to J. Grinnell and T. I. Storer, *Animal Life in the Yosemite* (Berkeley, University of California Press, 1924, xviii, 752 pp.), where on p. 39 "each species occupies a niche of its own." This statement was apparently overlooked by Elton (*in litt.* August 26, 1976).

3. A. N. Formosov was a distinguished Russian ornithologist and mammalogist with marked ecological interests. He undoubtedly was familiar with some American work, as is shown in his paper, The lake region of the forest-steppe and steppe of western Siberia as a breeding area of the water fowl. *Bull. Soc. Nat. Moscow, sect. Biol.*, 43:256–86, 1934. (Russian text, English summary). I cannot find that the paper on the ecology of terns, which Gause used, was ever published.

4. See chap. 4, notes 8 and 46.

terms of the inequalities given on p. 118, which are the standard didactic devices of any modern presentation of the matter, and which go further than either Volterra or Lotka.

History of competitive exclusion and the concept of the niche

Before we proceed further, it will be of interest to enquire how it came to be admitted that "two similar species scarcely ever occupy similar niches."

The idea that two similar species are not likely to live together was certainly held by some naturalists in the nineteenth and early twentieth centuries. The earliest specific reference so far known is to a remark of Hansmann[5] about warblers in the island of Sardinia. Darwin, as Hardin[6] points out, never published a succinct statement on the matter, though the idea is implicit in his writings. A remarkable example of what he thought is found in the recently published[7] part on natural selection of the "big species book," another part of which appeared as *Variation of Animals and Plants under Domestication*. He writes: "one is sometimes tempted to conclude, falsely as I believe, that nature has worked for mere variety: thus when we hear [ref. to A. R. Wallace, *Narrative of Travels on the Amazons*, 1850, p. 469] that Mr. Bates collected within a day's journey, in a quite uniform part of the valley of the Amazons, 600 different species of Butterflies (Gre[a]t Britain has about 70 species), one may at first doubt whether each is adapted to its own peculiar and different line of life; but from what we know of our own British Lepidoptera we may confidently believe that most of the 600 caterpillars would have different habits, or be exposed to different dangers from birds and hymenopterous insects."

Steere,[8] in 1894, indicated that among the land birds of the Philippine Islands, congeneric species do not occur together.

Grinnell,[9] in 1904, wrote: "two species of approximately the same food habits are not

5. A. Hansmann, Die Sylvien der Insel Sardinien. *Naumannia*, 7:404–29, 1857. Writing of the willow warbler (*Phylloscopus trochilus*), as contrasted with Bonelli's warbler (*P. bonelli*), he says (p. 417): "Die bei dem vorigen angegebenen Orte bilden auch seiner Lieblingsaufenthalt hierselbst, an den man ihn jedoch nicht auf demselben Punkte mit dem vorigen antrifft, sondern höchstens beide in derjenigen Entfernung von einander, welche die Reviere zwei so nahe verwandter Vögel zu trennen pflegt." The statement clearly indicates that the species keep apart, the areas occupied (Reviere is an area over which one has hunting rights; it here seems to imply a space occupied locally by a species rather than a territorial pair) by two such closely allied species being customarily separate. The passage was noted by Hartmut Walter and called to my attention by Martin L. Cody.

In the seventeenth century, Sir Thomas Browne clearly suspected that the kite (*Milvus milvus*) and the raven (*Corvus corax*) replaced one another as scavengers, but his observations were not published until 1902. See *Notes and Letters on the Natural History of Norfolk, More Especially on the Birds and Fishes, from the Manuscript of Sir Thomas Browne*. London, Jarrold, 1902, xxvi, 102 pp.

6. G. Hardin, The competitive exclusion principle. *Science*, 131:1292–97, 1960.

7. R. C. Stauffer, ed., *Charles Darwin's Natural Selection Being the Second Part of His Big Species Book Written from 1856 to 1858*. Cambridge University Press, 1975, xii, 692 pp. It is interesting to note that in the extract quoted, "line of life" is used as the exact equivalent of "niche" *sensu* Elton. I am indebted to my friend Stan Rachootin for calling my attention to this prior to Stauffer's publication.

8. J. B. Steere, on the distribution of genera and species of non-migratory land-birds of the Philippines. *Ibis*, 1894:411–20. This seems to be the first detailed and elaborate treatment of the phenomenon.

9. J. Grinnell, the origin and distribution of the chestnut-backed chicadee. *Auk*, 21:364–82, 1904; see p. 377.

Joseph Grinnell (1877–1939) was perhaps the greatest student of North American birds and mammals whom the continent has yet produced. He was born on February 27, 1877, 40 miles from Fort Sill in what was then Indian Territory, now Oklahoma, where his father was the agency physician. Later the family moved and, in 1880, were with the Oglaga Sioux in the Dakotas. Joseph Grinnell's early playmates were Sioux children, and later his wife believed that part of his keen capacity for observation was due to their influence. It is recorded that on one occasion, traveling in a wagon, he announced to a group of students that kangaroo rats were breeding; he had noticed, from the wagon, tracks in which the scrotum was indicated dragging between the imprints of the feet.

He was educated at Pasadena High School and, after graduating in 1893, entered Throop Polytechnic Institute, which later metamorphosed into the California Institute of Technology. In 1895, he contributed a list of the birds of Pasadena to Hiram A.

likely to remain long evenly balanced in numbers in the same region. One will crowd out the other." This is perhaps the simplest possible qualitative rephrasing of the Volterra competition equations.

A little later[10] Grinnell again discussed the matter: "competitive struggle between species has led to the adoption of remote and otherwise unexplainable habitats, temporary and constant. It has also led to the development of various and perfected means of food-getting. The geometric ratio of reproduction makes the population of a species an elastic quantity, expanding into any favorable food area presenting itself. And the masses of different species press against one another, like soap-bubbles, crowding and jostling, as one species acquires, through modification of food-getting powers and perfected adaptability to other conditions, some advantage over another."

A little reflection will show that in this soap-bubble analogy Grinnell is using an abstract space and not the actual physical space of the habitat. In his later papers, as we shall see almost immediately, he used the word *niche*, though without a completely clear indication of its meaning.

Ortmann,[11] in 1906, concluded that two allied species do not occupy the same range under identical ecological conditions, and the same idea was employed by Hilzheimer,[12] in 1909, to elucidate the distribution of the two European fossil species of *Bison*. Monard[13] concluded from his study of the benthic animals of the Lac de Neuchâtel that in a uniform environment there tends to exist only one species of each genus present, a conclusion identical to that of Steere. Finally, we may note that Cabrera,[14] in 1932, just as Gause was starting his investigations, published an excellent article, evidently influenced by both Grinnell and Ortmann, on what he called the ecological incompatibility of allied sympatric species.

Reid's *History of Pasadena*; Reid wrote of the 18-year-old Grinnell that "he seems to have a specimen of every species and variety of avian fauna ever found here all nicely preserved and neatly labelled with both its common and its scientific name."

He was as a student able to do much traveling, including taking part in the Alaska gold rush and doing a great deal of ornithology on the side. His more formal scientific career started at Stanford in 1900, and in 1908 he was already established in Berkeley, where the generosity of Miss Annie M. Alexander had established the Museum of Vertebrate Zoology, with which he was associated for the rest of his life. Grinnell's work consists of an enormous number of notes, papers, monographs, and books that would have been dismissed as merely descriptive by many of his laboratory colleagues of the time. Though elected a corresponding member of the Zoological Society of London and a Foreign Member of the British Ornithological Union, he was never admitted to the National Academy of Sciences or the American Philosophical Society. Some of his more far-sighted contemporaries, such as F. B. Sumner, writing in *Science* (Jan. 8, 1915) of Grinnell's work, indeed realized from it "that the field naturalist holds the key to some of the most important secrets of nature." However, neither the obituary written by his wife Hilda Wood Grinnell for the *Condor* (42:3–34, 1940), nor that by Jean M. Linsdale for the *Auk* (59:269–85, 1942), gives any hint of the scientific importance of his ideas about competitive exclusion and the niche, though this of course does appear in A. H. Miller's appreciation in *Systematic Zoology* (13:235–42, 1964), published 25 years after Grinnell's death.

10. J. Grinnell, The biota of the San Bernardino Mountains. *Univ. Calif. Publ. Zool.*, 5:1–170, 1908; see p. 26.

11. A. E. Ortmann, Facts and theories in evolution. *Science*, 23:947–52, 1906. Ortmann was particularly interested in the systematics and distribution of crayfishes.

12. M. Hilzheimer, Wisent and Ur im K. Naturalienkabinett zu Stuttgart. *Jahreshf. Ver. Vaterländische Naturkunde in Württemberg*, 65:241–59, 1909.

R. Lydekker, in his widely known *The Ox and Its Kindred* (London, Methuen, 1912, xi, 271 pp.), wrote (p. 258): "closely allied species of approximately equal bodily size are not found at the present day living actually in company, and it is accordingly suggested by Dr. Hilzheimer that while *B. bonasus* was, as it is at the present day, an inhabitant of the forests, *B. priscus* was a denizen of the open plains." Hilzheimer's ideas must thus have come to the attention, however abortively, of a wide circle of naturalists.

13. A. Monard, La faune profondal du Lac de Neuchâtel. *Bull. Soc. Neuchâteloise Sci. Natur.*, 44:65–236, 1920.

T. T. Macan, in his admirable *Freshwater Ecology* (London, Longmans Green; New York, John Wiley, 1963, x, 338 pp.), gives on pp. 101–102 a short discussion of other concepts, derived from Monard, and put forward by other European limnologists.

14. A. Cabrera, La incompatibilidad ecológica: Una ley biológica interesante. *Anal. Soc. Cient. Argentina*, 114:243–60, 1932.

The idea of the niche

The first known[15] use of the word niche as an ecological term occurs in a monograph on the evolutionary meaning of color pattern in lady beetles by R. H. Johnson,[16] which appeared in 1910. He wrote: "one expects

15. B. M. Gaffney, Roots of the niche concept. *Amer. Natural.*, 109:490, 1973.

16. R. H. Johnson, *Determinate Evolution in the Color-Pattern of the Lady-Beetles.* Washington, Carnegie Institution of Washington, publ. no. 122, 1910, 104 pp.; see p. 87.

Roswell Hill Johnson was born in Buffalo, N.Y., on October 9, 1877. He studied at Brown University, Harvard University, and the University of Chicago, where he took a B.S. in 1900, and later in 1903 received an M.S. at the University of Wisconsin. After some teaching experience in Wisconsin and in the state of Washington, he was appointed to the Carnegie Institution Station for Experimental Evolution at Cold Spring Harbor in 1905, and remained there until 1908, doing his work on Coccinellidae. He then became a consulting geologist and in this capacity was appointed Professor of Oil and Gas Production at the University of Pittsburgh in 1916. While at Pittsburgh he took a Ph.D. in sociology. His later career included the executive secretaryship of the Social Hygiene Association of Hawaii, and, from 1936, the Directorship of Counseling, American Institute of Family Relations, California. From this last position he retired in 1960, in his 83rd year. He died on January 17, 1967, at his home in Hollywood. His brief obituary in the Los Angeles *Times* (Jan. 19, 1967, Part II, p. 10, bottoms of cols. 7, 8) indicates that he developed the Johnson Temperament Analysis procedure widely used by marriage counselors throughout the United States and Europe, and that his work guided many families to happier lives. It is interesting that his intellectual life should have begun in playing down what he considered to be excessive competition, and he was able to stick to a career only when he found one that presumably promoted harmony.

In the same year as saw the appearance of his contribution on determinate evolution, Johnson had already published another work, a temperate and humane, but inevitably often erroneous exposition of eugenics. This essay, The evolution of man and its control, though printed in the *Popular Science Monthly* (76: 49–70) for January 1910, was, in spite of its brevity, conceived as a book, divided into 7 chapters. Considering natural selection, Johnson writes: "Sustentative selection, in the sense in which it depends on a supply of food and shelter insufficient for the population, has been considerably overvalued as an evolutionary factor. Very few species are affected directly by it, as is shown by the rarity of starvation among the lower animals, and in man it has practically disappeared, unless it be in India, Siam or a few savage and barbarous tribes." By sustentative selection Johnson clearly means competitive or density-dependent resource limitation of the less fit part of a population, while his non-sustentative selection involves density-independent elimination by changes in the physical parameters of the environment. As with his remarks about the niche, Johnson embarks on a sound analysis only to reject it. The whole essay clearly contains the germs of his later activity. It is remarkable in being totally un-Mendelian as well as fairly un-Malthusian. It appeared after two articles on the occasion of the centennial of Darwin's birth and the jubilee of the publication of the *Origin of Species.*

Later in his career, in R. H. Johnson, H. Randolf, and E. Pixley, *Looking towards Marriage* (Boston, New York, Allyn and Bacon, 1943, v, 99 pp.), which begins with a chapter on "dear" hunting, intellectual tension is abandoned and the resolution of all the conflict in nature takes place in the frontispiece on a flower-strewn cloud where a young man places an engagement ring on the finger of a young woman. I am much indebted to David L. Cox of Washington University, St. Louis, for information that led me to this and to the previously mentioned eugenic work of Johnson.

As an evolutionist studying lady beetles Johnson believed in what he described as determinate evolution, due to what we would now call mutations, occurring in a limited number of directions, but not in a strict temporal order, as his contemporary proponents of orthogenesis believed. Influenced by the apparently fraudulent work of W. L. Tower on the chrysomelid beetle, *Leptinotarsa*, he regarded a direct effect of the environment as a possible cause of the variation. He believed natural selection to be relatively unimportant in the Coccinellidae, as Bateson evidently had before him (see G. E. Hutchinson, Variations on a theme by Robert MacArthur. In: *Ecology and Evolution of Communities*, ed. M. L. Cody and J. M. Diamond. Cambridge, Mass., Belknap Press of Harvard University Press, 1975, pp. 502–03).

Johnson may be regarded as one of a number of evolutionists who late in the nineteenth and early in the twentieth century had become, for largely invalid reasons, rather disenchanted with natural selection and were groping for something to put in its place; even though as a eugenist he was inevitably something of a selectionist. The situation was most curious, for evolution had been widely accepted after 1858 because Darwin and Wallace had provided a mechanism which explained how it took place. Now the mechanism began to look old-fashioned and ricketty, though the doctrine of evolution had become an all but universally accepted biological dogma. This situation, which needs its historian, involved several highly disparate points of view, Lamarckian, orthogenetic, or mutationist. As Hans Gadow, with his heavy Wendish accent, put it to his classes, moving his hand forward in a spiral path, the index finger extended, "A sort of a kind of a something goes SO!"

the different species in a region to occupy different niches in the environment. This at least is a corollary of the current belief that every species is as common as it can be, its members being limited only by its food supply, a belief which is the result of the strong Malthusian leanings of Darwin. Major species of the Coccinellidae do not seem to be so distributed. With certain exceptions which we have given, the species of *Hippodamia* and *Coccinella* are in quite general competition. They are characterized for the most part by very wide distribution and extensive overlapping of other species."

I am inclined to view the use of the word niche in this statement as an ordinary example of the figurative employment of this architectural term, which had acquired the meaning, since early in the eighteenth century, of "a place or position adapted to the character or capabilities, or suited to the merits, of a person or thing" (*Oxford English Dictionary*). Since Johnson, who believed in the immediate effect of the environment in determining the direction of evolution, was doing little more than setting up what he thought was a straw man in this passage, it is unlikely that he believed himself to be establishing, in a technical sense, a theory of the niche. Nevertheless, it is most curious that he should have given so modern a statement of niche specificity, only to demolish it, if the idea, though perhaps not the word, had not been widely current when he wrote.

The next known use of niche in an ecological sense is by Grinnell[17] who, in discussing the mammals and birds of the lower Colorado Valley in a long paper published in 1914, wrote (p. 91): "A concurrent axiom is that if associational analysis is carried far enough, no two species of birds or mammals will be found to occupy precisely the same ecologic niche, though they may apparently do so where their respective associations are represented fragmentarily and in intermixture." Later in the same paper we read (p. 98), "The query presents itself: is the Colorado fauna full? Are all the ecological niches, which are available in this area and which have occupants in other regions, occupied here? Probably not, for the intervention of barriers has doubtless prevented the invasion of types, which if they could have gotten there, would have thriven and assumed a place as endemic elements in the fauna."

Again it seems possible that Grinnell used the word in a way common in ordinary parlance and initially without intending to introduce a new technical term. Conceivably Grinnell could have read Johnson; the Carnegie Institute publications were freely available and enjoyed considerable respect. Johnson's scientific style, however, is likely not to have appealed greatly to Grinnell, and there is at least evidence that Grinnell neither corresponded with Johnson nor owned his book.

In 1916, W. P. Taylor,[18] who had worked

Johnson was curiously right in associating the idea of the niche, as that thing in which two closely allied species cannot co-occur, with Darwin's Malthusian point of view, even though he tended to reject this position. Such antiselectionist ideas continued into the 1920s and 1930s, when the mathematical study of natural selection, largely initiated by Haldane and developed by Fisher and by Wright, reinstated the Darwinian position, suitably modified to accommodate Mendelian inheritance. A very interesting light on some aspects of the period can be obtained from A. D. Darbishire's almost forgotten book, *An Introduction to a Biology and Other Papers* (London, Cassell; New York, Funk and Wagnall, 1917, xiii, 290 pp.). I owe a knowledge of this book to my first biology teacher, H. W. Partridge ("P").

I have found no clear evidence that Johnson had visited California when working on Coccinellidae, but it is evident from his book that at least he had received material from and corresponded with Charles Fuchs in Berkeley.

17. Joseph Grinnell, An account of the mammals and birds of the Lower Colorado Valley. *Univ. Calif. Publ. Zool.*, 12:51–294, 1914.

In view of Johnson's previous use of the word niche, I have looked through all Grinnell's earlier major works and as many of his small contributions as were available at Yale without finding any use prior to 1914. Frank A. Pitelka most kindly has examined the collections of books that Grinnell had in 1910–14 and found that Johnson's work was not among them. There are, moreover, no letters from him in Grinnell's very complete file.

18. W. P. Taylor, The status of the beavers of Western North America, with a consideration of the factors in their speciation. *Univ. Calif. Publ. Zool.*, 12:413–95, 1916.

with Grinnell, wrote a systematic monograph on the beavers of western North America, in which there is a whole section entitled "How have different ecologic niches been filled?" Taylor's answer was that the process involved allopatric speciation and subsequent invasion of other areas, rather than a direct development of adaptation to a new unfilled niche by a species living in another niche, but in the same restricted area. Like Grinnell, he believed that a region could contain unfilled niches.

Taylor, by repeating the word in a context very like that of Grinnell two years earlier, began to make it more technical. This process was continued by Grinnell in his later writings. Thus in 1917 he published a paper[19] that has become famous with the title "The niche-relationships of the California thrasher."

In this work he explains the rather restricted range of the species, *Toxostoma redivivum*, as due to the "close adjustment of the bird in various physiological and psychological respects to a narrow range of environmental conditions." These are analyzed primarily in terms of life zones and plant associations. Actually, within the temperature range which defines the Upper Sonoran life zone, the bird is clearly primarily limited by the nature of the vegetation cover provided by the chaparral. This is an association of continuous cover, "open next to the ground, with strongly interlacing branch-work and evergreen leafy canopy close above"; two-thirds of the foraging of the bird is done on the ground beneath this cover. Within the climatic zone that it inhabits, the niche of the California thrasher could doubtless be defined by vegetation height diversity, as MacArthur and his associates[20] did later for various eastern North American birds. After giving a qualitative but quite convincing account of this niche, Grinnell adds: "It is, of course, axiomatic that no two species regularly established in a single fauna have precisely the same niche relationships."

In later papers[21] Grinnell described the niche as the smallest subdivision of the habitat, the "ultimate unit . . . occupied by just one species or subspecies." It is evident again that though Grinnell is thinking geometrically all the time, the space occupied by "just one species" is an abstract space that cannot be a subdivision of the ordinary habitat space.

In 1927, Elton published his immensely important *Animal Ecology*,[22] which he said, in the preface of the book, dealt with the sociology and economics of animals. In a chapter on Animal Communities there is a section simply headed "Niches." For Elton, the niche is initially an active concept. When an ecologist sees a badger, he tells us "that he should include in his thoughts some definite idea of the animal's place in the community to which it belongs, just as if he had said 'there goes the vicar.'" He qualifies this dynamic approach by saying that there are "all manner of external factors acting" on the animal and that these must be included in characterizing the niche as the place of an animal "in the biotic environment, *its relations to food and enemies*." Thus Elton, although defining the niche largely in terms of the role of any given species, also includes in its niche the relations of the animal to the external factors that act on it.

As has already been indicated, Elton, unlike Grinnell, initially paid no attention to the competitive significance of the niche. However, so long as we are concerned with animals, Elton's use of the word permits statements, such as those of Gause, on competitive exclusion, and Elton along with Lack and Varley was an important proponent of Gause's concepts at a symposium[23] held by the British Ecological Society in March 1944,

19. J. Grinnell, The niche-relationships of the California thrasher. *Auk*, 34:427–33, 1917.

20. R. H. MacArthur, J. W. MacArthur, and J. Preer, On bird species diversity: II. Prediction of bird census from habitat measurements. *Amer. Natural.*, 96:167–74, 1962.

21. J. Grinnell, Geography and evolution. *Ecology*, 5:225–29, 1924.

22. See note 2 of this chapter.

23. L. A. Harvey, Symposium on "The ecology of closely allied species." *J. Ecol.*, 33:115–16, 1945.

which did much to further among British naturalists ideas about competitive exclusion.

Throughout the 1940s and early 1950s Elton's meaning was the one usually given to the word,[24] though with a sidelong glance toward Gause.

When competitive exclusion is considered among unicellular organisms, particularly nonmotile unicellular phytoplankton, Elton's ideas seem to be somewhat strained, though perhaps this strain is more apparent than real. A rather different way of looking at much the same set of facts, but one that seemed more suitable for the members of the phytoplankton, was proposed in 1944 and subsequently developed in 1957 into a more general way of regarding the niche.[25]

It is assumed that at least ideally all the variables that affect a particular species are capable of being ordered linearly. If we consider an *n*-dimensional space any point in which is defined by some value of the variables x', x'', x''' . . . measured along rectangular coordinates, we can represent the conditions for the existence of a species, which requires that the values of x' be between x_1' and x_2', of x'' between x_1'' and x_2'', etc., as a hypervolume, which in the simplest case will have a rectangular projection onto any coordinate plane such as $x'x''$. This hypervolume is the *fundamental* or *preinteractive niche* of the species (figure 99). The two-dimensional case, of a planktonic animal limited solely by temperature and food size, is shown in figure 100. If a third variable such as oxygen concentration is also involved in deriving the niche, as would be likely, the reader can easily visualize a three-dimensional niche rising from the page or sinking into it. If a fourth variable such as pH is then needed, most readers will find further visualization[26] impossible and will have to be content with

24. Notably in E. P. Odum's *Fundamentals of Ecology* (Philadelphia and London, W. B. Saunders, 1953, vii, 384 pp.; see pp. 16–18), to name only the best textbook of the period.

25. G. E. Hutchinson, Limnological studies in Connecticut: VII. A critical examination of the supposed relationship between phytoplankton periodicity and chemical changes in lake water. *Ecology*, 25:3–26, 1944 (see note 5, p. 20).

G. E. Hutchinson, Concluding remarks. *Cold Spring Harbor Symp. Quant. Biol.*, 22:415–27, 1957.

See also, for the ecological definitions here employed, G. E. Hutchinson, *A Treatise on Limnology*, vol. II, *Introduction to Lake Biology and the Limnoplankton*. New York, London, Sydney, John Wiley, 1967, chap. 19.

There is an important discussion in R. H. Whittaker, S. A. Levin, and R. B. Root, Niche, habitat and ecotope. *Amer. Natural.*, 107:321–38, 1973. Further reference to this paper is made later. Almost all the work on the niche is made conveniently accessible in a most useful volume: R. H. Whittaker and S. A. Levin, eds., *Niche: Theory and Application*. Benchmark Papers in Ecology 3. Stroudsburg, Penn., Dowden, Hutchinson and Ross; Halsted Press, 1975, xv, 448 pp.

There is a perennial difficulty about the word *biotope*. Etymologically, since both o's in *τόπος* are short, the word should be spelled biotop and pronounced to rhyme with "stop." This usage seems to me to assimilate badly into English; as I believe that aesthetic considerations should have a slight priority over historic ones, I unashamedly use biotope.

Whittaker, Levin, and Root trace the origin of the multidimensional approach to the niche to the Russian investigator, L. G. Ramensky, whose most significant works are available in German, in abstract or in translation respectively, as:

Die Grundgesetzmässigheiten im Aufbau der Vegetationsdecke. *Botan. Cbl.*, n.s. 7: 453–55, 1926 (original paper published in 1924).

Zur Methodik der vergleichenden Bearbeitung und Ordung von Pflanzenlisten und anderen Objekten, die durch mehrere, verschiedenartig wirkende Faktoren bestimmt werden. *Beitr. Biol. Pflanzen*, 18:269–304, 1930.

The antecedent work influencing me most in framing my concept was E. J. Haskell, Mathematical systematization of "environment," "organism" and "habitat." *Ecology*, 21:1–16, 1940.

James R. Ziegler has drawn my attention to a passage in V. A. Kostitzin, *L'Evolution de l'atmosphère* (Paris, Hermann, 1935, p. 43), in which a formulation for scenopoetic factors identical to mine occurs. As I knew of this work in the early 1940s, it is almost certain that it had a direct influence, though I did not remember the passage.

Whittaker, Levin, and Root give several references to multidimensional approaches after 1944; such concepts were probably implicit in much botanical thought, but did not become widely used by zoological and general ecologists until Whittaker's earlier papers were available.

26. Certain investigators of the Euclidean geometry of four dimensions appear to develop a quite extraordinary capacity to see three-dimensional forms as the shadows or projections of four-dimensional figures. Though this capacity is doubtless greatly improved by practice, there is a suggestion of it having

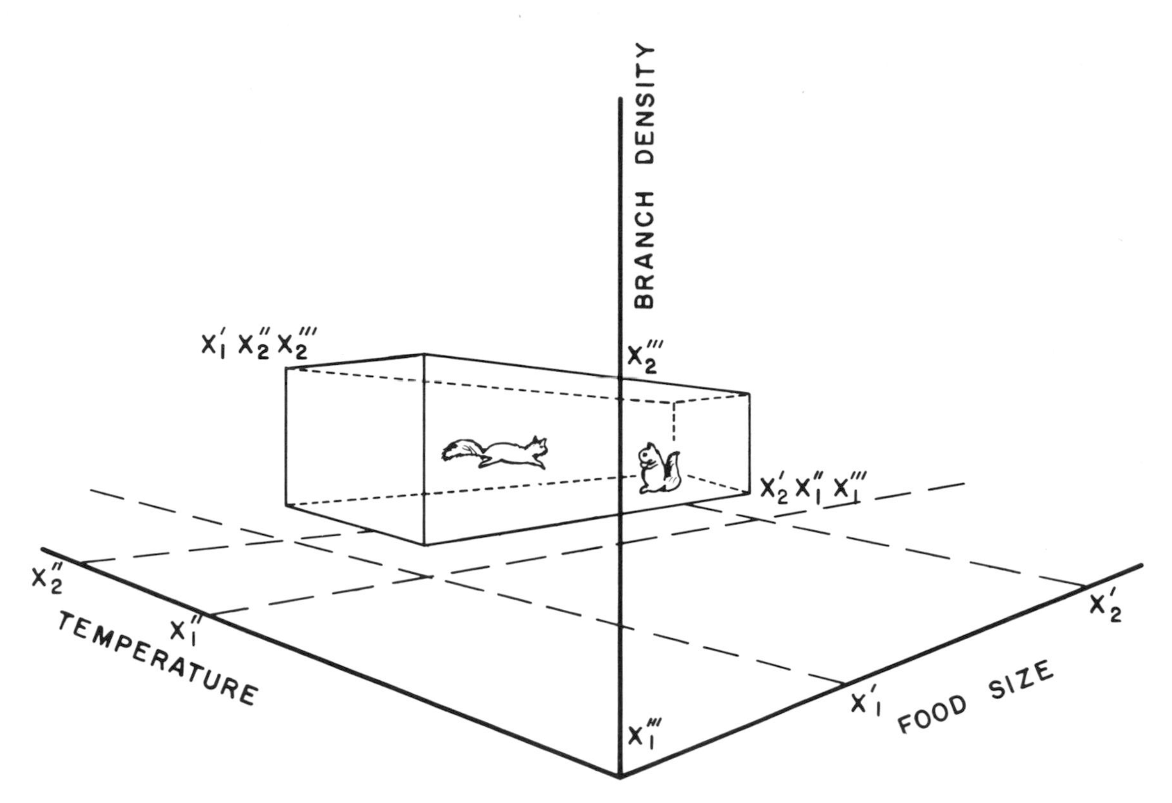

FIGURE 99. A three-dimensional orthogonal fundamental niche; x' might define food size, in this case the mean diameter of various seeds or fruits, notably acorns; x'' might define temperature tolerance, and x''' some measure of the density of branches, between certain diameters, in unit volume of physical space. Axis x' is therefore a bionomic axis, axis x'' is scenopoetic, while the status of x''' is not intuitively obvious, though the axis is clearly significant.

a purely verbal or symbolic statement, and so also for higher dimensionality.

If we now introduce a second species, its niche, as just defined, might not overlap that of the first species. If, however, it did overlap, we should expect that, in the part of the hyperspace where the overlap occurred, competitive exclusion would take place and the overlap would either be incorporated into the niche of one or the other species or be divided between the two, producing the *realized* or *postinteractive* niches of the two species (figure 100).

It is convenient at this stage in the argument to introduce the following definitions:

An area or volume will be called *homogeneously diverse* relative to a motile species if the relevant structural elements of the area or volume are small compared with the normal ranges of the individuals of the species, as would be the case of trees compared with the range of a deer. It will be called *heterogeneously diverse* if these elements are large relative to the ranges, as would be the case of trees relative to a bark-living insect or alga.

A *biotope* is any segment of the biosphere with convenient arbitrary upper and lower boundaries, which is horizontally homogeneously diverse relative to the larger motile organisms present within it; the diversity will be partly biological and not a prior physical property of the space. An *ecosystem*

an innate basis, as some people possessing it seem to have developed it rapidly and intuitively. Alicia Boole, later Mrs. Stott, the third of the five daughters of the inventor of Boolean algebra, was perhaps the most striking example of this, as she came on the capacity at the age of 18 when manipulating a number of wooden cubes (H. S. M. Coxeter, *Regular Polytopes.* New York and Chicago, Pitman, [1950?], xi, 521 pp.; see pp. 258–59).

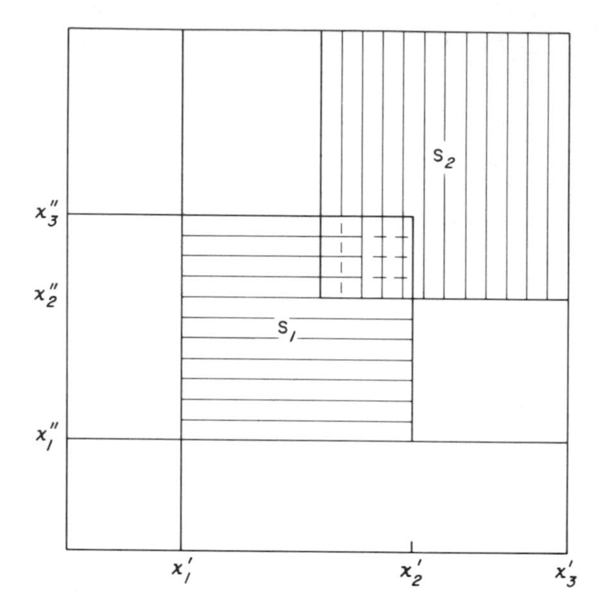

FIGURE 100. Two overlapping two-dimensional orthogonal fundamental niches. In the intersection, competitive exclusion occurs; either one or other species is eliminated or they divide the niche space between them, producing realized niches.

is the entire contents of a biotope. A *biocoenosis* is the totality of organisms living in a biotope, or the living part of an ecosystem.

Each niche can be regarded as a set of points, each one of which defines a possible set of environmental values permitting the species to live. The niche is clearly not part of the biotope since each point in the niche may correspond to many points (figure 101) in the biotope. It is, however, legitimate to speak of a biotope affording niches to the species living in it.

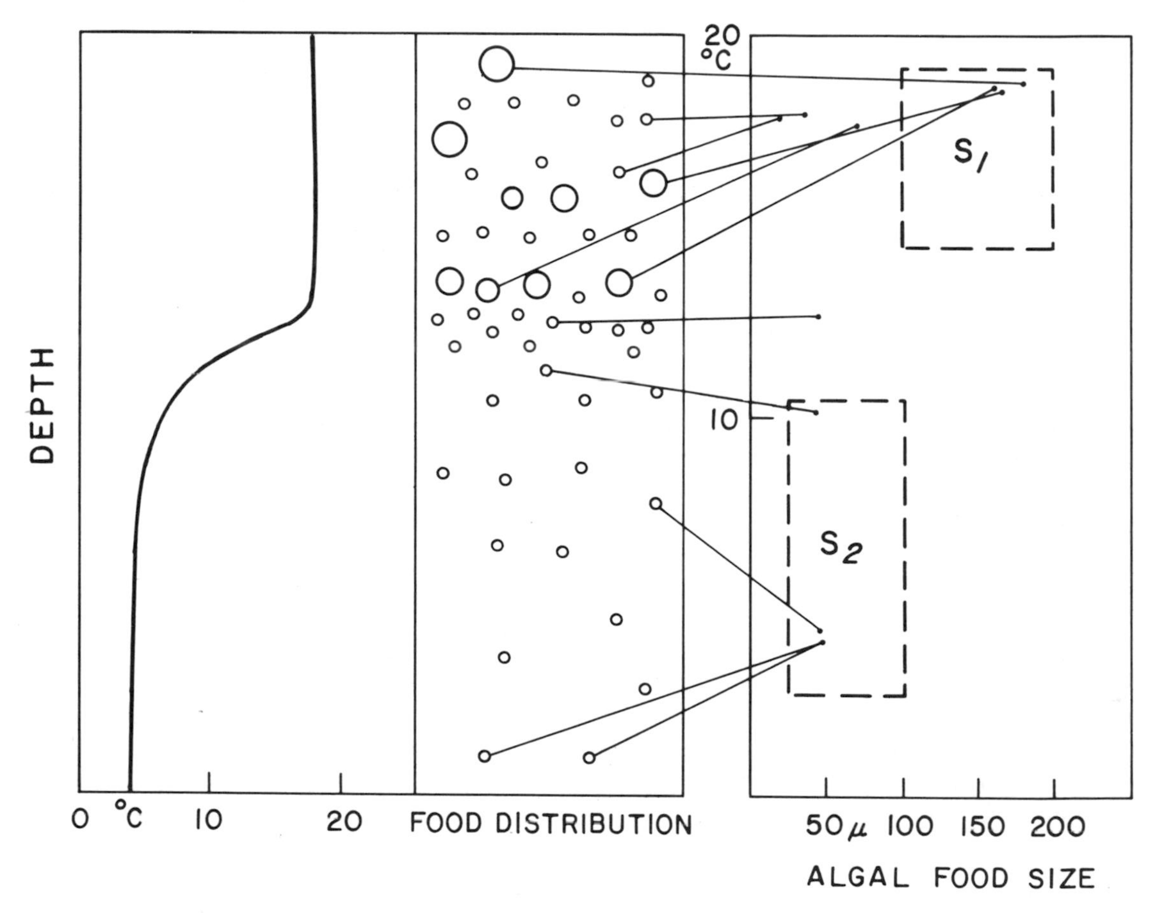

FIGURE 101. Niche space and biotope space in a very simple lacustrine system, involving variation in temperature and in algal food sizes. Any point in the niche space can correspond to many points in the biotope, but not all points are represented in any given biotope. The two nonoverlapping niches of species S_1 and S_2 contain areas that could be occupied by the species if they existed in the biotope, but in fact do not.

Though it is convenient to start the presentation with a rectangular two-dimensional niche, it is obvious that interaction between the different variables will modify this rectangular shape.

In the example of *Paramecium bursaria* studied by Gause, it is evident that there will be interaction of such a kind that lower oxygen contents are tolerated at high, but not at very low, light intensities, so providing a larger fundamental niche, part of which is not overlapped by the otherwise better competitors, *P. caudatum* or *P. aurelia*, as is roughly indicated in figure 102.

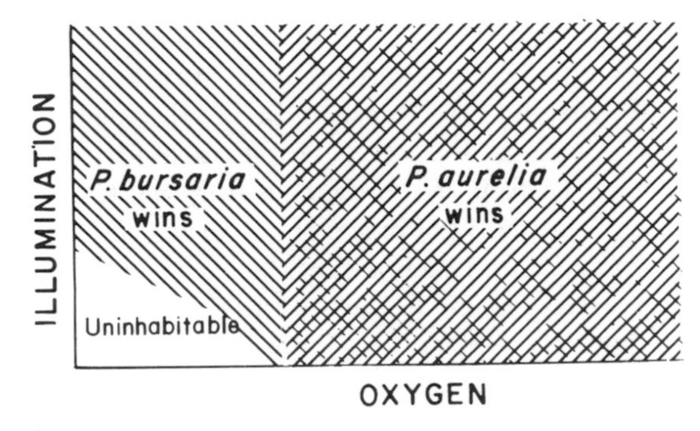

FIGURE 102. Ideal qualitative representation of the realized niches of *Paramecium aurelia* and *P. bursaria* projected onto the illumination-oxygen plane. At high oxygen concentrations *P. aurelia* displaces *P. bursaria*; but where the concentration is low, *P. bursaria* with its photosynthetic symbionts can live, provided the illumination is adequate, in a non-orthogonal or interactive niche. Fundamental niches are indicated by cross-hatching, realized niches by labeling.

It must be remembered that in this scheme the shape of any niche depends on the scales employed. A long rectangular niche can, by a suitable metric change, always be converted into a square one. Any volume not inside the boundaries of all the niches afforded by a given habitat is available without competition to a further species. In an old habitat we may expect niches to be filled, so that the average shape of a niche should be a space-filling figure. Without interaction this would give a structure of hyperspaces with rectangular projections onto planes. Interactions are, however, likely always to occur and one might expect there to be continual pressure to take over marginal parts of the realized niche of one species by a second species whose fundamental niche happens to trespass on the realized niche of the first species. In two dimensions the hexagon has the greatest area for a given perimeter, and the semiregular, space-filling tetrakaidecahedron the greatest volume for a given surface. If organisms extended their niches by competition and consequent natural selection, equally in both directions in a two-dimensional niche, or equally in three dimensions in a three-dimensional niche, following Grinnell's expanding soap-bubble analogy, these are the figures that we should get. Going beyond three dimensions may raise formidable difficulties; it is known, however, that any Euclidean *n*-space can be divided completely into orthotopes, the higher dimensional analogues of squares and cubes. In spaces of five or more dimensions, no regular *n*-dimensional figures other than orthotopes can fill space, but various kinds of semiregular figures may do so.[27]

The dimensionality of the niche

Before we proceed further, it is desirable to point out that the reader need not be frightened by the contemplation of an excessive number of dimensions. As the concept of the *n*-dimensional niche is developed, it will become increasingly clear that, although to define the tolerances and needs of a single species completely would indeed require a very large value for *n*, the study of the difference between two species can usually be conducted in a niche space of two or three dimensions.

The dimensionality of a niche is of some importance in considering the distribution of competition coefficients. In any stable community in which there are no negative coefficients, the values of all the α_{ij} will be positive numbers less than one. Species that have contiguous niches will have high values, species in widely separated niches will have values approaching zero. When only one niche axis is of significance, any species will

27. D. M. Y. Sommerville, *An Introduction to the Geometry of n-Dimensions*. London, Methuen, 1929, xvii, 196 pp.; see particularly chap. 10.

have two niche neighbors and most of the competitive limitations will involve interactions with them; outside them there will be two more and so on. In a plane, if the niche space is reasonably isotropic and there is an adequate species pool, one may expect on an average 6 nearest neighbors and 12 outside them, 18 in the next shell, and so on. If the variance of $\bar{\alpha}$, the mean competition coefficient, is computed for a distribution of α_{ij} dependent on distance from a given niche, in the first case this variance is found to be considerably greater than for the second. The variance of the competition coefficients thus may be expected to fall as the dimensionality of the niche increases.

In any given locality the number of species present is probably of the order of thousands or ten thousands. Elton[28] notes that 3,800 species of animals had been recorded from Wytham Wood near Oxford, a partly wooded but very diversified area of just over 5 km^2, an area small enough to permit most available species in the course of a century to invade any habitat in which they could live. As a number of the more obscure groups had not been studied, Elton suspected the total fauna to consist of about 5,000 species. The fauna thus is about one-fifth or one-sixth of that of Great Britain. There are about 60 breeding birds; this list includes species of very different ecological requirements and their presence certainly reflects the diversity of the area.

Tropical biotopes might well contain several times the number of species found in temperate areas. Lovejoy[29] found in one diversified station having an area of about 0.13 km^2, in the Amazonian rain forest near Belem, 161 species of resident birds. In the roughly defined ecological units of swamp forest (*varzea*), upland forest (*terra firme*), and an intermediate type, the number present was about 100. In all tropical rain forests, but most notably in South America, most species are very rare, so that intense study is needed to provide a relatively complete list. At times this rarity must give rise to local extinction. Lovejoy notes that a single adult male rufous-capped ant thrush, *Formicarius colma*, was repeatedly caught and released during 1967 and 1968, but no other specimen of the species was living in the area. This bird evidently can have left no descendants during that time.

For most invertebrates the data on species abundance are much less satisfactory. Bates,[30] as we have seen, found hundreds of species of butterflies in an area comparable to that worked by Lovejoy. For the whole Lepidoptera the number of species must be well into the thousands and the same would certainly be true of the Coleoptera, Diptera, and Hymenoptera. A total fauna of between 10,000 and 20,000 species would seem reasonable.

In terms of niche specificities the observed local faunas of say 5,000 to 20,000 species could be accommodated by a hyperspace

28. C. S. Elton, *The Pattern of Animal Communities.* London, Methuen; New York, John Wiley, 1966, 432 pp.

29. T. E. Lovejoy, Bird diversity and abundance in Amazon forest communities. *Living Bird*, 13:127–91, 1974.

30. H. W. Bates, *The Naturalist on the River Amazons.* London, John Murray, 1863, vol. I, ix, 351 pp.; vol. II, vi, 434 pp. Darwin, quoting Wallace (see p. 153), gives 600, but Bates wrote (vol. I, p. 102), "about 700 species of that tribe are found within an hour's walk of" Para.

If the unicellular algae of a temperate aquatic habitat are considered, quite high numbers may be obtained. N. Foged (On the diatom flora of the Funen Lakes. *Folia Limnol. Scand.*, 6:1–75, 1954), found up to 81 species of diatoms at any one season in the plankton of Braendegård Sø, while the study of chrysophycean cysts in small samples of sediments of very limited thickness suggest that over 50 species of these flagellates may have occurred together in some soft-water lakes in the past (G. Nygaard, Ancient and recent flora of diatoms and chrysophyceae in Lake Gribsø. In: *Studies on the Humic Lake Gribsø*, ed. K. Berg and I. C. Peterson. *Folia Limnol. Scand.*, 9:32–94, 1956; E. A. Leventhal, Chrysophycean cysts. In: *Ianula: A Study of the History of Lago di Monterosi. Trans. Amer. Phil. Soc.*, 60, pt. 4, 123–42, 1970). J. W. G. Lund (The marginal algae of certain ponds with special reference to bottom deposits. *J. Ecol.*, 30:245–340, 1942) found 196 species of various algae largely unicellular in ponds on Richmond common in England, but they were not all present simultaneously. There may be some reason to regard these associations of small rapidly reproducing species as rather different from associations of larger organisms (see pp. 213–14).

with four to five dimensions, the coordinates divided into 10 lengths, each representing the possible minimum and maximum values for tolerances or requirements of a single species. This is obviously far too formal a way of trying to put nature together, but indicates the kinds of magnitudes that might be involved.

In any given case even if we confine our attention to animals, the number of dimensions needed to provide environmentally different niches for a group of organisms small enough to handle, such as a class of vertebrates or an order of insects, is usually likely to be less than four.

In all cases it is probably impractical to deal with all organisms in a single niche space. In too many cases complicated properties of organisms, defined as one set of axes, are liable to turn up ordered along another set of axes. Trees would be determined by physically derived niches and competitive relationships. Most terrestrial animals are determined by the existence of trees and in some cases by the properties of quite small parts of such organisms. This will give to the niche space an involuted character, the existence of one axis being dependent on other axes.

There is no reason, however, not to use the concept of the multidimensional niche in plant ecology, even though it has rarely been done. In the present context a particularly illuminating case has been examined by Watson,[31] who found (figure 103) that in the mosses of the family Polytrichaceae, three members of the genus *Polytrichum*, namely, *P. commune*, *P. piliferum*, and *P. juniperinum*, exhibited considerable separation in a two-dimensional space defined by light and pH; the three sympatric species of the very closely allied *Polytrichastrum*, namely, *P. alpinum*, *P. formosum*, and *P. pallidisetum*, however, were not separated in this way. In such a case two

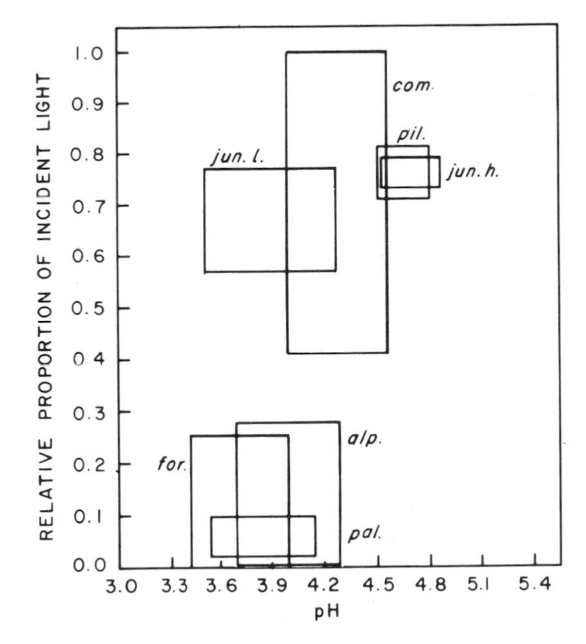

FIGURE 103. Realized niches relative to illumination and soil pH of three species of *Polytrichum* (com., *P. commune*; *jun. h.*, *P. juniperinum* at high elevations; *jun. l.*, *P. juniperinum* at low elevations; *pil.*, *P. piliferum*) and of *Polytrichastrum* (*alp.*, *P. alpinum*; *for.*, *P. formosum*; *pal.*, *P. pallidisetum*). Note the separation of the two genera on the illumination axis and the apparent greater niche specificity in *Polytrichum* than in *Polytrichastrum*. *P. piliferum*, being very short-lived (see Figure 66), probably differs much in its biology from the other two species of the genus (Watson).

very similar groups of plants are apparently separated in quite different ways, and at least three dimensions would be needed to effect a good differentiation in the family. What are essentially projections of niche space have proved convenient in considering the chemical ecology of water plants (figure 104).

The included niche

Miller[32] has pointed out that there is a common arrangement in which the niche of one species lies inside that of another. Usually the

31. M. A. Watson, *The Population Biology of Six Species of Closely Related Bryophytes* (Musci). Ph.D. Thesis, Yale University, 239 pp., 1974.

32. R. S. Miller, Ecology and distribution of pocket gophers (Geomyidae) in Colorado. *Ecology*, 45:256–72, 1964.

See, however, A. E. M. Baker. Interspecific aggressive behavior of pocket gophers *Thomomys bottae* and *T. talpoides*. *Ecology*. 55:671–73. 1974.

R. S. Miller, Conditions of competition between redwings and yellow-headed blackbirds. *J. Anim. Ecol.*, 37:43–62, 1968.

G. M. Branch, Mechanisms reducing competition in limpets: Migration, differentiation, and territorial behaviour. *J. Anim. Ecol.*, 44:575–600, 1975.

R. G. Jaeger, Competitive exclusion: Comments on the survival and extinction of species. *Bioscience*, 24: 33–39, 1974.

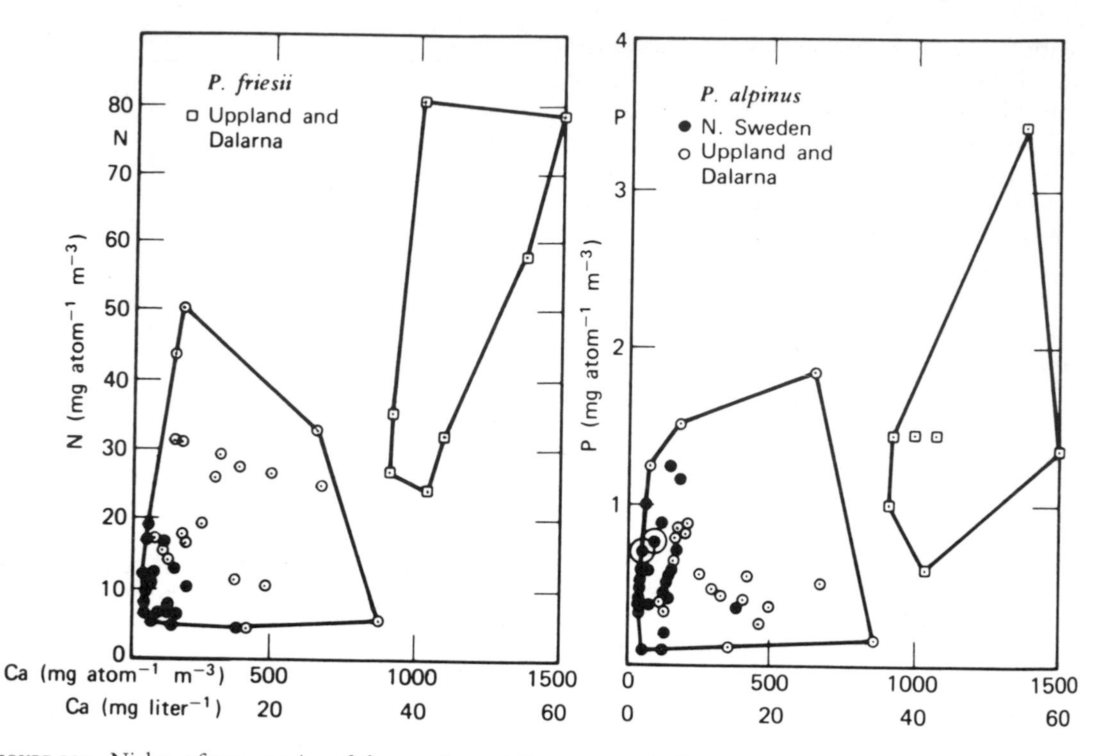

FIGURE 104. Niches of two species of the pond weed *Potamogeton*, *P. friesii* and *P. alpinus*, in Sweden, projected on the mean total dissolved nitrogen-calcium, and mean total dissolved phosphorus-calcium planes (data of Lohammar, diagrams after Hutchinson).

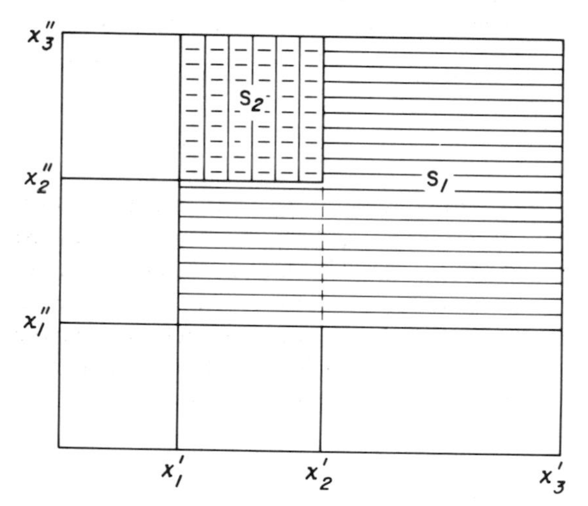

FIGURE 105. Miller's included niche. The more adaptable species S_1 has a fundamental niche $x_1'x_1''$, $x_1'x_3''$, $x_3'x_1''$, $x_3'x_3''$, but the more specialized species S_2 occupies the best part of this $x_1'x_2''$, $x_2'x_2''$, $x_1'x_3''$, $x_1'x_3''$ from which it excludes S_1, so that the realized niche of the latter is concave.

species share part of two niche boundaries and one vertex between them. The species with the included fundamental niche if it is to survive must be the competitive superior within at least part of that niche. The other species survives by having a larger niche. This in itself means that it can make better adjustments to variation in the environment than is usually possible for its competitor, which is superior only under a limited range of conditions. Miller put forward this idea with regard to four species of pocket gophers found in Colorado. The optimal conditions for all species are provided by deep light soils, but when the species are in competition they can be arranged in a series, *Geomys bursarius*, *Cratogeomys castanops*, *Thomomys bottae*, and *T. talpoides*, the first tending to displace all species in favorable habitats, the last none. The sizes of the fundamental niches are, however, inversely related to competitive ability in an optimal habitat; so *G. bursarius* occupies the best corner of the niche and the other species are arranged around it (figure 105). The idea is essentially a geometrical interpretation of Gause's concept[33] of an antithesis

33. G. F. Gause, Problems of evolution. *Trans. Conn. Acad. Arts Sci.*, 37:17–68, 1947.

between the adaptable species, able to use a very wide range of environments, and the adapted species specialized for a particular part of this range. Though the biological picture of the Colorado pocket gophers just given may be oversimplified, Miller's concept is clearly a valid one. He has further applied it to the niche of the yellow-headed blackbird (*Xanthocephalus xanthocephalus*) included within the fundamental niche of the redwing blackbird (*Agelaius phoeniceus*). Other investigators have also used the idea, notably Branch (see p. 202) in his discussion of South African limpets and Jaeger in his treatment of the niches of salamanders of the genus *Plethodon*.

By far the most dramatic examples of species having niches that include those of other species in Miller's sense are provided by birds studied by Diamond[34] in the islands off New Guinea. He records 13 species which he designates as supertramps, 6 being pigeons and the other 7 belonging to 6 genera in as many different passerine families. The supertramp species are mainly found on small outlying islands where they may form large populations. Where an island has been sterilized by volcanic action during the relatively recent historic past and has regained a good deal of vegetation but not the complete avifauna, the 13 species are found to be highly adaptable to all conditions except that of having a large number of competitors. They are also clearly very easily dispersed. Diamond concludes that they, and a number of less extreme tramp species, are ultimately kept out of the islands with a really rich avifauna by the presence of species that he designates as "companions in starvation"; these can exhaust resources down to a level at which they can just survive, but at which the supertramps cannot. This method of competing by the species in the included niche must lead to the evolution of low, on occasions to dangerously low, populations, as with Lovejoy's rufous-crowned ant thrush. It must be characteristic of many tropical environments.

The adaptation of the species occupying the best, or included, part of the niche is likely to involve innate behavioral dominance. This is clearly indicated in the case of the two woodrats, *Neotoma lepida* and *N. fuscipes*. Both prefer *Quercus turbinella* shoots as food when they are living separately, but when sympatric, *N. lepida*, known to be less dominant in experimental situations, is likely to switch its food to *Juniperus californica*.

The separation of niches by dominance relationships can be quite complicated. Heller, working with chipmunks of the genus *Eutamias* in the Sierra Nevada, found that *E. alpinus* and *E. amoenus monensis*, both of which use terrain supplying moderate amounts of defensible food and a good deal of cover against predators, are aggressively territorial, while *E. speciosus frater* and *E. minimus scrutator*, which live respectively between and below the two first-named taxa and are probably limited by predators rather than food, are not markedly territorial. In such species territoriality is of no great advantage and may in fact be dangerous if it involves conspicuous behavior. In *E. m. scrutator* living in the hot dry sagebrush zone, it is probable that significant aggressive behavior would not be metabolically feasible. The limits of the niches of these animals appear to be largely determined by differences of this sort.

It is interesting to note that the realized niches of all species except the most successful one in the best habitat are *concave*, in the sense that points exist in the realized niche which can be joined only by lines passing outside that niche.

Measurement of competition coefficients

As a feasible approximation in the field, MacArthur and Levins,[35] who published

34. J. M. Diamond, Assembly of species communities. In: *Ecology and Evolution of Communities*, ed. M. L. Cody and J. M. Diamond. Cambridge, Mass., Belknap Press of Harvard University Press, 1975, pp. 342–444. A very impressive work.

G. N. Cameron, Niche overlap and competition in woodrats. *J. Mammal.*, 52:288–96, 1971.

H. C. Heller, Altitudinal zonation of chipmunks (*Eutamias*): Interspecific aggression. *Ecology*, 52:312–19, 1971.

35. R. H. MacArthur, The theory of the niche. In: *Population Biology and Evolution*, ed. R. C. Lewontin. Syracuse University Press, 1968, pp. 159–76.

R. Levins, *Evolution in a Changing Environment:*

separately in 1968, but seem to have worked out the theory together, proceeded as follows.

In a set of microhabitats chosen in the overall habitat under investigation, the frequency of occurrence of each species of some taxocene is determined by a suitable technique of observing or trapping. In large diurnal animals such as birds the fraction of the total time of observation of the species spent in each microhabitat is an excellent measure of frequency. Since competition necessarily involves co-occurrence, the product of these relative frequencies p_{ih} and p_{jh} of the i-th and j-th species in the h-th microhabitat is taken as a measure of competition in that habitat, and is expressed relative to p_{ih}^2.

At first sight it may appear that this begs the question of competitive differences between individuals of different species. MacArthur, however, believes that closely related, morphologically similar species, living in coarse-grained environments from which appropriate pieces of food are picked out individually, need not differ greatly in their capacities to search and pursue effectively. Such differences as occur, he suspects, are just enough to make each species superior in its own characteristic type of environmental patch. Moreover, the species contributing the larger factor to the product will presumably do so because it is effective and not because it happens to be there for no reason at all.

We therefore write as an expression for the competition coefficient

$$\alpha_{ij} = \Sigma p_{ih} \cdot p_{jh} / \Sigma p_{ih}^2. \qquad (5.1)$$

We note that $1/\Sigma p_{ih}^2$ is regarded by Levins as the best expression of the *niche breadth* of species i in habitat h.

A very instructive example is given by MacArthur and concerns four of five species of New World warblers of the family Parulidae, living in spruce trees in northern New England. MacArthur divided the typical spruce tree into zones (figure 106) and recorded the times, relative to the total observation times for the species, spent in each zone by the four kinds of warblers studied.

The mean number of pairs, (N_i) per unit area (in this case 5 acres or about $2.10^4 m^2$) is known, as well as the competition coefficients from (5.1). We may therefore write for each species, assuming equilibrium so $dN_i/dt = 0$;

$$\frac{dN_i}{dt} = N_i r_i (K_i - N_i - \alpha_{ij} N_j) = 0 \qquad (5.2)$$

or evaluating K_i from the different values of N_i and α_{ij} that we have obtained, for

the myrtle warbler, *Dendroica coronata*, $\overline{N_1} = 2$

$$6.190 - N_1 - 0.490 N_2 - 0.480 N_3 - 0.420 N_4 = 0$$

the black-throated green warbler, *D. virens*, $\overline{N_2} = 5$

$$9.082 - 0.519 N_1 - N_2 - 0.959 N_3 - 0.695 N_4 = 0$$

the Blackburnian warbler, *D. fusca*, $\overline{N_3} = 1$

$$6.047 - 0.344 N_1 - 0.654 N_2 - N_3 - 0.363 N_4 = 0$$

the bay-breasted warbler, *D. castanea*, $\overline{N_4} = 3$

$$9.014 - 0.545 N_1 - 0.854 N_2 - 0.654 N_3 - N_4 = 0$$

the first, absolute terms in the equations being the K_i.

In the mathematical treatment of a number of simultaneous equations, matrix notation is convenient, the equations (5.2) being written

$$\mathbf{K} = \mathbf{AN} \qquad (5.3)$$

Some Theoretical Explorations. Princeton University Press, 1968, ix, 120 pp.

See also R. H. MacArthur, Population ecology of some warblers of northeastern coniferous forests. Ecology, 39:599–619, 1958.

The concept of a guild is due to R. B. Root, The niche exploitation pattern of the blue-grey gnatcatcher. *Ecol. Monogr.*, 37:317–50, 1967.

The guild is not the same as the taxocene or assemblage of species belonging to a particular classificatory unit or higher taxon, such as a family, found in a given habitat, irrespective of niche similarities. The guild may be, but often is not, a taxocene.

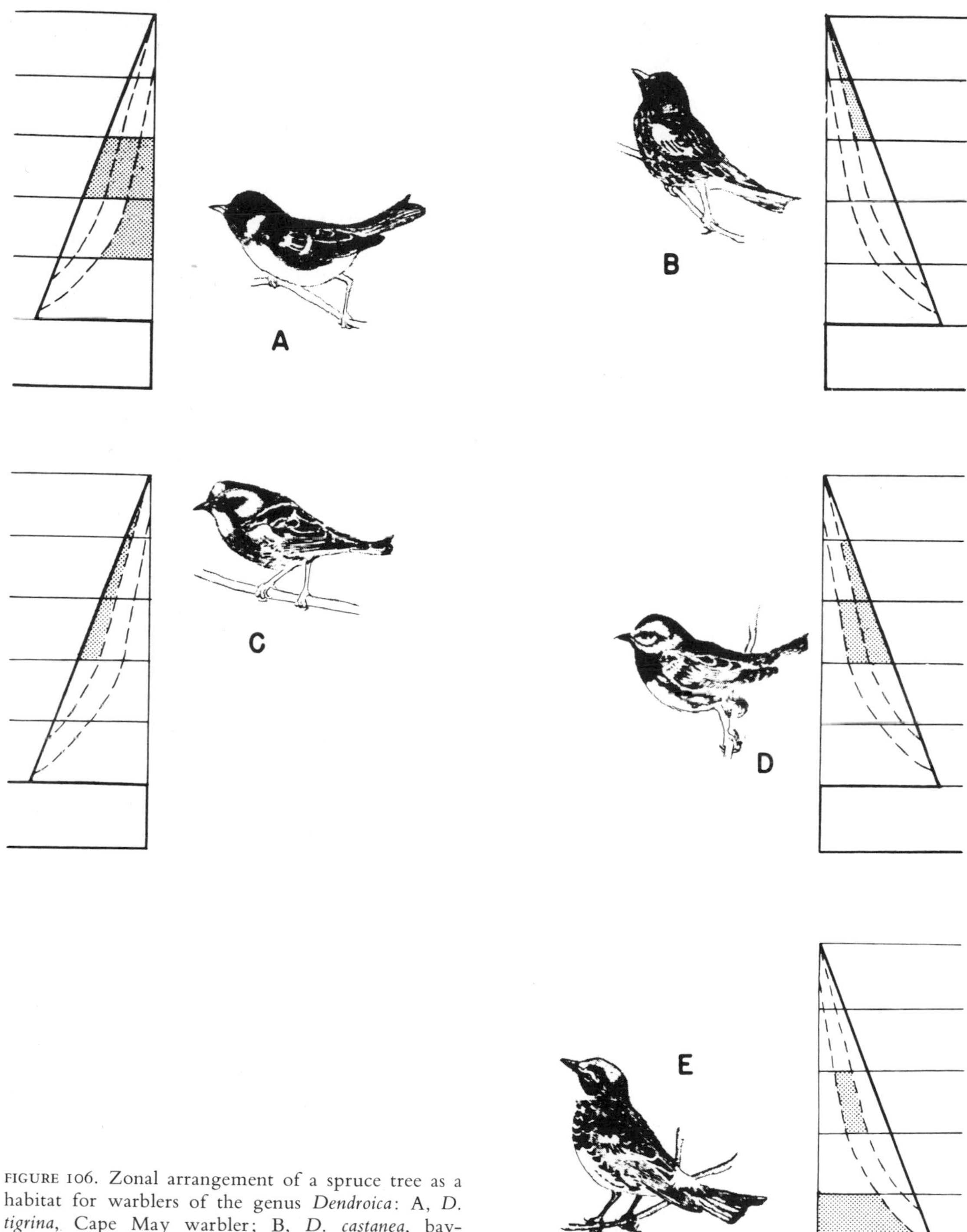

FIGURE 106. Zonal arrangement of a spruce tree as a habitat for warblers of the genus *Dendroica*: A, *D. tigrina*, Cape May warbler; B, *D. castanea*, bay-breasted warbler; C, *D. fusca*, Blackburnian warbler; D, *D. virens*, black-throated green warbler; E, *D. coronata*, myrtle warbler. The time each species spent in each of the defined zones was determined and from it the values of the competition coefficients estimated. In this diagram each species spent more than half its time in the stippled zones (Hutchinson, from MacArthur).

where **K** and **N** are merely the columns giving the saturation values K_i and the observed populations N_i, while

$$\mathbf{A} = \begin{vmatrix} \alpha_{11} & \alpha_{12} & \alpha_{13} & \alpha_{14} & \cdots \\ \alpha_{21} & \alpha_{22} & \alpha_{23} & \alpha_{24} & \cdots \\ \alpha_{31} & \alpha_{32} & \alpha_{33} & \alpha_{34} & \cdots \\ \alpha_{41} & \alpha_{42} & \alpha_{43} & \alpha_{44} & \cdots \end{vmatrix}$$

For the four warblers

$$\mathbf{A} = \begin{vmatrix} 1 & 0.490 & 0.480 & 0.420 \\ 0.519 & 1 & 0.959 & 0.695 \\ 0.344 & 0.654 & 1 & 0.363 \\ 0.545 & 0.854 & 0.654 & 1 \end{vmatrix}$$

The term *community matrix* is ordinarily now used for **A**. Since its theoretical foundations involve assumptions about similarities in mode of life even if there are small but important limiting differences, the term community is unfortunate and indeed incorrect in this context. The usual word for a group of animals with comparable roles, such as insectivorous birds living in trees, plant-feeding insects mining in leaves, or carnivorous water beetles swimming among plants, all having similar though not identical Eltonian niches, is *guild*. The matrix that we have been discussing is therefore most legitimately thought of as a *guild matrix*.

A little mathematical manipulation makes it possible to determine under what conditions a species can invade the guild. In the present case it is possible to show, for instance, for the bay-breasted warbler, that invasion would be possible when only the other three species were present if the environment provided resources for the invading species of such a kind and quantity that

$$K_4 > 0.1392K_1 + 1.0775K_2 - 0.4461K_3,$$

where K_1, K_2, and K_3 are the carrying capacities of the first three species and K_4 of the invading bay-breasted warbler. It will be noted that the presence of Blackburnian warblers actually increases the possibility of the existence of bay-breasted warblers. Reference to the equations shows that the Blackburnian has a smaller competitive effect on the bay-breasted than on the myrtle or black-throated green, and a greater effect on the latter than on other species. In turn, the black-throated green is a strong competitor of the bay-breasted, so that anything depressing the former would encourage the latter. This example shows how ideally the matrix can give a good deal of insight into the structure of a biological association.

Culver[36] has concluded that where interference rather than exploitation is taking place, the proper form of the equation for α_{ij} is

$$\alpha_{ij} = \frac{\Sigma p_{ih} p_{jh}}{\Sigma p_{ih}{}^{\star 2}} - \frac{\Sigma p_{ih} p_{jh}}{\Sigma p_{jh}{}^{\star 2}} \qquad (5.4)$$

in which $p_{ih}{}^{\star}$ and $p_{jh}{}^{\star}$ are frequencies in different but comparable microhabitats where the two species are present by themselves.

Culver was interested in the aquatic crustacea of caves in West Virginia. Four species were studied, the isopod, *Asellus holsingeri*, and the three amphipods, *Stygonectes emarginatus*, *S. spinatus*, and *Gammarus minus*. Food appears to be present in excess and the animals are limited by competitive displacement of each other when hiding under stones in a current (figure 107).

The interference competition matrix obtained for these animals is:

	Ss	*Gm*	*Se*	*Ah*
Ss	1	0.07	0.09	0.01
Gm	0.08	1	1.13	0.73
Se	0.08	0.84	1	0.52
Ah	0.18	0.98	0.95	1

$$- \begin{vmatrix} 1 & 0.22 & 0.01 & 0.10 \\ 0.22 & 1 & 0.07 & 0.17 \\ 0.01 & 0.05 & 1 & 0.10 \\ 0.10 & 0.18 & 0.13 & 1 \end{vmatrix}$$

36. D. C. Culver, Analysis of simple cave communities: Niche separation and species packing. *Ecology*, 51:949–58, 1970.

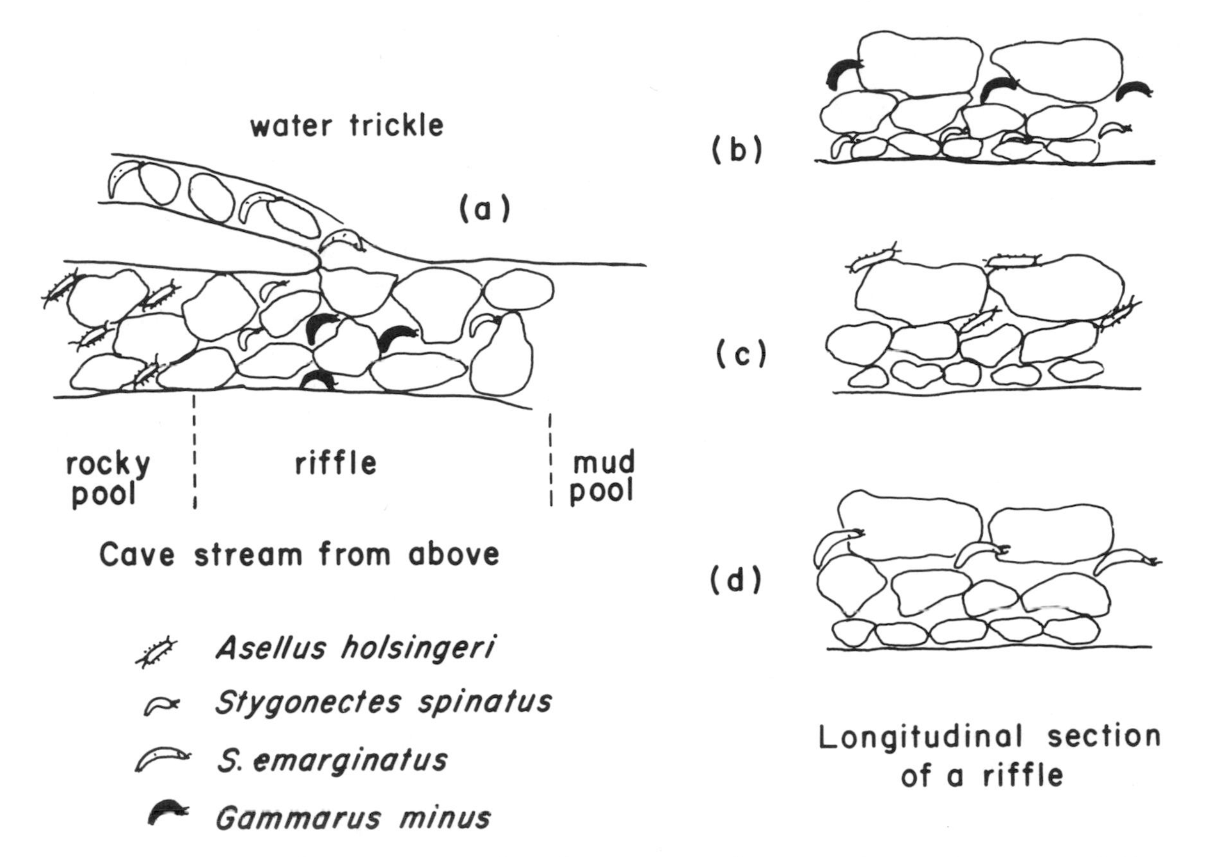

FIGURE 107. (a) Portion of a cave stream, seen from above, containing four species of peracarid crustaceans. *Asellus holsingeri* occurs primarily in pools with a rocky bottom, *Stygonectes emarginatus* among stones in small trickles, *S. spinatus* deep in riffles, and *Gammarus minus* above the *S. spinatus*; the latter arrangement is shown in (b), in section. When only one species is present, it will tend to be in the riffles. *G. minus* and *S. spinatus* singly occupy the positions that they have in the full community, while *A. holsingeri* occurs around the upper stones (c), and *S. emarginatus* in the region (d) which would contain *G. minus* if it were present (Culver, modified).

$$= \begin{vmatrix} 1 & -0.15 & -0.08 & -0.09 \\ -0.14 & 1 & 1.06 & 0.56 \\ 0.07 & 0.79 & 1 & 0.42 \\ 0.08 & 0.80 & 0.82 & 1 \end{vmatrix}$$

The negative values in the final results are spurious and the first and third matrices are not significantly different. Culver believes that food is of little importance in this case. Study of the matrices along the lines given by MacArthur for the bay-breasted warbler suggests the community is saturated with peracarid crustacea when four or five species are present.

Niche breadth and niche size

The concept of niche breadth is initially due to Levins,[37] who used the term to indicate the length of that part of any niche axis that included all the points defining viable values of the variable that is measured along that axis. Thus if we are dealing with temperature, the niche breadth is the range permitting permanent survival of the organism; if with food size, the range that the organism can eat. Since near the ends of the range the conditions are likely to be suboptimal, it is best to consider the matter stochastically. For any species S_i we write p_{ih} for the proportion of the whole surviving at temperature h, or eating food of size h, the niche breadth may then be expressed in several ways. Levins considers two:

$$\log B' = -\Sigma p_{ih} \log p_{ih} \qquad (5.5)$$

37. R. Levins, see note 35.

and

$$B = \frac{1}{\Sigma p_{ih}^2} \qquad (5.6)$$

The first of these is the familiar Shannon-Wiener approximation to information-theoretic diversity, to be used again in discussions of the relation of species diversity and habitat diversity. The second measure, B, which Levins uses, has appeared before in equation (5.1). In both cases when there are n intervals or sampling points and the individuals of the species are equally viable or equally well-fed or equally common at all of them, the value of p_{ih} will be $1/n$ and the niche breadth will be n. Since one often needs to make comparisons between different sets of data, the breadth can be normalized by division by n; or in a given locality, if a taxocene or guild is being studied, the mean value ($\bar{B}$) of the B_i can be determined and each value divided by this mean. Since if the B_i have been divided by n, $\bar{B}$ is also divided by n, the width relative to the mean ($B/\bar{B}$) is the same if calculated from B or from B/n. In general, the strict determination of niche breadth along a given axis is difficult in natural communities. Levins gives some examples in *Drosophila* based on seasonal occurrence; these are doubtless partly based on temperature tolerances and on food preferences, which are hard to order linearly.

Of greater interest are the measurements that can be made by observing the distribution of species in the various parts of a biotope or habitat either directly or by sampling at a number of stations within the habitat by trapping of some sort. In such cases we do not know what niche axes are involved, though it is probably always safe to assume that a single axis is unlikely to be isolated accidentally in the sampling process. If the process of gathering the data is efficient and unbiased, we may hope that the result approximates to some measure, not along some single known axis, but of the realized niche as a whole, whatever largely unknown axes are significant. The measure may be appropriately called *niche size*.[38]

The same methods of computing the value of B or B' just given may be used. Colwell and Futuyama[39] have developed an elaborate mathematical technique by which a number of weaknesses in the original formulation are removed. Anyone working in the subject will need to consult their paper, but for the purpose of general exposition Levins's simpler original treatment seems satisfactory. It can be used, for instance, on the data of figure 106 for the four species of warbler that MacArthur used in his treatment of the competition coefficient matrix, the numbers in the diagrams being taken as p_{ih}. The habitat is divided into 16 pieces which have qualitatively different properties relative to the birds' behavior, but are obviously of different volume. Using the simple Levins approach, we obtain the results of table 12.

There are also some data for the Cape May warbler (*D. tigrina*) in MacArthur's earlier paper, which gives a niche size of 0.20.

In a study of this sort it is reasonably certain that the whole of that part of the habitat

	B/n	N pairs in 5 acres	K pairs in 5 acres
Myrtle (*Dendroica coronata*)	0.45	2	6.19
Black-throated green (*D. virens*)	0.48	5	9.08
Blackburnian (*D. fusca*)	0.33	1	6.05
Bay-breasted (*D. castanea*)	0.58	3	9.01

TABLE 12. Niche size, actual and maximal populations of four species of warbler studied by MacArthur.

38. This expression was used, though in an undefined way, by P. H. Klopfer and R. H. MacArthur, On the causes of tropical species diversity: Niche overlap. *Amer. Natural.*, 95:223–26, 1961.

39. R. K. Colwell and D. J. Futuyama, On the measurement of niche breadth and overlap. *Ecology*, 52:567–76, 1971.

See also M. D. Sabath and J. M. Jones, Measurement of niche breadth and overlap: The Colwell-Futuyama method. *Ecology*, 54:1143–47, 1973.

used for feeding and all reproductive activities has been under observation. A great deal is also known about differences in feeding and breeding behavior, though it is not possible to identify a minimum number of independent niche axes. The values of B/n, however, must be reasonable approximations to a measure of niche size. When numbers of individuals are given so that K can be calculated, it appears that there is some, but by no means a perfect, correlation of niche size and the maximum possible or actually realized populations. Such a correlation would be expected since the larger the niche, the more points it contains, each of which corresponds to volumes of habitat space where the conditions defined by the niche points are realized. Levins also inferred from some observations on Puerto Rican species of *Drosophila* that the relative niche size and K were related, concluding that the different species differed in the ranges of their resources more than in the efficiency with which they utilized them. This to me seems a reasonable conclusion, though not without interest.[40]

When the habitat is large and is sampled by local trapping so that most of the individuals present are unnoticed and unenumerated, the relationship of the observations to the structure of the niche is by no means obvious, and Lovejoy,[41] who has studied the matter extensively in the tropical rain forest, prefers the term *habitat tolerance* to niche breadth or size.

Lovejoy determined the frequency of birds in various parts of a small area of rain forest near Belem in Brazil by means of capture and release of birds in mist nets over a period from the beginning of October 1967 to the end of August 1969. The bird fauna of the area is one of the richest in the world, as has earlier been indicated. Lovejoy gives data for B and indicates that n is 76. The maximum possible value of B is therefore 76. There is an enormous spread of abundance and of habitat tolerance, the two being of course related. A considerable number of species are too rare to give any valid figure for habitat tolerance; a bird found only 10 times cannot have a tolerance of more than $B/n = 0.13$. Five species, however, have habitat tolerances of more than 0.65 ($B = 50$) and 23 of more than 0.39. Of these 23, 5 are nectivores, 7 frugivores, 3 follow army ants, feeding on whatever animals, mainly insects, are fleeing from them, and 8 are mainly fairly specialized insectivores, often feeding in peculiar ways, but in ways that are possible in any of the forest types represented near Belem. The very habitat-tolerant and very abundant birds in this extremely rich fauna are thus primarily fairly specialized for living in ways that are possible in a wide variety of habitats. Their specialization gives them niche separation; in a full model one would presumably need a separate axis for army-ant invasion frequency, and would find that the limited parts of the hyperspace, where this was not zero, constituted the niches of the plain-brown woodcreeper (*Dendrocincla fuliginosa*), the white-shouldered fire-eye (*Pyriglena leucoptera*), and the black-spotted bare-eye (*Phlegopsis nigromaculata*). All of the 23 species of habitat-tolerant birds eat food that, when found, is easily taken and does not have to be pursued through the forest. This is obvious for the nectivorous and frugivorous species, feeding on conspicuous food, but less so for the insectivores, which in one case, that of the yellow-billed jacamar (*Galbula albirostris*) may be an ambush predator.[42] Active long pursuit of a given prey may well involve sensory adaptation to particular ranges of light intensity or contrast, characteristic of different forest types. Lovejoy's analysis clearly shows that the species with large niches may be specialists that can live in a variety of habitats provided these habitats give opportunities for their specialities.

Scenopoetic and bionomic niche axes

The axes of a multidimensional niche are of

40. J. M. Emlen, Niches and genes: Some further thoughts (*Amer. Natural.*, 109–472–76, 1975), gives a model in which niche breadth is measured as K.

41. T. Lovejoy, see note 29.

42. A term suggested by J. Gerritsen and R. Strickler, Encounter probabilities and community structure in zooplankton: A mathematical model, *J. Fish. Res. Board Canada* 34:73–82, 1977.

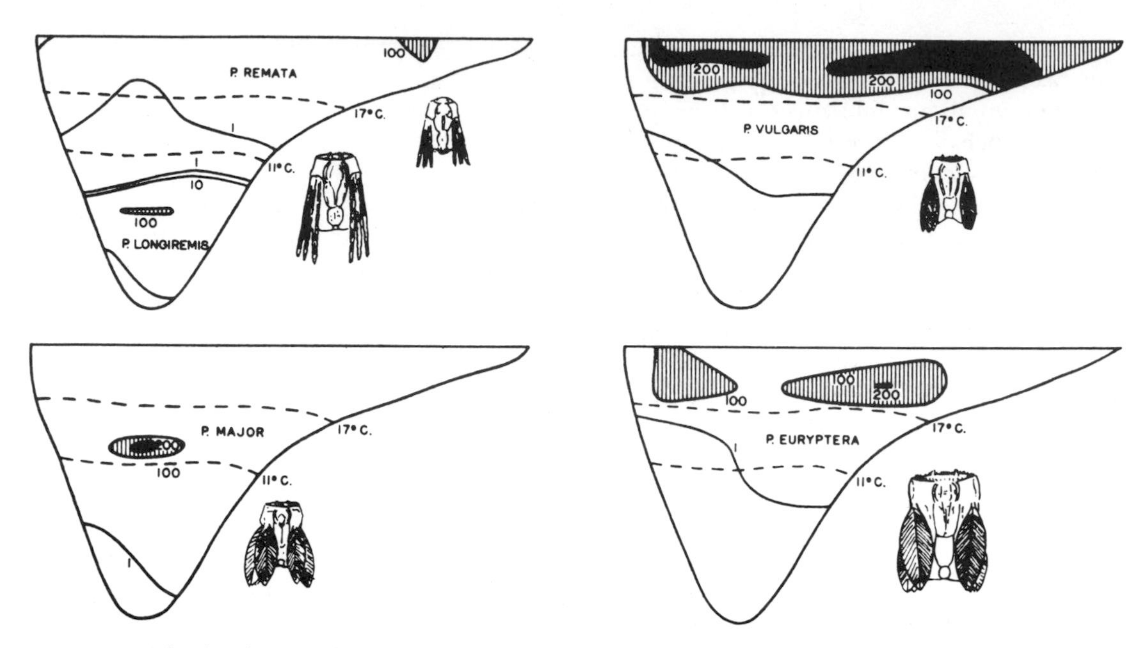

FIGURE 108. The distribution of the various species of the planktonic rotifer, *Polyarthra*, in Skarshultsjön, a small lake in Sweden, showing separation of species partly by size, probably indicating different food, and partly by tolerance of physical conditions (Bērziņš, modified by Hutchinson).

two sorts. In one set of cases the niche of an organism is determined by the tolerance limits to certain intensive physical and chemical variables, temperature, illumination, and humidity being obvious examples. If the species is to live in a habitat in which a certain range of values is expected, the species must tolerate these values. In the simplest case the mere tolerance of a particular temperature range by a species makes no difference to the existence of other species so far as temperature tolerance is concerned; there is no direct competition for a temperature of say 20°C. Such niche axes set the stage and may be called *scenopoetic*. The other axes relate to variables that are directly involved in the lives of organisms, in general, as resources for which there may be competition; these may conveniently be termed *bionomic* axes. A niche space could in theory be divided up into niches by the variables of the scenopoetic axes alone, without any competition taking place, provided a biotope existed in which a sufficient variety of values of the physicochemical variables were present simultaneously. This situation is a commonplace in limnology, where one may often find in small stratified lakes in summer in temperate regions, within the swimming distance of any but the most minute organism, a range of temperatures of from 5° to 25°C, of oxygen concentrations from 0.0 to 8.0 mg per liter, of pH from 6.8 to 7.6, with comparable variations in other chemical variables. Sufficiently restricted tolerances might make possible a number of niches in such a biotope. It is, however, unlikely that this would be a stable situation; natural selection would continually act to extend the tolerances, and so the niche sizes, of the various species and in this process the species would continually come into competition with each other for resources that had previously been allocated solely by a variety of physicochemical factors defining the niches, and so the habitats, of the species. In a case such as that of the rotifers of the genus *Polyarthra* studied by Bērziņš[43] in a small lake in Sweden (figure

scenopoetic

bionomic

43. B. Bērziņš, Ein planktologisches Querprofil. *Inst. Freshw. Res. Drottningholm Rep.*, 39:5–22, 1958. See also G. E. Hutchinson, *The Ecological Theater and the Evolutionary Play* (New Haven and London, Yale University Press, 1965, xiii, 139 pp.), in which the matter is discussed on pp. 43–45.

108), some direct determination by physico-chemical factors may have occurred. The very common *P. vulgaris* in the warm, well-oxygenated waters of the epilimnion must come into contact with the two other epilimnetic species, *P. euryptera* and *P. remata.* The large size of *P. euryptera* (160–210 μ long) may indicate that it feeds on larger flagellates than does *P. vulgaris* (100–145 μ long); separation between the latter and the rather more littoral *P. remata* (80–120 μ long) is unlikely on the basis of food size as there is much overlap in the sizes of the two rotifers. The other two species, *P. major* and *P. longiremis*, may be separated, at least partly, by temperature preferences; *P. longiremis* presumably is tolerant of lower oxygen concentrations than the other species, but this, while permitting it to enter deeper colder water, would be unlikely by itself to prevent it from living in more oxygenated water; obligate microaerophil animals are rare. When a situation is found where two species are apparently sharply separated by physico-chemical factors, it is always wise to enquire, though usually hard to decide, whether the separation is primary or whether the physico-chemical factors are acting, not directly, but by the regulation of competition. When regulation of competition occurs by the interaction of factors, plotted on the scenopoetic axes, with resources plotted on the bionomic axes, the regular rectangular form of the niche space is modified; in *n* dimensions it has ceased to be orthotopic.

Species packing

While there is no difficulty in packing as many species as may exist into a limited part of a scenopoetic axis, there is obviously a problem involved in packing along a resource axis. The problem is, moreover, a very important one because on it depends the total number of species that can be present in a region of niche space corresponding to a given habitat.

The matter has been examined in a characteristically ingenious and significant way by May and MacArthur.[44] It is supposed that *n* species share the resources along a resource axis, along which is ordered the quantity k_i

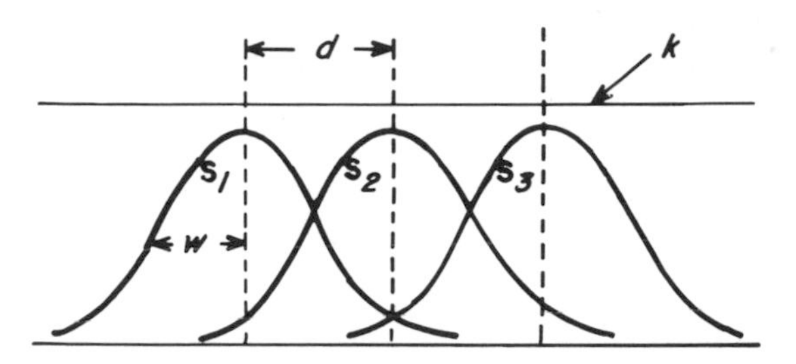

FIGURE 109. Resource continuum plotted as the amounts of food (k) available along a food-size axis, comparable to the x' axis of figure 99, and the utilization functions or measures of availability of food of various sizes, distributed normally about the optimal size for three different species. The distribution has a standard deviation w and the optimal sizes are separated by distances d (May and MacArthur).

of foods, of varying size or other quality, optimal for each of the S_i species. Each species can, however, use, with less efficiency or preference, resources on either side of k_i, according to a utilization function of such a kind that the variance about the preferred position is normal and equal to w. Let there be n species arranged along the axis, with distances between the peaks of their utilization functions equal to d (figure 109). May and MacArthur establish that as long as the k_i are constant, the packing of the species can be arbitrarily close. However, it is extremely unlikely that the k_i would not fluctuate. Suppose that the value of the food supply for the i-th species at any time t be given by

$$k_i = \overline{k}_i + \gamma_i(t)$$

where $\overline{k}_i$ is the mean value over time and $\gamma_i(t)$ is "white noise" normally distributed, with variance σ^2.

May and MacArthur now prove that provided n is reasonably large, the ratio d/w lies between about 0.9 and 1.6 for any value of $\sigma^2/\overline{k}$ between 10^{-4} and 10^{-1}. Even when $\sigma^2/\overline{k} = 0.5$ the ratio d/w is still not quite 2.

With a smaller number of species the limits are closer, but with $n = 4$ the possible values of d/w over the range of $\sigma^2/\overline{k}$ from 10^{-4} to 10^{-1} is about 0.4 to 1.1. Only when $\sigma^2/\overline{k}$ is approaching unity, so that there is a real

44. R. M. May and R. H. MacArthur, Niche overlap as a function of environmental variability. *Proc. Nat. Acad. Sci. U.S.A.*, 69:1109–13, 1972.

possibility of almost complete elimination of a species by its food supply being drastically reduced, does the packing distance respond strikingly to environmental variability (figure 110). This might happen in extremely rigorous but variable environments such as the arctic, which does in fact have a restricted faunal diversity, or when small variations in the seasonal meteorological cycle can have great environmental effects, as in the semiarid marginal areas around deserts. On the whole the variance of the utilization function defining the species capacity to eat a particular kind of food is evidently more important than is the variance in the quantity of that food, in determining how many species can be accommodated along a single resource axis.

Extremely specialized species will have a low value of w, striking generalists a high value. Since most adjacent competitors will be members of the same guild, having structural and behavioral similarities which may be due to phyletic affinity or to convergence, it is likely that some parts of a fauna will be characterized by low w and high packing and other parts by high w and low packing.

In a good many cases it appears that allied species in competition differ in linear dimensions by a factor of 1.2–1.4, implying one species is about twice as heavy as the other Bossert[45] has obtained some evidence from computer models of character displacement, making reasonable biological assumptions, that the process goes quickly until a difference of about this amount has developed, after which the rate of divergence rapidly declines. The result depends on an assumption as to the initial genetic variability of the organism which, for a mean value of some character of 100, is taken as given by $\sigma = 10$. Bossert believes that this is a reasonable figure. May and MacArthur believed that the little em-

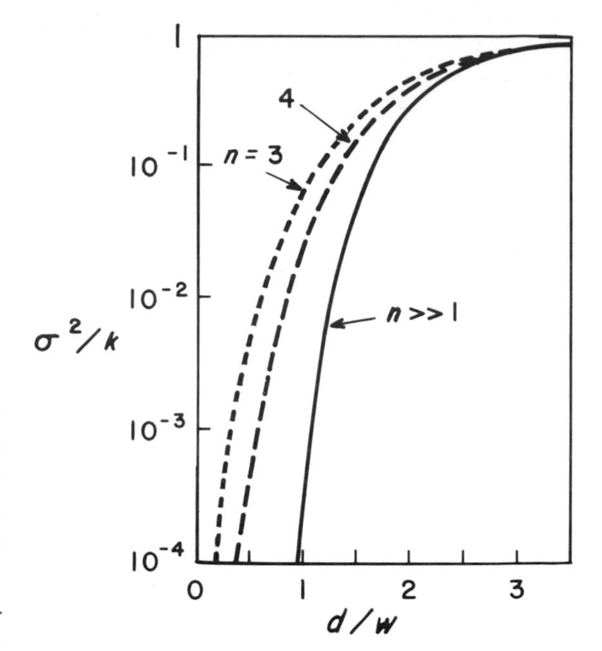

FIGURE 110. Relationships of d/w of the previous figure to the relative variance of k_i, the amount of food of size i. In a community of more than a few species, d/w is about unity for a large range of this variance (May and MacArthur).

pirical data that they considered tended to support their theoretical conclusion.

The magnificent study of the feeding ecology of sympatric seabirds breeding on Christmas Island in the Pacific Ocean by N. P. and M. J. Ashmole[46] provides data that look roughly in agreement with May and MacArthur's theory. It is, however, complicated by the fact that not merely size of food, but the distance of the feeding ground from the island also provides a method of specific separation.

It is apparent that an additional aspect of the matter needs to be considered. When dealing with size, the distribution of the utilization function is generally likely to be asymmetrical, as D. S. Wilson[47] has emphasized. An animal of any given size will be

45. W. H. Bossert, *Simulation of Character Displacement in Animals*. Ph.D. Thesis, Harvard University, 1963, iii, 771 pp.

46. N. P. Ashmole and M. J. Ashmole, Comparative feeding ecology of sea birds of a tropical oceanic island. *Bull. Peabody Mus. Nat. Hist. Yale Univ.*, 24; 1967, v, 131 pp.

N. P. Ashmole, Body size, prey size, and ecological separation in five sympatric tropical terns. *Syst. Zool.*, 17:292–304, 1967.

47. D. S. Wilson, The adequacy of body size as a niche difference. *Amer. Natural.*, 109:769–84, 1975.

The case of the co-occurrence of species of *Anisops* is discussed in G. E. Hutchinson, *The Ecological Theater and the Evolutionary Play* (see note 43), pp. 59–60.

unable to eat food that is too big, for purely mechanical reasons. Such food, however, might be taken by a somewhat larger species. Going in the opposite direction, smaller food rather below the optimal size will be quite easily handled, and the limitation on its use will be set primarily by the amount of work that must be done by a large animal in gathering up small pieces when large, more easily gathered pieces are available. This gives the large species an absolute advantage, in the simplest possible case, as a competitor. A smaller species cannot eat the optimal food of the larger species, while the latter can eat the optimal food of the smaller species, though less efficiently so.

Wilson thinks that the kind of size difference in trophic structures, which has so frequently been seen as a corollary to niche separation, is probably found only in the top predators of food chains, which, on land, will generally be mammals, birds, and lizards. This seems to be somewhat restrictive, as appears in the next paragraph. Wilson quotes several authors who have unsuccessfully looked for the phenomenon in insects. There are, however, cases such as the co-occurrence of the small backswimmers of the genus *Anisops* which would be very hard to explain in any other way, though admittedly the observations are quite inadequate.

The clearest case so far published is probably the guild of fruit-eating pigeons discussed by Diamond[48] in his extraordinary study of the ecology of the birds of New Guinea. Eight sympatric species living in lowland rain forest form a series, the mean weight of any species being about one and a half times that of the next smallest species, the five species of *Ptilinopus* ranging in mean weight from 49 g to 245 g, the three species of *Ducula* from 414 g to 802 g. Here not only do the large species have a capacity to eat larger fruits than the small, but because the small species can perch on much smaller branches than the large, the asymmetry noticed by Wilson is counterbalanced by these smaller species having a food supply not available to the larger members of the guild.

One species in more than one niche

Emlen[49] has considered the nature of the variance w^2, which he concludes can be divided into two parts. One part depends on

48. J. M. Diamond, see note 34.

49. Emlen, see note 40. It should not be supposed that no phenotypic adaptation exists in small animals, nor genetic polymorphism in large and educable species. Even in man it is quite conceivable that behavioral as well as biochemical differences, insofar as they have a genetic basis, are adaptive. This is not a racist point of view, but rather the reverse, indicating that in a population near equilibrium each individual has his or her unique worth.

If two sympatric species of mammals, differing in size and in food habits, are studied, it is usually found that the standard deviation of measurements of structure is much less than the difference between the means. This is, for instance, true of the skull lengths of the sympatric shrews, *Sorex araneus* (19.65 ± 0.45 mm) and *S. minutus* (15.65 ± 0.65 mm) or of the field mice, *Apodemus sylvaticus* (22.96 ± 0.76 mm) and *A. flavicollis* (26.09 ± 0.84 mm), given in G. S. Miller, *Catalogue of the Mammals of Western Europe* (Europe excluding Russia). London, British Museum (N.H.), 1912, xv, 1,019 pp. In such cases we may suppose that Emlen's phenotypic variation permitting a fair amount of difference in the way individual pieces of food, of different sizes, are handled, is of much greater importance than any genetically determined structural polymorphism.

Parenthetically, it may be pointed out that the dichotomy between genetic and phenotypic adaptation, even if synergistic, does not exhaust the possibilities presented by animals having large brains. In such animals it has often been recognized that the problem of genetic coding for the development of the nervous system could be formidable. We may suppose that coding on a practical scale produces a nervous system, constructed within certain tolerance limits, which permits its possessor to function adequately in most of the circumstances likely to be encountered in its environment. Within the tolerance limits there will be a certain amount of structural variation, which may produce various minor differences in behavior among the individuals. This expression of constitutional variation is neither phenotypic nor genetic in the conventional meanings of those words, though the possible range of the variation will be determined genetically. It is in fact conceivable that in social species the range will not be as narrow as it could be, so allowing for an adaptively significant set of random behavioral differences. If this happened at all, it would be most likely to happen in man and would explain a great deal of the sporadic incidence of behavioral variation, otherwise so hard to understand.

the ability of the individual organism to adapt to conditions on either side of the optimum, the other is due to the population consisting of a variety of genotypes with slightly different optima. Emlen's analysis leads to the conclusion that selection pressure on a population in which the variance is mainly phenetic leads to it becoming more phenetic, while where it is more genetic it tends to become still more genetic. One should therefore get two types of evolutionary processes involved in widening a niche, one leading to more polymorphic, less individually adaptable organisms, the other to less polymorphic, more individually adaptable organisms. Large motile animals generally have larger nervous systems and show greater adaptability than do small. Small animals can divide the potential biomass of the species up into more individuals and so exhibit more genetic variability. In general, we might expect a high phenetic capacity to maintain a wide niche in the higher vertebrates, as indeed man obviously does to an unparalleled degree, while the insects would be expected to employ polymorphism more extensively. It is, however, likely that in very few cases can genetic polymorphism do as much for a species as can intelligence.

The most striking and widespread type of polymorphism is sexual. In the majority of species most females leave offspring, but only a limited number of males do so. As Fisher[50] showed over 40 years ago, the existence of two parents implies that in a population, whichever sex is in the minority, each individual of that sex will leave an average number of offspring greater than that of the other sex. This means that divergences from equality in a population will be automatically corrected insofar as they have a genetic basis. At the same time most of the males could be removed without reducing the fecundity of the females. There is therefore a genetic tendency toward equality, which is ecologically unfortunate as many individuals are doing nothing for the fitness of the species.

This can be corrected in various ways. Parthenogenesis or hermaphroditism can replace bisexuality. This may slow evolutionary processes by removing the mechanism for genetic recombination.[51] Males may be reduced in size, so that they use little of the potential resources, as in many spiders; this may raise problems in the synchronization of maturity in the two sexes. The minute parasitic males of the echiuroid worm, *Bonellia*, the complemental males of some barnacles, and the parasitic males of the oceanic anglerfish, in which synchrony may be achieved hormonally, all combine the advantages of hermaphroditism and outbreeding. Males may be put to work, as in many vertebrates and a few insects such as the termites, or perhaps the belostomid water bugs, the females of which lay their eggs on the backs of the other sex. Finally, the male may live in a niche somewhat different from that of the female. This is best known in birds,[52] particularly in the birds of prey, where the

50. R. A. Fisher, *The Genetical Theory of Natural Selection*. Oxford, Clarendon Press, 1930, xiv, 272 pp. The slight excess of males at birth in man, compensated by a greater male mortality, giving an approximately equal distribution of the sexes in the reproductive period, was discovered by John Graunt (see chap. 1, note 6). It was early considered, as by Sir Matthew Hale, as a providential adaptation to compensate for the additional mortality of men in warfare and navigation. The ratio at birth is subject to a great deal of variation among different populations, which in part might be due to compensation, by Fisher's mechanism, for a culturally determined tendency to poorer female survival in childhood in some population that in others (U. M. Cowgill and G. E. Hutchinson, Sex-ratio in childhood and the depopulation of the Peten, Guatemala. *Hum. Biol.*, 35:90–102, 1963. U. M. Cowgill and G. E. Hutchinson, Differential mortality among the sexes in childhood and its possible significance in human evolution. *Proc. Nat. Acad. Sci. U.S.A.*, 49:425–29, 1963. R. Cook and A. Hanslip, Nutrition and mortality of females under 5 years of age compared with males in the "Greater Syria" region. *J. Trop. Pediat. Afr. Child Hlth*, 10:76–81, 1964).

51. G. C. Williams (*Sex and Evolution*. Princeton University Press, 1975, x, 200 pp.), in a most interesting discussion, concludes that the biological meaning of sex is still rather mysterious.

Where hermaphrodites are self-sterile, hermaphroditism, of course, does not imply a reduction of genetic recombination. The conditions favoring hermaphroditism have recently been analyzed by E. R. Charnov, J. Maynard Smith, and J. J. Bull, Why be an hermaphrodite? *Nature*, 263:125–26, 1976.

52. A. L. Rand, Secondary sexual characters and

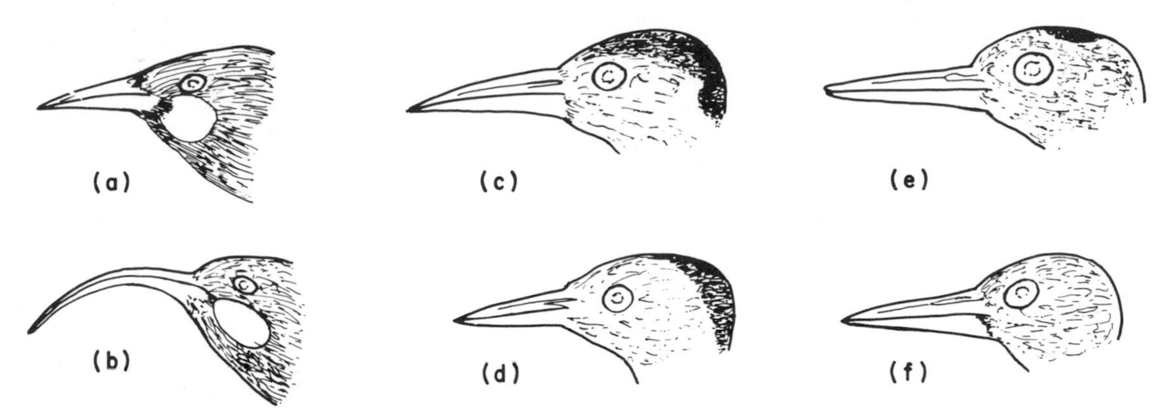

FIGURE 111. Birds in which the food habits of the male and female are different and strikingly reflected in anatomical sexual dimorphism; (a) the huia, *Heteralocha acutirostris*, an extinct starling formerly found in New Zealand, ♂; (b) the same ♀; (c) the Hispaniolan woodpecker, *Melanerpes striatus* ♂; (d) the same ♀; (e) the allied Continental *M. aurifrons*, for comparison, ♂; (f) the same ♀. (In part after Selander.)

male or tiercel, reduced by one-third, follows the ordinary pattern of size difference among sympatric species of predators; comparable cases are known in a few fishes and lizards. The most extreme avian case[53] was doubtless the extinct New Zealand huia, *Heteralocha acutirostris*, in which the male had a short stout bill while the female had a long curved one. The bird fed largely on wood-boring insect larvae; the male's beak was adapted to hammering open partly decaying wood, the females to probing into the galleries made by the larvae. It is often believed that the two sexes used their two different tools in a co-operative way, but the evidence is not quite adequate (figure 111, a, b).

A less extreme, but very interesting case described by Selander[54] is provided by the

ecological competition. *Fieldiana, zool.*, 34 (no. 6): 65–70, 1952.

R. W. Storer, Variation in the resident sharp-shinned hawks of Mexico. *Conder*, 54:283–89, 1952.

A comparable situation is found in owls (C. M. Earhart and N. K. Johnson, Size dimorphism and food habits of North American owls. *Condor*, 72:251–64, 1970). The larger species show the phenomenon more strikingly than do the smaller insectivorous species: in some owls, the female, though larger-bodied, does not have a longer wing than the male.

The phenomenon is also known in some fishes (A. Keast, Trophic interrelation shifts in the fish fauna of a small stream. *Univ. Mich. Great Lakes Res. Div. Publ.*, 15:51–79, 1966. A. Feduccia and B. H. Slaughter, Sexual dimorphism in skates (Rajidae) and its possible role in differential niche utilization. *Evolution*, 28:164–68, 1974), and in lizards (T. W. Schoener, The ecological significance of sexual dimorphism in size in the lizard *Anolis conspersus*. *Science*, 155:474–77, 1967; The anolis lizards of Bimini: Resource partitioning in a complex fauna. *Ecology*, 49:704–29, 1968).

53. The most complete account appears to be W. L. Buller, *Manual of the Birds of New Zealand*. Wellington, New Zealand, Colonial Museum and Geological Survey Department, George Didsbury, Govt. Printer, 1882, ix, 107 pp.

Buller writes of a captive pair, on p. 32, "What interested me most of all was the manner in which the birds assisted each other in their search for food. . . . The male always attacked the more decayed portion of the wood, chiselling out his prey after the manner of some woodpeckers, while the female probed with her long pliant bill the other cells, where hardness of the surrounding parts resisted the chisel of her mate."

W. Colenso, A description of the curiously deformed head of a huia (*Heteralocha acutirostris*, Gould), an endemic New Zealand bird (*Trans. Proc. N.Z. Inst.*, 19:140–45, 1887 [for 1886]) gives a description of the head of a female specimen with a spirally deformed maxilla; after quoting Buller, he suggests that the specimen survived by being helped by its mate.

See also D. Lack, *Darwin's Finches*. Cambridge University Press, 1947, x, 208 pp.; see pp. 154–155.

W. J. Phillipps, *The Book of the Huia*. Christchurch, etc., N.Z., Whitcombe and Tombs, 1963, 159 pp.

54. R. K. Selander, Sexual dimorphism and niche utilization. *Condor*, 68:113–51, 1966.

See also F. A. Pitelka, Geographical variation and the species problem in the shore birds of the genus *Limnodromus*. *Univ. Calf. Publ. Zool.*, 50:1–108, 1950.

R. A. Norris, Comparative biosystematics and life-history of the nuthatches *Sitta pygmaea* and *Sitta pusilla*. *Univ. Calif. Publ. Zool.*, 56:119–300, 1958.

woodpecker, *Melanerpes striatus*, on the island of Hispaniola. Unlike the mainland *M. aurifrons*, from Texas to Nicaragua, *M. striatus* is social and its populations are often equivalent in size to those of all the different species of woodpecker occurring in a comparable mainland area. The male bill averages 121.3 percent that of the female and the tongue and its muscles are rather better developed (figure 111, c, d). The male spends 34.8 percent of its feeding effort in probing, 37.5 percent in pecking and excavating, and only 25.0 percent gleaning, while for the female the figures are 8.6 percent, 24.8 percent, and 58.1 percent. Here the longer male beak is clearly used in a rather different way from the shorter female beak. No such differences are observed in *M. aurifrons*. Selander concludes that on the island, with no immediate competitors, the niche is extended, not merely by the birds using their feeding mechanisms on a wider selection of foods, as often happens in the absence of competitors on islands, but by the development of sexual dimorphism that promotes such wider utilization.

A number of comparable, but less striking cases have been described. Perhaps the most interesting is that of the two species of dowitcher, *Limnodromus griseus*, a smaller bird living largely on tidal flats, with the length of the male bill 108.4 percent that of the female, and the larger *L. scolopaceus*, living along freshwater streams, with the male bill 117 percent of that of the female. The latter species is thus able to compensate, by its longer legs and sexual dimorphism, for the more restricted freshwater habitat, in which tidal variation does not continually change the depth of water in which the bird can feed.

A good many cases of purely behavioral difference are known in birds, chiefly related to the height at which they feed, and providing an intraspecific analogy to the warblers studied by MacArthur. The male indigo bunting,[55] *Passerina cyanea*, normally forages higher in bushes and trees than does the female. In the nuthatch, *Sitta pusilla*, the male forages below the female, but in *S. pygmaea* there is no sexual difference in feeding behavior. A number of other instances of the same kind are cited by Selander.

Though a fair minority of mammals have slightly larger females than males,[56] the most striking dimorphism in size is in the opposite direction from that found in the birds of prey, though it does occur in a group of carnivores. In the weasels of the genus *Mustela* the male tends to be much larger than the female. In the common North American *M. frenata*, he is about twice the weight of his mate. In the European weasel, *M. nivalis*, the two sexes seem in the past to have been taken to be two species by country people familiar with the animal. Brown and Lasiewski conclude from metabolic studies that the long, thin form of these animals, which permits them to hunt in the burrows of a quite small prey, is very disadvantageous in cold weather owing to the heat loss that it entails. They suggest that at such times niche diversification might increase the potential food supply, so offsetting the disadvantage of the form. There is some evidence that both sexes may take part in bringing food to the weaned but not fully independent young.

All cases of mimicry in which the mimetic form is confined to the female sex imply a slightly different niche, or subniche as Selander would say; and in the cases in which

55. The male is largely occupied with singing high in the tree and the female with the nest, which is set relatively low, so that the different disposition of feeding grounds follows as naturally from other activities as does that of a suburban housewife at lunchtime when her husband is in the city. See Thompson, in Selander, note 54.

56. K. Ralls, Mammals in which females are larger than males *Quart. Rev. Biol.*, 51:245–76, 1976. A significant paper with a rich bibliography.

J. H. Brown and R. C. Lasiewski, Metabolism of weasels: The cost of being long and thin. *Ecology*, 53:939–43, 1972.

The belief that the common weasel of Europe, *Mustela nivalis*, might consist of two species is discussed by G. B. Corbet, *The Terrestrial Mammals of Western Europe*. London, G. T. Foulis, 1966, xi, 264, see p. 141.

The feeding of young *Mustela frenata* by both male and female is noted by W. J. Hamilton, Weasels of New York. *Amer. Midl. Natural.*, 19:289–344, 1933; see p. 328.

polymorphic mimicry occurs, the different female forms are similarly in slightly different niches. In at least one case, the beetle, *Phytodecta variabilis*, it seems probable that the male, emerging earlier, on a broom plant of which the foliage is not fully developed, is ordinarily a mimic of coccinellid beetles, while the female emerging later is usually cryptically colored. The genetic situation is rather complicated, but apparently permits easy and rapid adaptation of the proportions of the color patterns in a population to the phenology of any particular locality.[57]

It has been found by Freeman, Klikoff, and Harper[58] that in a number of dioecious plants growing in dry regions, the male plants tend to occur in exposed and more arid sites from which pollen is easily blown, while the females grow preferentially in moister places, which no doubt promote seed production.

Recently cases have been described in fishes in which nonsexual polymorphism involving dentition has been supposed to exist. If these are established, it means that a single species occupies niches that are as discretely different as is usually the case among validly different species.

Roberts[59] concluded that in two species of the genus *Saccodon*, a characoid fish living in streams in Panama and northern South America, there are five dental morphs, three of which occur both in *S. wagneri* in Ecuador

57. This interpretation is given in G. E. Hutchinson, Variations on a theme by Robert MacArthur. In: *Ecology and Evolution of Communities*, ed. M. L. Cody and J. M. Diamond. Cambridge, Mass., Belknap Press of Harvard University Press, 1975, pp. 492–521.

The observational data are derived from W. Bateson, On the colour-variation of a beetle of the family Chrysomelidae, statistically examined. *Proc. Zool. Soc. Lond.*, 1898:851–60.

L. Doncaster, On the colour-variation of the beetle *Gonioctena variabilis*. *Proc. Zool. Soc. Lond.*, 1905 (2): 528–36.

The genetics are due to A. de Zuelueta, La herencia ligada al sexo en el coleóptero *Phytodecta variabilis* (Ol.). *Revist. Espan. Entom.*, 1:203–31, 1925.

In addition to the mechanism of the same locus being present in both X and Y chromosomes, but with no crossing-over in the male, as seems to be the case in *P. variabilis*, rapid changes in the nature of the most abundant phenotype, in response to selection when there is an environmental change, are also possible when the gene involved is dominant in one sex and recessive in the other. This may happen in the spittlebug, *Philaenus spumarius*, as is indicated in my discussion of what is now known of that very variable plant bug, in the same paper in the volume dedicated to the memory of Robert MacArthur.

In the two species, *Mustela erminea* in Britain and *M. frenata* in eastern North America, there are populations which are probably polymorphic for genes controlling winter whitening. These genes appear to promote whitening much more in the female than in the male. Again the different expression of the polymorphism in the two sexes would permit rapid natural selection in response to slight changes in the intensity of the winter. The matter is discussed by G. E. Hutchinson and P. Parker in a forthcoming paper (Sexual dimorphism in the winter whitening of the stoat, *Mustela erminea*. In press in *J. Zool.*). It is odd that the phenomenon should be so rare.

A few cases are known among insects in which a morph, occurring only in the female, has physiological or behavioral properties, as well as recognizable morphological characters, which cause the morph to live in a slightly different niche, no mimicry apparently being involved.

R. F. Bretherton and G. G. de Worms (Butterflies in Corsica. *Entom. Rec.* 75:93–104, 1963; see p. 97) and A. L. Panchen and M. N. Panchen (Notes on the butterflies of Corsica, *Entom. Rec.*, 85:149–53, 198–202, 1973) both note that in Corsica the ♀ f. *valesina* of the fritillary, *Argynnis paphia immaculata*, in which the orange-brown background of the wings is replaced by a pearly greyish green, behaves differently from the nominotypical female and the male, by frequenting mossy stones, while the ordinary females and males are feeding at flowers. F. *valesina* is said to pay more attention by day to the beech trees in which the species rests by night and during cloudy weather. It is possible that some thermal preferendum is involved in this case.

However, in the *alba* forms of the females of many sulfur butterflies of the genus *Colias*, which occur at greater frequencies in relatively cool climates, there is evidence of no direct thermal advantage for the white form, but rather a suggestion of a more efficient nitrogen metabolism of the pupa directing the precursors of the yellow pigment into other channels of greater value to the insect. There must be some balancing counterselection. See W. B. Watt, Adaptive significance of pigment polymorphisms in *Colias* butterflies: III. Progress in the study of the "alba" variant. *Evolution*, 27:537–48, 1973.

58. D. P. Freeman, L. G. Klikoff, and K. T. Harper, Differential resource utilization by the sexes of dioecious plants. *Science*, 193:597–99, 1967.

59. T. R. Roberts, Dental polymorphism and systematics in *Saccodon*, a neotropical genus of freshwater fishes (Characoidci). *J. Zool.*, 173:303–21, 1974.

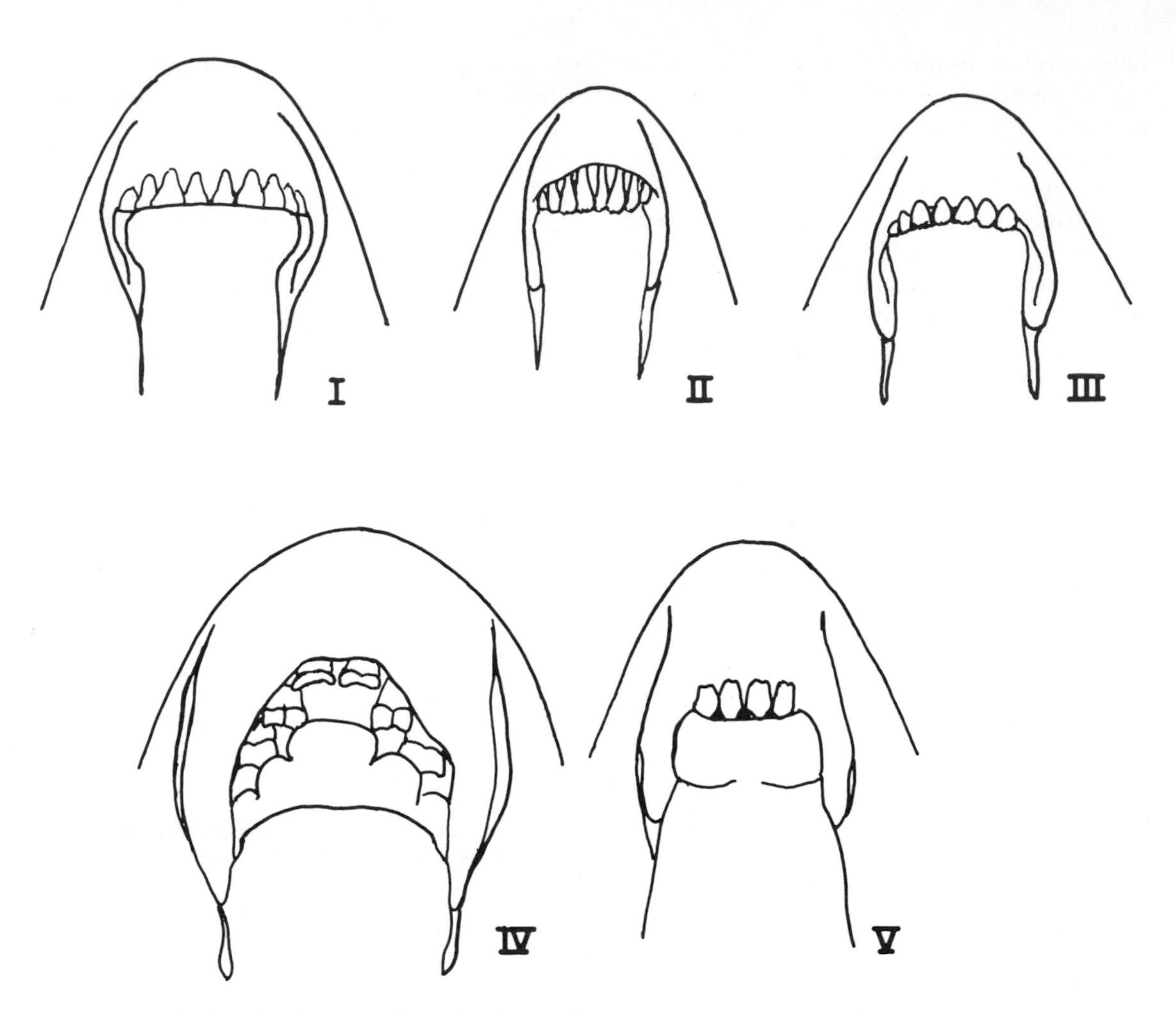

FIGURE 112. The five dental morphs found in *Saccodon dariensis*. Some intermediates between I and IV are known (Roberts).

and *S. dariensis* in Panama and Colombia, while the latter species has two more. The tooth form is very different (figure 112) in the different morphs. The evidence of conspecificity is primarily sympatry with no character other than the teeth to separate the morphs, while the two species differ clearly in many ways. Roberts supposes that the possibility of exploiting several niches is adaptive in uncertain environments. A genetic analysis would be of great importance in this example.

A very extraordinary case in the cichlid genus, *Cichlasoma*, has been discovered in the Cuatro Cienegas basin of Coahuila, Mexico, by Taylor and Minckley,[60] who believed that a flock of three species was present, one feeding on snails and with molariform teeth (figure 113, a) and a short gut, one on algae and detritus with papilliform teeth (figure 113, b) and a long gut, and one fish-eating fusiform species. Later a second fish-eating species was recognized, either tooth form occurring with a fusiform body. The snail-eating and algal-feeding species were found to lack any different proteins as demonstrated by electrophoretic mobility and were regarded by Kornfield and Koehn as excep-

60. D. W. Taylor and W. L. Minckley, New world for biologists. *Pacific Discovery*, 19:18–22, 1966.

I. L. Kornfield and R. K. Koehn, Genetic variation and speciation in New World cichlids. *Evolution*, 29:427–37, 1975.

R. D. Sage and R. H. Selander, Trophic radiation through polymorphism in cichlid fishes. *Proc. Nat. Acad. Sci. U.S.A.*, 72:4669–73, 1975.

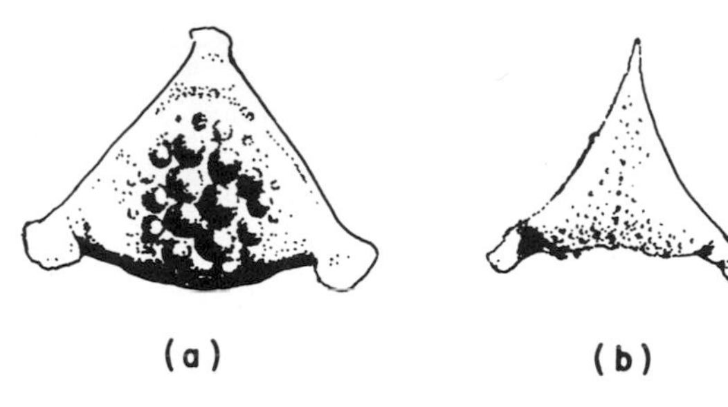

FIGURE 113. The teeth of a species of *Cichlasoma* apparently composed of three morphs, one of which, with (a) molariform teeth, feeds on molluscs, and another with (b) papilliform teeth, on allgae.

tional in showing speciation without any incidental biochemical differention. Sage and Selander, however, finding three populations which exhibited between themselves concordant variation in all biochemical loci studied, concluded that each population, consisting of more than one morph, is unispecific. The matter clearly needs much further study.

Fine-grained and coarse-grained species

An insight into the possible ways by which species may replace one another, or continue to exist in stable combinations, may be obtained from the work of MacArthur and Levins.[61]

It is assumed that two species S_1 and S_2 co-occur and feed on two renewable resources R_1 and R_2. If both species eat both foods, though not necessarily in the same proportions,

$$\frac{dN_1}{dt} = N_1[l_1(R_1 - m_1) + l_1'(R_2 - m_1')]$$

$$\frac{dN_2}{dt} = N_2[l_2(R_1 - m_2) + l_2'(R_2 - m_2')]. \qquad (5.7a, b)$$

Here the terms m_1, m_1', m_2, and m_2' represent the quantities of resources just able to support life, but not to permit reproduction. No increase in the populations is possible unless the expressions in the brackets are greater than 1. The terms l_1, l_1', l_2, and l_2' represent efficiency of conversion of resources into new individuals when this is possible. When two species eat the same food, though not necessarily in equal quantities, they are called *fine-grained*.

Taking two axes for the two resources R_1 and R_2, we can plot (figure 114) the isocline

$$\frac{dN_1}{dt} = N_1[l_1(R_1 - m_1) + l_1'(R_2 - m_1')] = 0$$

or

$$l_1(R_1 - m_1) + l_1'(R_2 - m_1') = 0. \qquad (5.8)$$

This line gives the lower limit of the mixed resources for the increase of species S_1. Any mixture of resources above and to the right of the line permits S_1 to increase. The line cuts the two axes at

$$R_1 = m_1 + \frac{l_1'}{l_1}m_1', \qquad R_2 = 0$$

$$R_1 = 0, \qquad R_2 = m_1' + \frac{l_1}{l_1'}m_1.$$

We now consider a second species S_2, and draw its isocline

$$l_2(R_1 - m_2) + l_2'(R_2 - m_2') = 0. \qquad (5.9)$$

If this line falls wholly above (in the all-positive quadrant) that of the first species, there is a range of resources represented by the set of points between the two lines, permitting S_1 but not S_2 to increase. S_1 therefore cannot be displaced by the newcomer, which fails to become established. If the isocline of S_2 lies entirely below that of S_1, the reverse holds and S_2 displaces S_1. If the isoclines cross, as at P, coexistence is possible, there being always some mixture of R_1 and R_2 superior for S_1 and another mixture superior for S_2.

If we add a third species S_3, so that its isocline intersects that of S_1 at Q and that of S_2 at T, the two intersection points imply a greater necessary consumption if increase is to occur, than is implied by P, so that S_1 and S_2 remain coexisting and S_3 cannot establish itself. If, however, another species S_4 is introduced with an isocline lying between P and the origin, S_4 will persist and an equilibrium mixture with either S_1 or S_2 will result.

61. R. H. MacArthur and R. Levins, Competition, habitat selection, and character displacement in a patchy environment. *Proc. Nat. Acad. Sci. U.S.A.*, 51:1207–10, 1964.

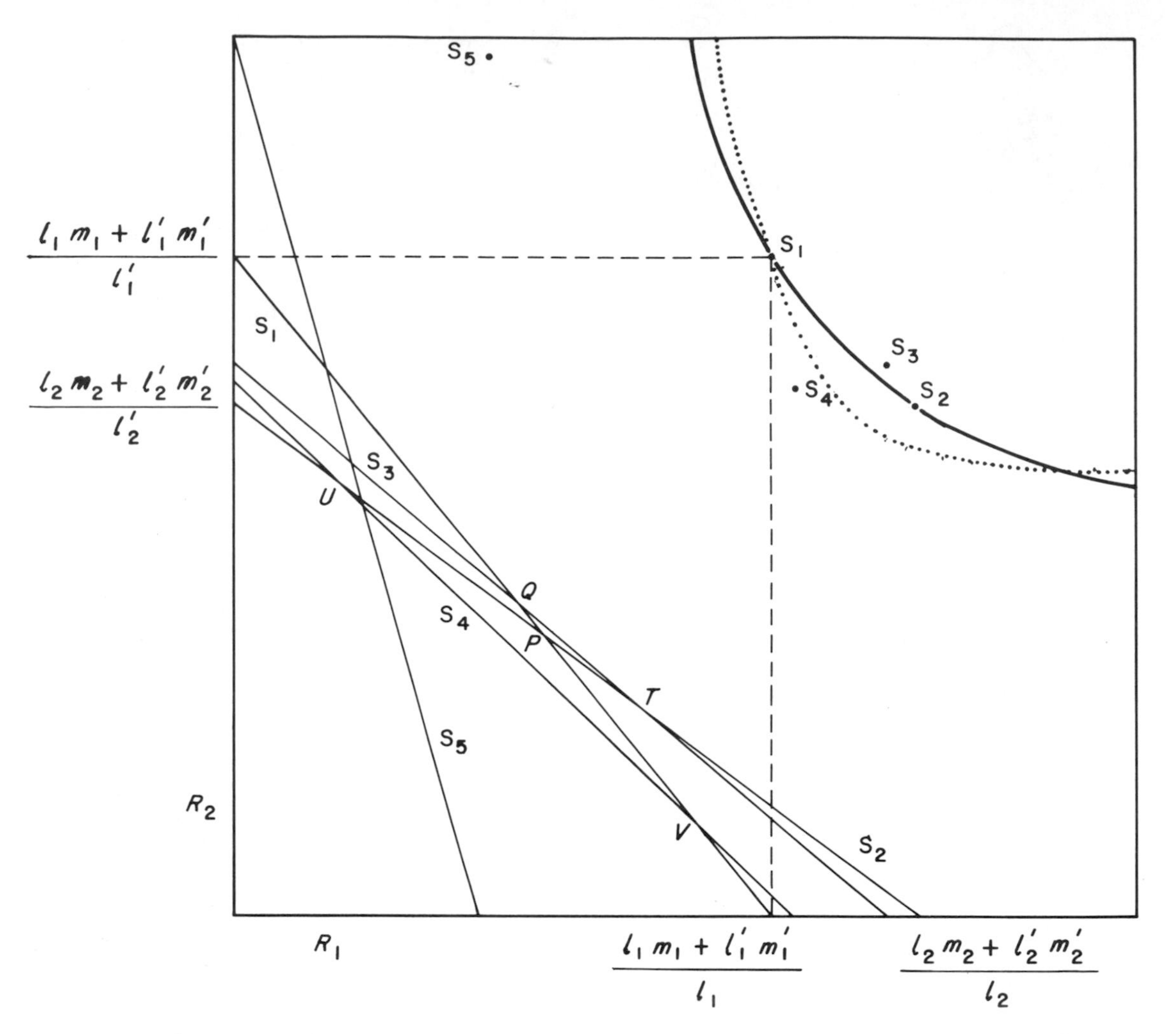

FIGURE 114. The geometrical treatment of MacArthur and Levins of coarse-and fine-grained species. For explanation consult the text with care.

For any species for which an isocline can be drawn intersecting the R_1 and R_2 axes, the values of R_1 and R_2 for which the isocline cuts the two axes define a point, e.g., S_1 in figure 114. If a number of isoclines pass through a single point P, which is very unlikely in practice, they define a set of representative points for a set of species of equal competitive ability; such points S_1 and S_2 lie on an equilateral hyperbola. Any species with a representative point S_3 lying inside the hyperbola is a poorer competitor than the species whose isoclines intersect at P, and will not be able to displace any of the set that happen to occupy the habitat. Any point lying outside the hyperbola can represent an invading species S_4 that has better competitive capacity than the set represented by the hyperbola. Since in practice it is most unlikely that three lines would intersect exactly in the same point, one may assume that ordinarily the hyperbola represents but two species, one of which can be replaced by the invader. In the part of the diagram that lies near the diagonal of the all-positive quadrant, the species represented are feeding on commensurate amounts of the two resources. Relatively small changes in the habitat will presumably cause changes in the l's and m's and so in the intersection points of the isoclines. If the competitor has a representative point near the greatest curvature of the hyperbola, it is likely under some circumstances, in some habitats, to be inside the latter and so fail to become established in such habitats, but to be outside and to become established in others. The ultimate result will be species feeding on approximately the same mixture of foods, with different potential competitors establishing themselves in slightly different environments.

If, however, the invading species has a representative point outside the hyperbola of its competitors but far from the greatest curvature, so that the isocline is very steep or very flat, and the values of R_1 and R_2 where it crosses the axes are very different, selection might favor avoidance of competition by giving up the food that is used in very small proportions and specializing more effectively in the utilization of the other food. This would ultimately lead to a *coarse-grained species*.

MacArthur and Levins believe that in some cases the representative points of a species pool might be concentrated as a bulge down the diagonal (dotted curve of figure 114), and in other cases be distributed more at right angles to the latter. In the first case, the evolution would be toward fine-grained, morphologically similar species, specialized in seeking optimal habitats but feeding in much the same way as their competitors. In the second case, the evolution would be toward coarse-grained, morphologically specialized species, each feeding in an efficient way on a limited range of food and seeking this food over a wide range of habitats. This postulate about the kinds of species pool may to some extent beg the question as it is neither intuitively obvious nor seemingly empirically demonstrated. Possibly different groups may have different patterns of representative points on a resource plane, but if this were so it would probably already involve their categorization as fine-grained or coarse-grained. What is apparent, however, is that such types of difference do exist. As MacArthur and Levins point out, animals which spend much time searching for food, but can eat it as soon as they find it, tend to belong to the fine-grained category. The various warblers studied by MacArthur provide an excellent example. Most small rodents and insectivores doubtless also belong here.

Animals that spend much time pursuing prey that they have easily located belong in the coarse-grained categories; they are often specialized in size categories appropriate to particular prey. This is true of most falcons and various carnivora among the mammals.

The idea can probably be extended to the enormous number of plant-specific phytophagous insects that fundamentally depend on green leaves or on sap, but use the chemical and biological properties of different plants to recognize environments in which they can live but competitors cannot. Carnivorous insects, like carnivorous mammals, are more likely to trespass over a wide variety of habitats. If, as in the Anthocoridae (Hemiptera), they are both carnivorous and plant-specific, they may have originated from herbivores that took to sucking aphids, coccids, and the like, which, due to the material that they had ingested, may have tasted like their host plants.

The classification is not absolute and the incidental characteristics of the categories may vary. Thus in the genus *Conus*[62], where every species has a restricted range of food, the species show practically no morphological adaptation to their resources, but are quite selective with regard to the habitats in which their specific prey may be found.

Foliage height diversity and other aspects of niche dimensionality

MacArthur and his associates[63] have studied the correlation of the occurrence of various birds in eastern North America with the structure of the plant cover of the areas on which they hold territory. The plant cover was measured by estimating the leaf area projected against a disc sighted over a standard distance. The vegetation was divided into three layers; the first, 0–2 feet high, is essentially herbaceous; the second, 2–15 feet high, bushy; the third, over 15 feet, may be regarded as arboreal. The proportion of the total foliage that is in each layer over a particular plot can be plotted in an equilateral

62. A. J. Kohn, The ecology of Conus in Hawaii. *Ecol. Monogr.*, 29:47–90, 1959.

63. See note 20.

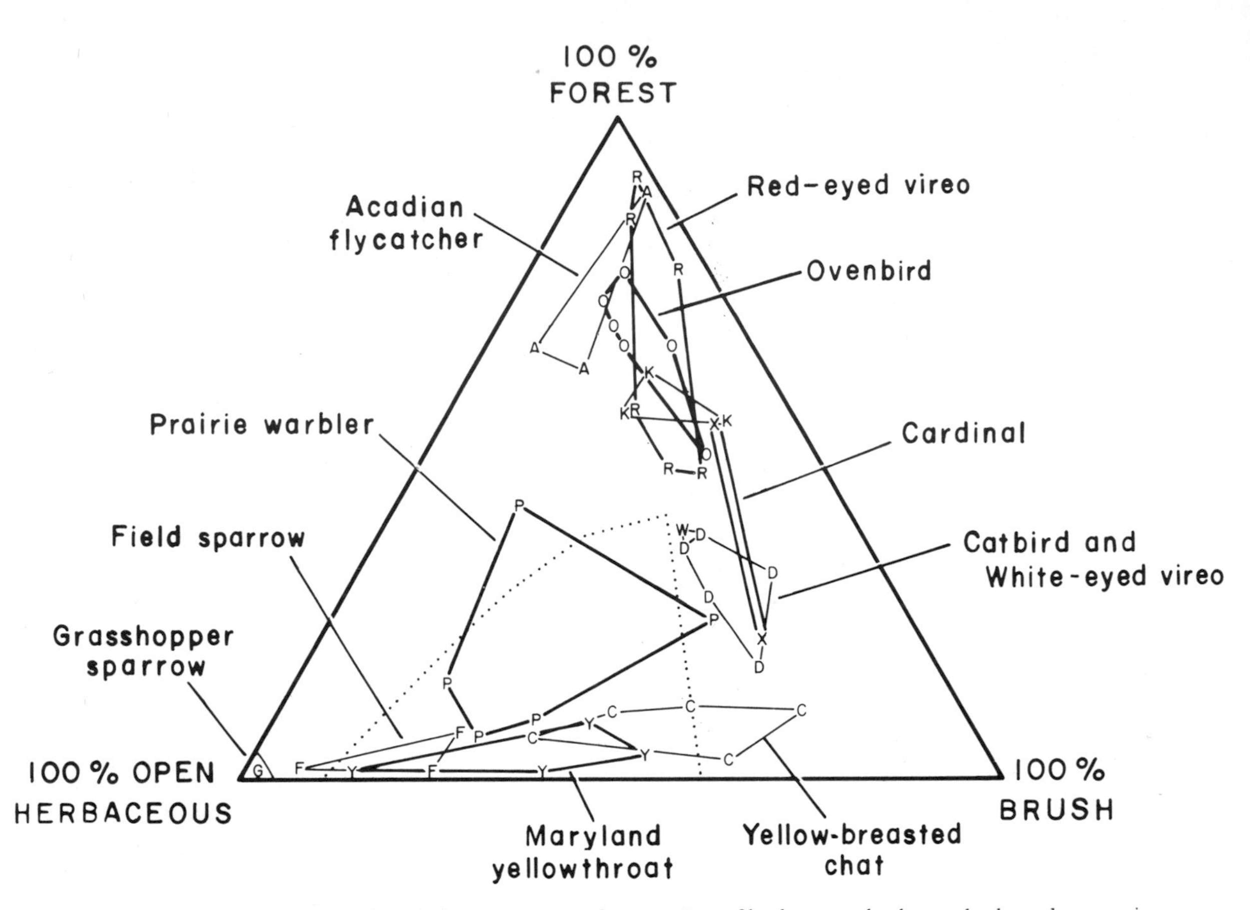

FIGURE 115. Triangular plot of bird habitats in terms of proportion of herbaceous, bushy, and arboreal vegetation. The dotted line encloses an area, not used in preparing the diagram, for which a census was also made, as given in the text (MacArthur, MacArthur, and Preer, modified).

triangle; any point may be defined by three perpendiculars on the three sides, the lengths being proportional to the amounts of herbaceous, bushy, or arboreal foliage. A large number of territories of twelve representative passerine birds in eastern Pennsylvania was studied, an area of 100 square feet in the most used part of each territory being measured for foliage diversity. The results of this survey are plotted in figure 115. It will be apparent from this diagram that most of the species can be separated by their preferences for different types of vegetation profile. The two pairs of species most nearly identical in their requirements are the white-eyed vireo (*Vireo griseus*) and the catbird (*Dumatella carolinensis*) in patches that were largely bushy, and the red-eyed vireo (*Vireosylva olivacea*) and the ovenbird (*Seiurus aurocapillus*) in the most wooded patches studied. Both the vireos feed primarily on insects and their larvae on leaves. The ovenbird, though insectivorous, feeds primarily on the ground; while the catbird, which feeds on insects and berries, doubtless takes most of its insect food in different places and in different ways from that of the insectivorous vireos.

If the three dimensions of the vegetation profile are now determined in another relatively uniform area not involved in preparing the diagram, a prediction can be made as to which of the twelve species will be present. Such an area is enclosed by the dotted line in figure 115. Of the species entered in the figure, one would predict that the field sparrow, yellowthroat, prairie warbler, and yellow-breasted chat would be common and that possibly catbirds, white-eyed vireos, and cardinals might also occur as their areas are close to the right-hand limit of the area enclosed by the dotted line. The actual census within that area was:

prairie warbler	3 pairs
yellowthroat	3 pairs
yellow-breated chat	2 pairs
field sparrow	1 pair
cardinal	1 pair
catbird	1 pair
towhee	1 pair
blue-winged warbler	1 pair
robin	1 pair
yellow-billed cuckoo	1 pair

The last four species, not being plotted in figure 115, could not have been predicted.

The three measures of foliage area define the vegetation profile adequately, though perhaps arbitrarily; it is reasonably obvious that two measures would be inadequate. These three may then be regarded in practice as niche dimensions, though their biological meaning is far from clear. The addition of a fourth dimension related to trophic behavior, possibly the mean height at which insects are caught by any species feeding on them, would probably give complete separation of all twelve niches. It is important to remember that these twelve are not the whole of the terrestrial avifauna of eastern Pennsylvania. The general success at three-dimensional separation would suggest that for the whole fauna four or five dimensions would be enough, while in a local biotope, with a uniform vegetation profile, perhaps one or two.

MacArthur and MacArthur[64] in an earlier paper made another use of the same sort of measurement that permits the prediction of the occurrences of birds from vegetation profiles. In any habitat if the total number of birds be N_s and there be N_i individuals of the i-th species, the information-theoretic diversity (see p. 227) is given approximately by $-\Sigma p_i \log p_i$ where $p_i = N_i/N_s$. It is also possible to consider for each observation the amount of foliage at the three observation levels (F_i) and to calculate from this the foliage diversity where $p_i = F_i/F_s$, F_s being the total of the leaf cover at all three levels. We now can plot the species diversity against the foliage diversity in any habitat. Provided the vegetation profile is comparble to that studied by MacArthur in Pennsylvania, the bird species diversity appears to be linearly related to the foliage height diversity, following the same line (figure 116) whether the observations are made in eastern North America or in Australia. However, when coniferous forest is considered, the outside and the inside of the cone formed by the branches of the tree may form separate habitats, as in the case of the Blackburnian and myrtle warblers of figure 106. In such cases the bird diversity may be higher than would be expected. Lovejoy,[65] moreover, believes that as a number of the commoner birds in the Amazonian rain forest have rather special food requirements, but ones that are met by several types

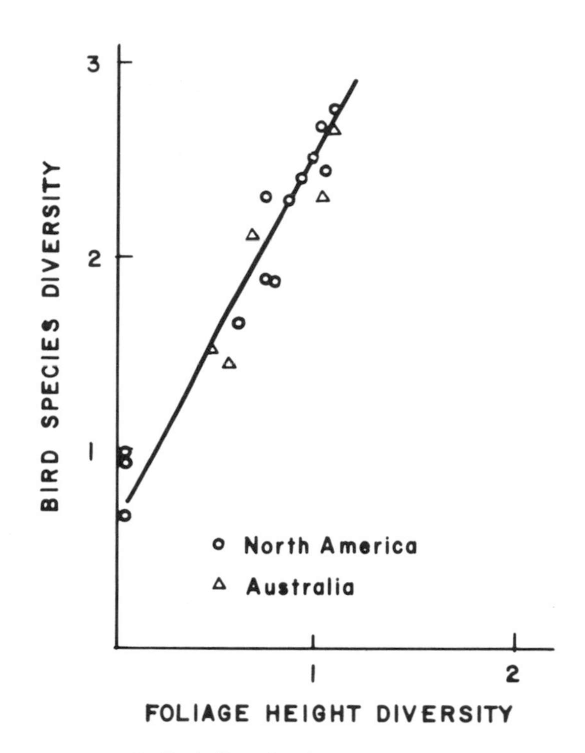

FIGURE 116. Bird diversity in angiospermatous forest plotted against foliage height diversity in eastern North America and in Australia (Recker).

64. R. H. MacArthur and J. W. MacArthur, On bird species diversity. *Ecology*, 42:594–98, 1961.
H. Recker, Bird species diversity and habitat diversity in Australia and North America. *Amer. Natural.*, 103:73–80, 1969.

65. See note 29.

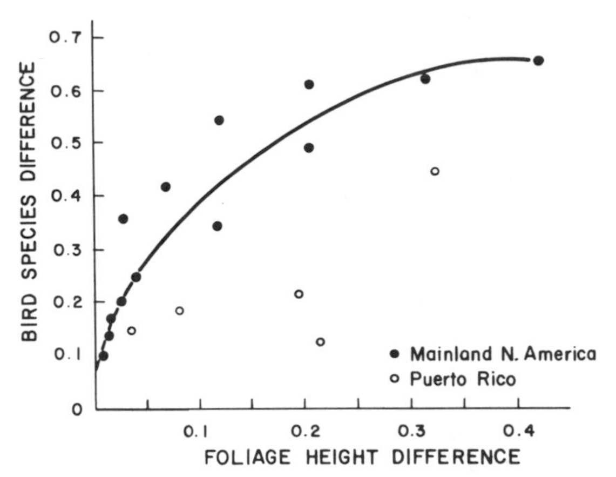

FIGURE 117. Difference in bird species composition between areas differing in foliage profile, plotted against foliage profile diversity. The same change in habitat causes far less change in bird fauna in Puerto Rico than in continental North America (MacArthur and Recker).

of forest, foliage height diversity, even with several more levels than are needed in a temperate forest, would not adequately characterize the habitat.

The other niche dimensions making up the seven or eight that might be needed would include the incidence of army ants, because some quite common species find their food primarily by association with the ants, as they move along striking terror into the smaller inhabitants of the forest, which are then easily eaten. Another dimension would have to be assigned to the availability of fruits, which can be obtained in some parts of the forest at all seasons and again determine the existence of a number of species of birds.

On islands where the species pool is likely to be lower than on the adjacent mainland, increasing the foliage height diversity causes a smaller increase in bird species diversity. MacArthur and Recker,[66] who devised a standard way of measuring the differences, show this clearly for Puerto Rico compared with the continental United States (figure 117).

Cody[67] has made an elaborate study in which the ecological separation of birds living in grassland, and so mainly in the lowest of MacArthur's three strata, is analyzed in detail.

Separation is likely to be achieved in three different ways, by behavioral reactions to horizontal differences in the biotope, by reactions to vertical differences, which will be much more limited than in most of MacArthur's cases, and by structural and behavioral differences in the species of birds which lead to the use of different foods at any given place in the habitat.

In defining a three-dimensional niche (figure 118) Cody considers not the properties of the species per se, but rather the differences between every possible pair of species. If there were methods of measuring habitat preference in terms of numbers expressing reactions to horizontal and vertical environmental diversities and the variation from species to species of the trophic structure and behavior, any species pair could be represented by three numbers, expressing the difference in horizontal habitat preference, the difference in vertical habitat preference, and the difference in mode of feeding. Any of these three might give sufficient niche separation to enable the species to coexist. Coexistence thus might be possible, even though the vertical and trophic components were identical, if the horizontal difference were great enough or, in figure 118, equal to OA. Similarly, if only the vertical component varied, coexistence would be possible if it were OC, and if only the trophic specialization varied if it were OB. These three points define a triangular plane ABC in space.

If the scales of the three coordinates are suitably graduated, all the differences being expressed as percentages of maximum values, OA = OB = OC = 100, the sum of the coordinates of any point on the triangle will equal this distance. Any pair of species outside the triangle will be more different than they need be to achieve separation, and so another species might come between them. If the species differed by less than 100, they would lie inside the tetrahedron ABCO and would not differ enough to coexist. The only

66. See R. H. MacArthur, Patterns of species diversity. *Biol. Rev.*, 40:510–33, 1965.

67. M. L. Cody, On the methods of resource division in grass-land bird communities. *Amer. Natural.*, 102:107–47, 1968.

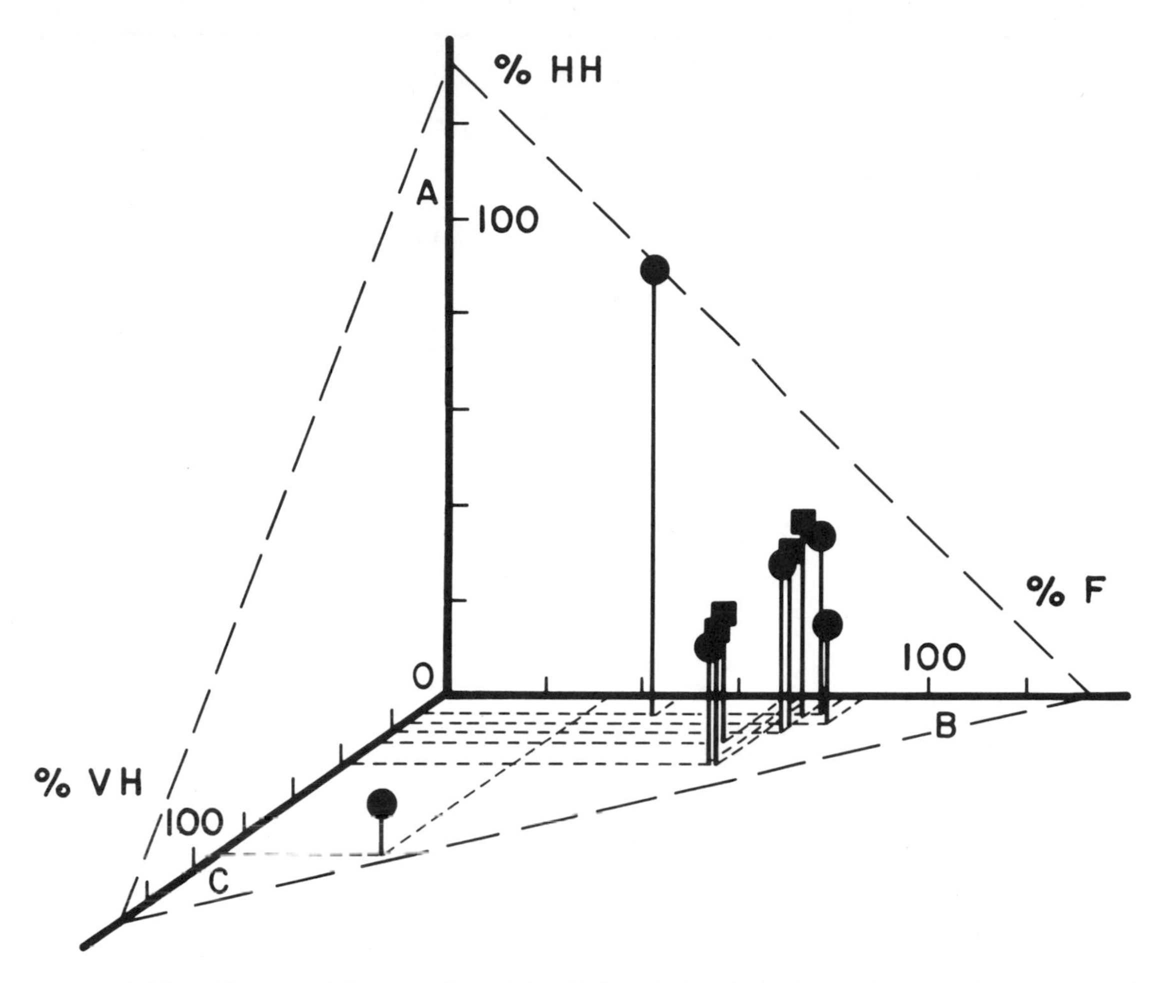

FIGURE 118. Three-dimensional diagram of the niche of a bird defined by horizontal habitat preference, vertical habitat preference, and mode of feeding, as indicated in the text (Cody).

difficulty lies in measuring the three kinds of difference. The easiest one to determine is vertical separation. The amount of time spent when feeding in three layers, 0–3″, 3″ to 1′, and above 1′ (roughly 0–7.5 cms, 7.5 to 31 cms, and over 31 cms), was determined for each of fourteen species or subspecies in various grasslands in North America. The percentage of time that any two species spent in the same layer was determined and its difference from 100 was taken as the vertical separation (table 13.)

The total separation in the habitat would be the mean of 0, 11, and 11, giving 7.3. The horizontal separation is much more complicated. Four environmental variables were

	0–3″	3″–1′	>1′	vertical separation	
Chestnut-collared longspur (*Calcarius ornatus*)	100	0	0	0	11
Lark bunting (*Calamospira melanoconys*)	100	0	0	11	
Western meadowlark (*Sturnella neglecta*)	89	11	0		

TABLE 13. Percentage of feeding time spent in each layer for the birds of a grassland habitat in Saskatchewan (Cody).

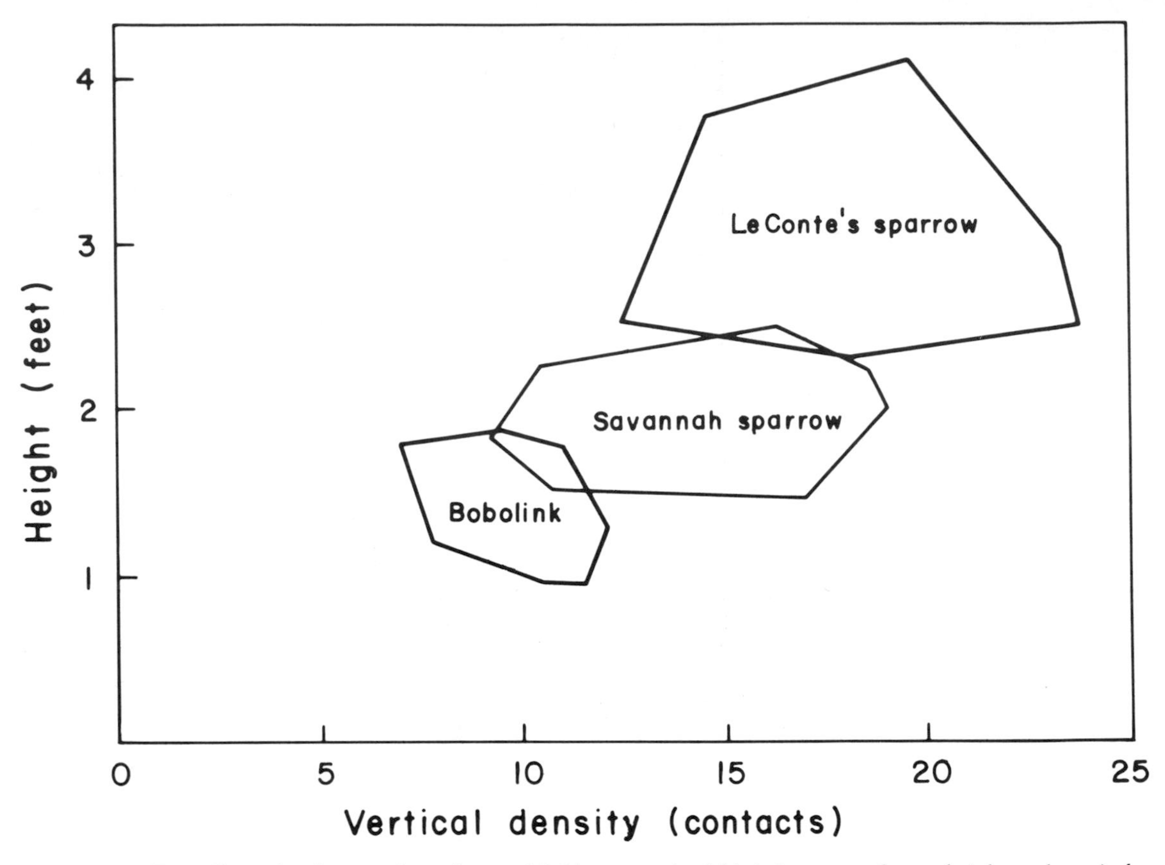

FIGURE 119. Two-dimensional separation of some Michigan grassland birds in terms of grass height and vertical vegetation density. Other North American species tend to be crowded into the region between the bobolink and the origin (Cody).

measured in the territories of the various birds studied. Vertical density (v) is measured as the number of leaves or stems over a point. Horizontal density (d) is measured as the average number of leaves or stems touching a line of standard length when placed at a standard number of heights. Vegetation height (h) is the height at which the horizontal density is less than one contact in 2 feet. The area (A) within a curve relating horizontal vegetation density to height is also obtained. For any habitat, a niche using these four variables can be constructed. In figure 119, the relations of three species of birds on the h–v plane for an area of grassland in Minnesota is shown. In this particular case, very good separation is obtained; but for other areas, notably one in Colorado with four species, all the points are crowded into a small region near the origin.

The four variables can be expressed as a discriminant function of the form $D = k_1h + k_2v + k_3d + k_4A$. This is a statistical device which permits maximum separation when the species are plotted along a single axis. From this their mean distance is calculated.

The third axis, of tropic specialization, is obtained from two morphological and three behavioral characters of each bird. The morphological characters are bill length and bill depth relative to length. The behavioral characters can be understood with reference to figure 120. This gives the mean behavior of the bird when traveling. The mean slope of the line gives the speed of movement, the sum of the horizontal parts as a percentage of the whole gives the percentage stationary time while the bird is feeding, the mean absolute length in seconds of the horizontal

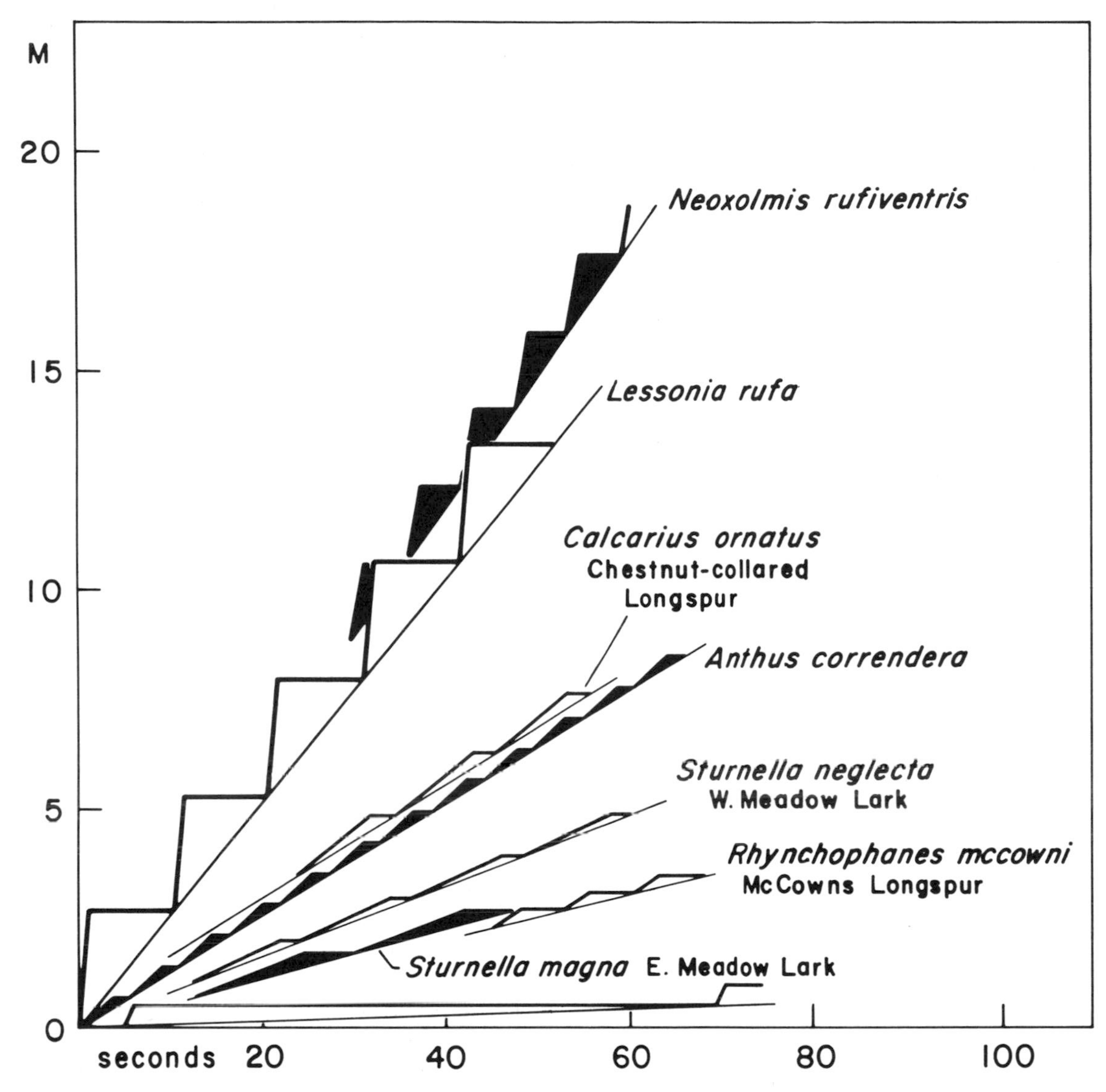

FIGURE 120. Diagrammatic representation of foraging behavior of grassland birds in North America (species with English names) and South America, selected from Cody's examples to show the range and the difference in behavior, as in the two longspurs and two meadowlarks, between very closely allied species. The unbroken lines give the overall rate of movement, the horizontal sections of the teeth the duration of an average stop. Note that the two most rapid birds are both South American. The bottom line refers to a number of very slowly moving species in North America, but only one in South America (Cody, modified).

parts gives the mean length of the feeding stops. All these data on feeding can be combined into a single figure.

The differences between pairs of species for each habitat are now calculated. Since the results are expressed as percentages, any pair differing only in one character should have coordinates 100, 0, 0. Actually, even the most extreme specialist will not, except for the character of its specialization, be exactly like any other species, and we should expect the sum of the three differences to be over 100. The differences are now averaged for each locality (table 14), and except for a single point representing a marine *Juncus* meadow in Maryland, all fall very close together. If a similar analysis is performed on grassland communities in Chile, the resulting points are also very close to the North American ones. A similar result can

Study Area	Horizontal Habitat Separation	Vertical Habitat Separation	Relative Food Specialization	Sum of Ecological Differences
NORTH AMERICA				
Saskatchewan	36.5	7.7	83.0	127.2
Minnesota	93.3	12.7	48.0	154.0
Colorado	21.2	14.3	85.2	120.7
Kansas	36.5	22.0	81.7	140.2
Maryland				
Spartina	25.2	40.0	74.5	139.7
Maryland				
Juncus	9.5	92.0	33.5	135.0
SOUTH AMERICA				
Phalaris				
field	37.5	16.7	79.6	133.8
Irrigated field	40.0	11.1	79.2	130.3
Cerro Castillo	27.8	27.7	71.5	127.0
Punta Delgado	28.5	41.0	75.9	145.4
Mean Values	35.6	28.5	71.2	135.0

TABLE 14. Mean Ecological Differences between Species in Ten Grassland Communities (Cody).

be obtained from the data in other ways. This analysis therefore suggests that with many different environmental resources, species tend to cope with their problems in a quite limited number of ways.

The meaning of foliage profile diversity and many of the other parameters that might be measured to provide satisfactory niche separation in vertebrates is quite likely to be found in the use that higher animals can make of differences in the appearance of a habitat. As a bird watcher knows, by looking at a limited area in a familiar terrain, what species to expect, so a bird may know what sort of area in that terrain provides it with a suitable home. Both bird watcher and bird may judge by roughly estimating a foliage height profile, but neither is primarily interested in such a finding; the bird watcher wants to study a particular species of bird, the bird wants a particular set of resources. The parameters measured by the ecologist, or intuitively seized upon by the bird or bird watcher, may primarily be indicators of the presence or absence of these resources.

A rather intensive study of two sympatric species of woodpecker, *Melanerpes e. erythrocephalus*, the red-headed woodpecker, and *Centurus carolinus zebra*, the red-bellied woodpecker, made in Kansas by Jackson,[68] illustrates this. Both have similar courtship and mating calls, nest at about the same height, and are similar in their brooding behavior, though there is some difference in the timing of the reproductive cycle. The red-headed is rather less dependent on trees for its food than is the red-bellied, and though both must nest in holes in trees, they would certainly be placed in different parts of the triangular MacArthur diagram. This is correlated with the insect food of the two species being quite different. The red-headed in its rather open habitat feeds more by catching flying insects or by flying to prey on the ground. The main groups eaten are grasshoppers and beetles, which together make up 48 percent of its

68. J. A. Jackson, A comparison of some aspects of the breeding ecology of red-headed and red-bellied woodpeckers in Kansas. *Condor*, 78:67–76, 1976.

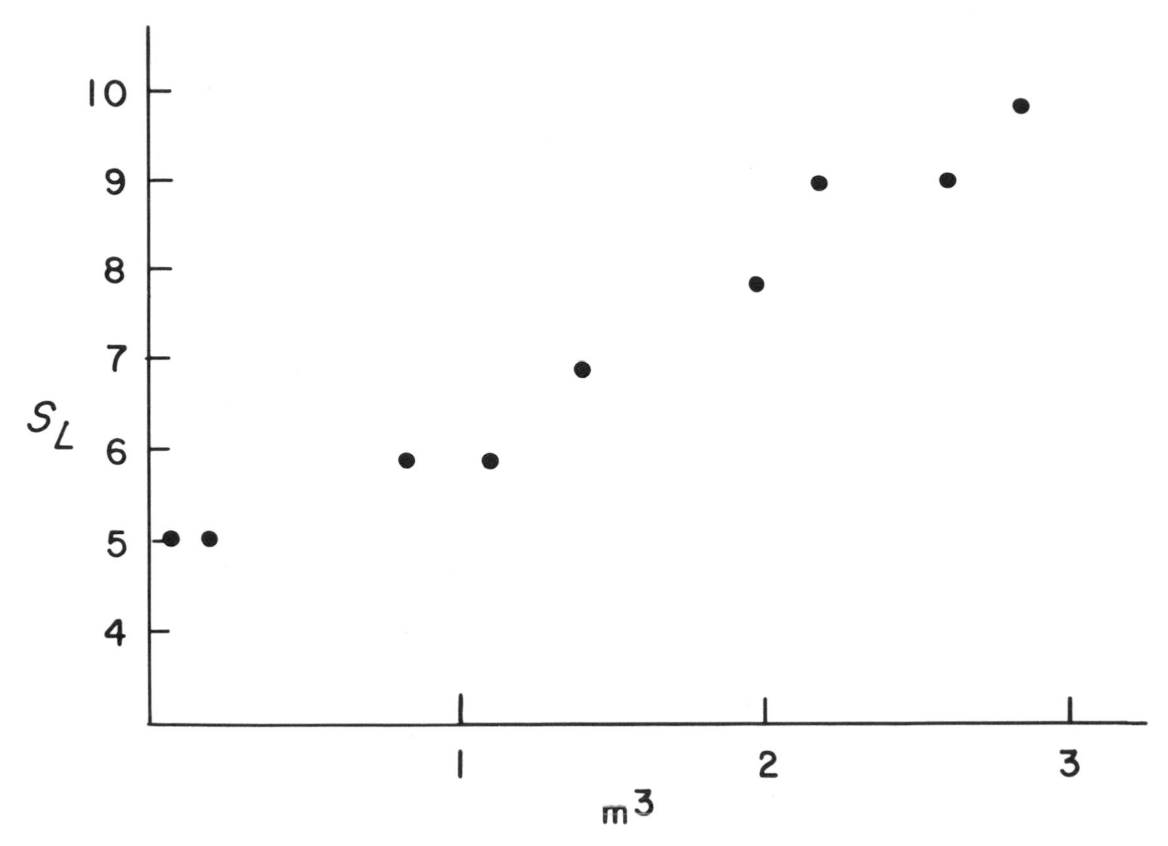

FIGURE 121. Number of lizard species (S_L) plotted against volume of major shrubs in m^3 (MacArthur after Pianka).

food. The red-bellied is more apt to feed by gleaning insects from trees. The main group eaten consists of wood roaches (Blattidae). The red-headed in its more open habitat nearly always excavates its nests in dead branches and starts digging out the hole at a place where there is a natural crack. The red-bellied is less particular about the kind of branch and does not need to start from a crack.

At least the large difference in food is clearly correlated with the foliage height profile preference. A more open nesting site means more grasshoppers and beetles in the diet, a more wooded one, more blattids. The most easily seen and so most easily measured characteristics of a habitat may provide the most suitable indicators to the bird, and axes which are at first sight scenopoetic are thus perhaps quite often competitive axes in disguise.

Pianka[69] has developed a method of analysis, comparable to that of MacArthur and MacArthur, of the habitats and niches of lizards, largely in terms of the openness or bushyness of the terrain (figure 121). Similar methods could probably be used in the study of other animals living in an environment composed of higher plants and the spaces between them, whether these spaces were filled with air or water. Casual records of aquatic insects show that some species prefer weed beds and others may be found on an adjacent open mud bottom. An example[70] is provided by the small corixid bug, *Micronecta dorothea*, living in the Transvaal, which was present in *Potamogeton*, while the allied *M. scutellaris* occurred close by, on an open bot-

69. E. R. Pianka, The structure of lizard communities. *Ann. Rev. Ecol. Syst.*, 4:53–74, 1973.

70. G. E. Hutchinson, A revision of the Notonectidae and Corixidae of South Africa. *Ann. S. Afr. Mus.*, 25:359–474, 1929.

tom between reeds. A quite complicated analysis of the insect fauna of a pond could probably be made in terms of foliage height diversity.

Examples of littoral fish in lakes or on reefs, showing similar differences, would presumably come to the minds of any ichthyological readers.

Multivariate statistical methods for defining niches

Various methods, comparable to the use of discriminate functions, introduced in the last few decades, provide ready-made approaches to the description of multidimensional niches. Like some other modern conveniences, they must be applied with care and thought. The principle involved is to determine points in an *n*-space and then by mathematical manipulation reduce the number of dimensions so that all are orthogonal, independent, and significant. The statistical techniques used may be obtained from the references given in the papers to be discussed. The main intellectual difficulty comes in evaluating the meaning of the final dimensions.

Green[71] has made an elaborate study of the bivalve molluscs of the lakes of central Canada. Each occurrence of a species is regarded as a point defined by nine environmental variables. These nine are reduced to five which have the following interpretation:

Axis I. Concentration of calcium relative to total alkalinity

II. Depth and depth-related sediment particle size

III. Organic content related to sediment particle size

IV. NaCl related to other noncalcareous inorganic compounds

V. A variable that appears as an interaction between depth and alkalinity. It probably represents the degree of wind disturbance, as the alkaline lakes are mainly shallow wind-swept prairie lakes.

With the use of these five variables, moderate niche separation can be obtained. A very interesting example is provided by a comparison of Lake Winnipeg and Lake Manitoba (figure 122). The potential niche space is smaller in Lake Manitoba largely owing to the shallowness of the latter, with a maximum depth in the southern basin of 6 meters rather than 13 meters as in Lake Winnipeg. The organic content of the sediment is also greater. Four species are common in Winnipeg; the two species of *Pisidium* have almost identical niches and are widely overlapped by *Sphaerium striatinum*. *Lampsilis radiata* overlaps the latter, but not either species of *Pisidium*. In Lake Manitoba only *L. radiata* and *P. casertanum* occur, and they show some overlap in the absence of *P. lilljeborgi*. The niche of *P. striatinum* clearly is essentially absent in Lake Manitoba, and in its absence *L. radiata* and *P. casertanum* extend into each other's niches.

It is to be noted that all the parameters initially measured are physicochemical, so all the axes inevitably appear at least superficially as scenopoetic. There is no representation of either food resources or the fish that would be needed to act as hosts for the *glochidium* larvae of the Unionidae. The lack of any certain bionomic axes presumably explains the considerable overlaps observed. The method, however, is clearly as promising in animal studies as in ecological botany, where it is much more often used.

Miracle,[72] studying the zooplankton of the Lake of Banyoles near Gerona in Spain, uses a formally comparable, but technically different approach. The initial space is constructed by considering each sample of zooplankton as a point determined by values along 17 axes which indicate the abundance, if present, of 17 species of zooplankton. From this an ordination of the species in 3-space in terms of three principal components is found to explain 58 percent of the observed variability.

71. R. H. Green, A multivariate statistical approach to the Hutchinsonian niche: Bivalve molluscs of Central Canada. *Ecology*, 52: 543–56, 1971.

72. M. R. Miracle, Niche structure in freshwater zooplankton: A principle component analysis. *Ecology* 55: 1306–16, 1974.

The attention of the interested reader may also be called to B. Seddon, Aquatic macrophytes as limnological indicators. *Freshw. Biol.*, 2: 107–30, 1972.

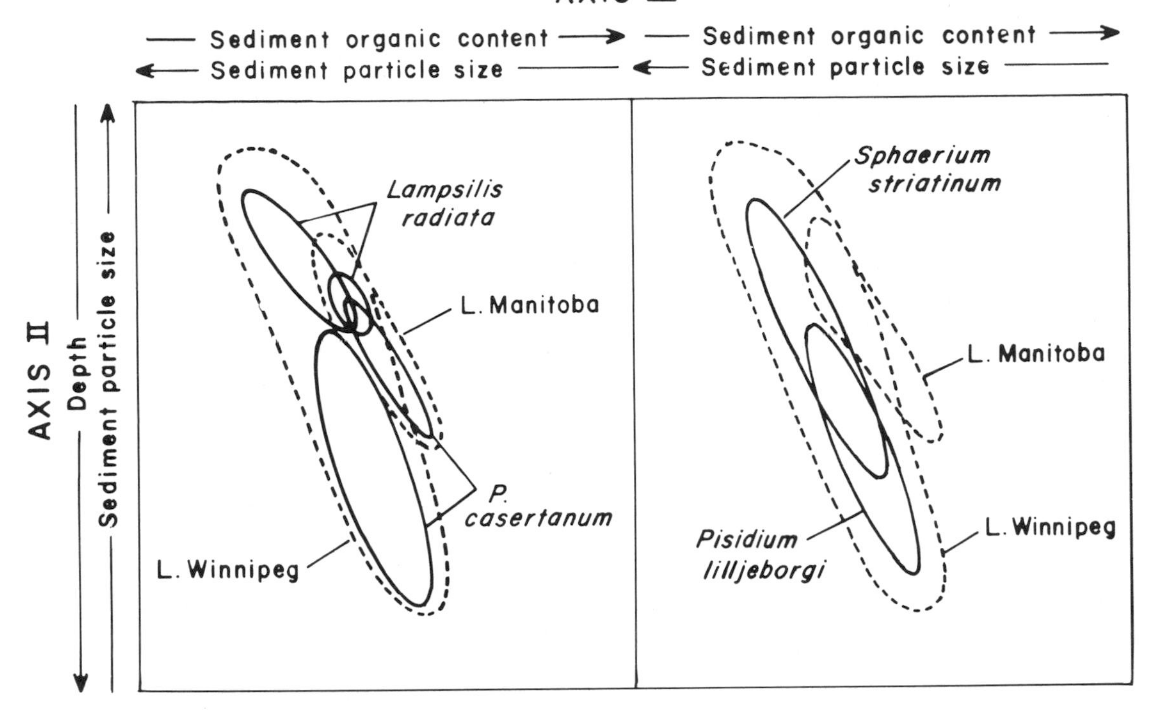

FIGURE 122. Scenopoetic niche structure of Lake Winnipeg and Lake Manitoba along two discriminant function axes, one (II) related to increasing depth and decreasing sediment particle size, the other (III) to increasing organic content of the sediment and decreasing particle size. The possible range for each lake is indicated by dotted lines. (a) The ellipses enclose 50 percent of the localities for *Lampsilis radiata* and *Pisidium casertanum*. (b) The two ellipses enclose 50 percent of the localities for *Sphaerium striatinum* and for *P. lilljeborgi* which do not occur in Lake Manitoba. Note that though *P. lilljeborgi* and *P. casertanum* seem, relative to the variables considered, to have comparable niches in Lake Winnipeg. *P. lilljeborgi* probably has a smaller fundamental niche not extending into the area occupied by any habitat in Lake Manitoba.

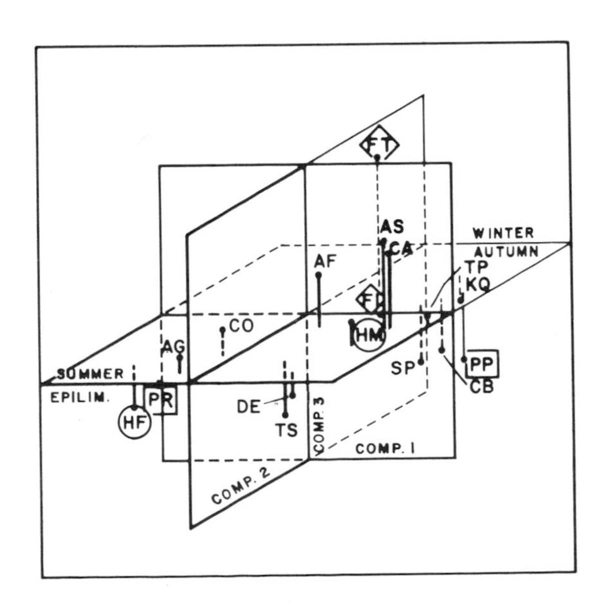

FIGURE 123. Correlation of occurrence of zooplankton at all stations and seasons in Banyolas Lake with three principal components. The three pairs of congeneric species, all rotifers, are indicated by circles for *Hexartha*, diamonds for *Filinia*, and squares for *Polyarthra*; all are well separated. AF, *Anuraeopsis fissa*; AG, *Asplanchna girodi*; CA, *Cyclops abyssorum*; CB, *Diacyclops bicuspidatus*; CO, *Collotheca* sp.; DB, *Diaphanosoma brachyurum*; AS, *Arctodiaptomus salinus*; FL, *Filinia limnetica*; FT, *Filinia terminalis*; HF, *Hexarthra fennica*; HM, *Hexarthra mira*; KQ, *Keratella quadrata*; PP, *Polyarthra platyptera*; PR, *Polyarthra remata*; SP, *Synchaeta pectinata*; TS, *Trichocerca similis*; TP, *Tropocyclops prasinus*.

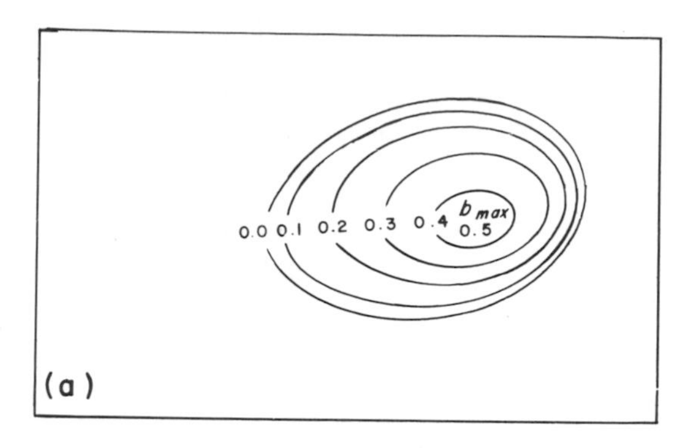

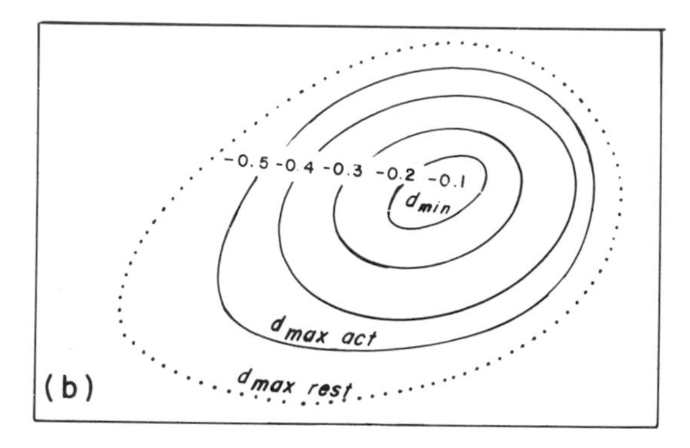

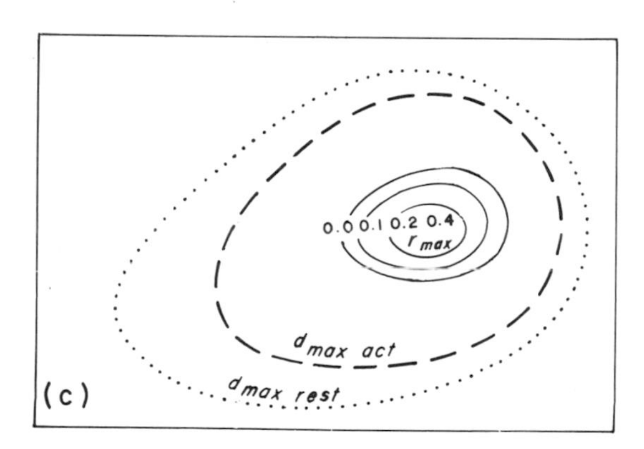

FIGURE 124. A niche space defined by two interacting variables x and y. (a) $\mathbf{b}_{max}$ is the representative point for all parts of the biotope permitting a maximum rate of reproduction 0.5; contours give the representative points for a series of lower birthrates from 0.4 to 0.0, beyond which birth does not occur. (b) A comparable figure in which $\mathbf{d}_{min}$ is the representative point for all parts of the habitat permitting a minimum death rate of -0.1. Contours define death rates up to -0.5 at which all active organisms die rapidly, though organisms in diapause may survive up to $\mathbf{d}_{max}$ rest. (c) A comparable plot $r = \mathbf{b} - \mathbf{d}$ showing the outer limit at which the species can be self-perpetuating, though active individuals may survive without reproducing at points up to the broken line and passive individuals to the dotted line (Maguire, modified).

The three axes are not easy to interpret. The lake has a peculiar thermal structure owing to warm springs in its floor; a permanent temperature inversion (crenogenic meromixis) is maintained by the sediment load of this incoming subterranean water. In the autumn there appears to be more upwelling and mixing into the freely circulating part of the lake than at other seasons.

The distribution of the vertical and temporal variation of the correlation of the logarithm of the zooplankton numbers on the three components are shown in Figure 123. Component 1 accounts for most of the seasonal variation; it probably involves the upwelling process. Component 2 is clearly primarily related to temperature, while component 3 is probably related to the phytoplankton as food for the zooplankton. This mode of proceeding is perhaps the purest type of niche analysis available.

It is significant that both the examples discussed are the result of limnological studies. It has already been noted that scenopoetic variables are more easily studied in lakes than elsewhere; it is probable that at least some information about bionomic variables can also be simply obtained from inland bodies of water.

Maguire's extension of niche theory

A considerable extension of the multidimensional niche has been suggested by Maguire.[73] In two dimensions two factors x and y are plotted along the two axes (figure 124). The effects of these are taken to interact, producing a niche with an elliptical or egg-shaped boundary ($\mathbf{b}_{lim}$). This boundary sets the limit outside of which the species cannot survive; in such a case $\mathbf{b} = \mathbf{d}$ or $r = 0$. We may take a point inside the niche ($\mathbf{b}_{max}$) where the birthrate is maximal and draw around it

73. B. Maguire, Jr., Niche response structure and the analytical potentials of its relationship to the habitat. *Amer. Natural.*, 107:213–46, 1973.

The terminology and symbolism used in this important paper have been modified a little to conform to the usage of the present book. This paper and the next are doubtless the most important contributions to the concept of the niche since the death of Robert MacArthur.

contours of decreasing birthrate; as the optimal conditions at the point are left behind, the limiting contours are ultimately reached. Similarly, we can take a minimum death rate at $\mathbf{d}_{\min}$, which may well fall on or very near the point representing $\mathbf{b}_{\max}$, and draw contours around it, until finally the limiting contour, at which death exactly balances birth is reached. Where resting stages exist there may be a zone around the active niche in which such stages can persist passively. The difference between **b** and **d** gives a further series of contours for *r*, which will in general have a maximum value in the region near $\mathbf{b}_{\max}$ and $\mathbf{d}_{\min}$, and a value of zero at the limiting contours.

Any uniform part of a habitat is represented by a point, the *representative point*, inside the niche. As the habitat changes, this point will move in the niche space, either as the result of external changes or as that of the effect of the organism on the habitat. A set of such changes will correspond to a trajectory along which the representative point of a habitat moves. If the species under consideration is coming into equilibrium, the trajectory, starting at a point defined by the initial state of the habitat, ends at a point on the limiting contour for *r* where in effect $N = K$. This is shown in an example calculated by Maguire for an alga plotted on the nitrogen-phosphorus plane of its niche space (figure 125). Here as the alga grows, the nutrients are exhausted; the habitat is consequently changed, and the trajectory of the representative points moves from some point of higher nutrient content to the limiting contour. The population meanwhile exhibits essentially logistic growth, but at a rate dependent on the part of the niche from which the trajectory takes its origin.

If we have two species competing, the trajectories can be calculated as the two species reproduce, according to the contours giving *r*, until the nutrients are exhausted. If one species, for example, requires more nutrients to produce a given amount of growth and

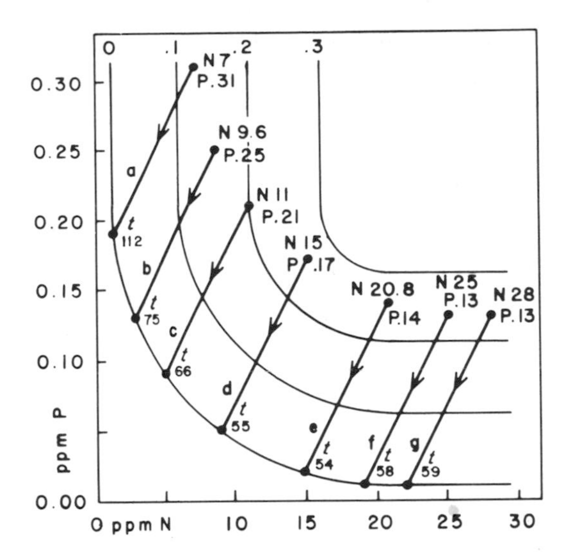

FIGURE 125. Computer simulation of changes in concentration of nitrogen and phosphorus in an algal culture using the elements in a ratio N:P = 10:1, starting with a variety of concentrations indicated at the tops of the trajectories and ending when the concentration is reduced to a level at which further growth cannot occur, at the points on the contour 0. The figures marked *t* indicate time taken to arrive at this contour (Maguire, modified).

starts with a higher maximum rate of growth, it can be shown that whatever trajectory is considered the more thrifty species with the lower growth rate will always displace its competitor. In doing so the trajectories may bend suddenly when they pass through the limiting contour of one species, but are still within the limiting contour of the other (figure 126). The potentialities of this model, some of which are further developed in the original paper, seem to be considerable.

Habitat, niche, and ecotope

A further analysis of the concepts that have so far been used in this chapter was made by Whittaker, Levin, and Root.[74] They consider the various possible variables acting on any species and group them into two categories. The first of these defines the *niche* space in the sense in which this term has been used throughout the discussion in this chapter.

74. R. H. Whittaker, S. A. Levin, and R. B. Root, Niche, habitat and ecotope. *Amer. Natural.*, 107:321–38, 1973. See also M. Rejmanek and J. Jenik, Niche, habitat, and related ecological concepts. *Acta Biotheoretica*, 24:100–07, 1975.

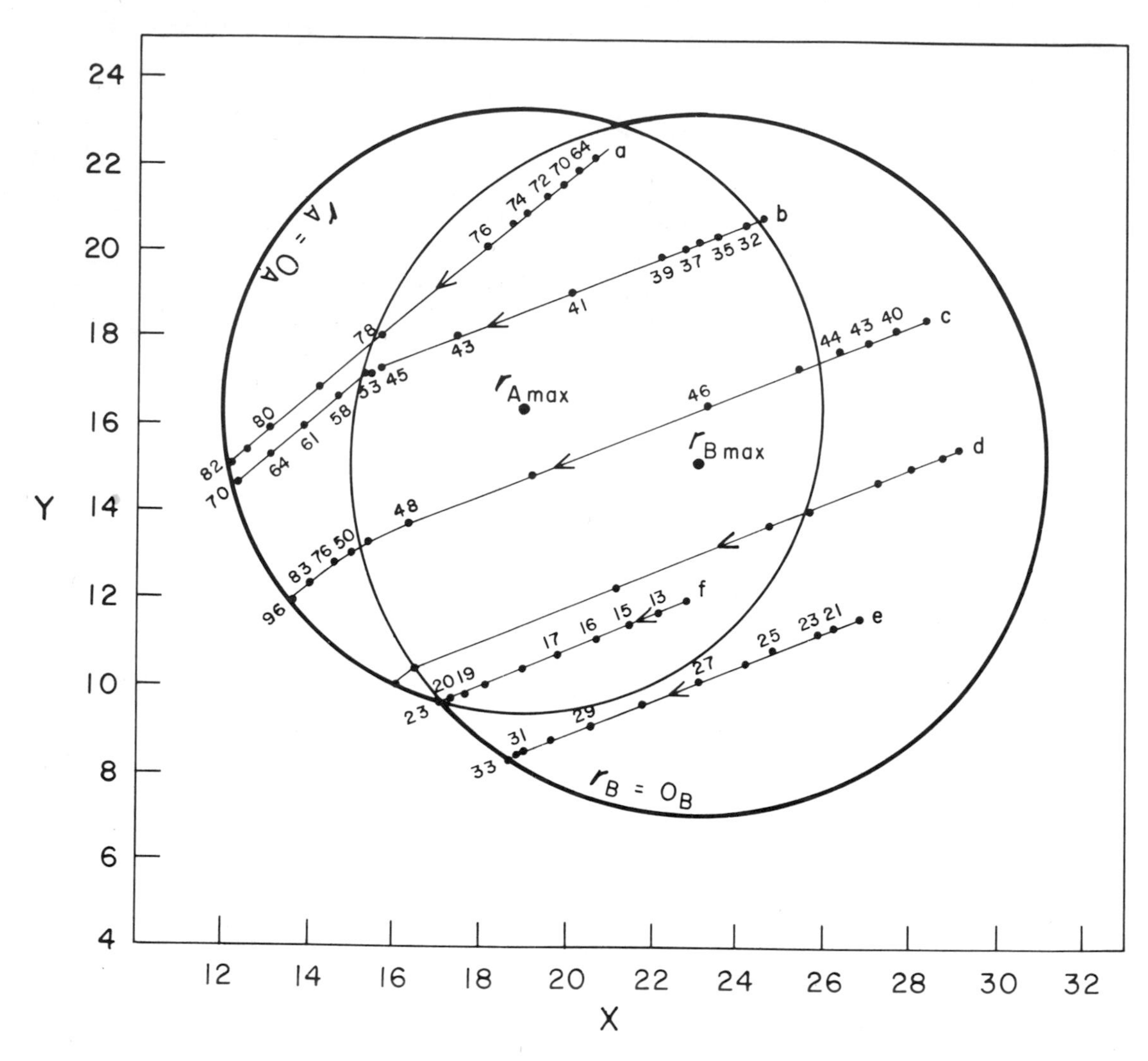

FIGURE 126. Trajectories of change in two limiting interacting nutrients Y, in a culture of two competing species S_A and S_B with different positions for r_{max}. Growth stops when the concentration of limiting nutrients lies on the contour O_A or O_B for no growth, lying nearest the axes. At this point only one species will be present. For trajectories a–d, this will be species A; for trajectory e, it will be species B; for f, it may be indeterminate (Maguire, modified).

All the niche variables are considered over a range defined by the biotope space or physical space actually containing a recognizable community, a homogeneously diverse three-dimensional space as defined on p 158. The second category is defined by variables taking sets of values as one passes from community to community. The axes of these variables define a second space, which they designate a *habitat* space. If both niche and habitat, in this special sense, are to be considered together, the resultant hyperspace is called an *ecotope* space. The distinction is best understood by considering the examples given in Whittaker, Levin, and Root's paper. The California thrasher considered by Grinnell (see p. 157) has its niche partially defined by nesting in dense foliage 60 to 200 cms above the ground. Outside the chaparral community, at lower elevations in the desert country, the Crissal thrasher (*Toxostoma crissale*) nests in low bushes near the ground. For each species the requirements for nesting and the nest height could be described most simply when the two communities are considered separately. If one needs to describe the ecotope in which both species are living, a habitat axis describing a specific kind of

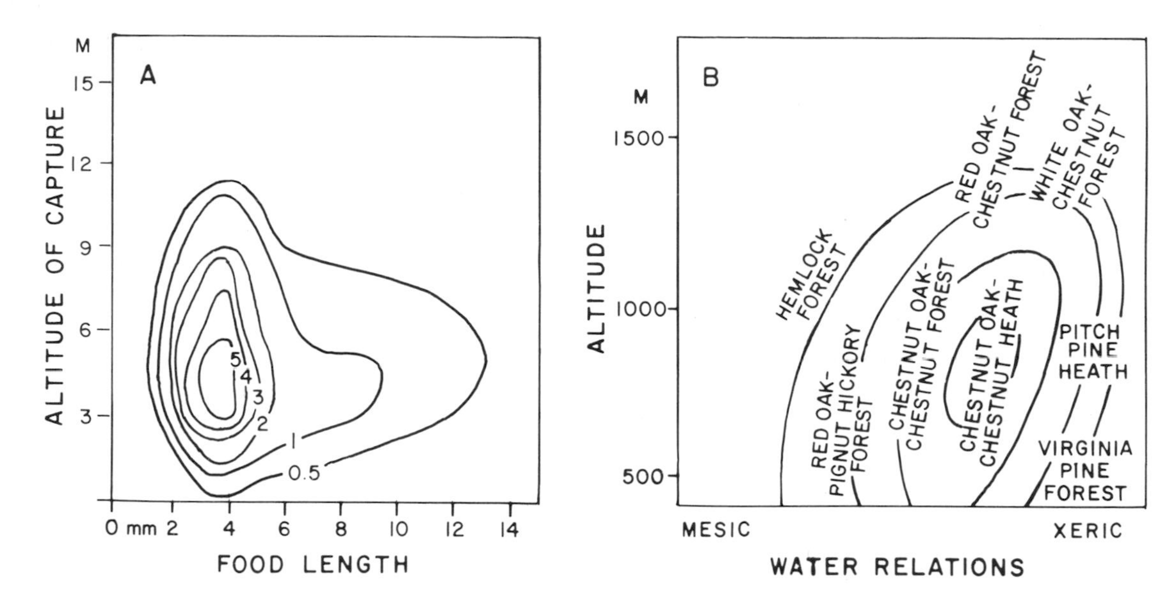

FIGURE 127. Diagram to illustrate Whittaker, Levin, and Root's conception of niche and habitat. A, Projection onto a two-dimensional space, defined by axes giving the percentage frequency of the food lengths and the percentage frequency of the various heights of capture, of the niche of the blue-grey gnatcatcher (*Poliophila caerulea*); B, Habitat of the chestnut oak (*Quercus prinus*) in the Great Smoky Mountains of Tennessee, in terms of altitude and of a scale of mesic to xeric regimes (Whittaker, Levin, and Root).

geographical variation of the vegetation relative to nest building is required.

In figure 127 another example of the distinction between niche space and habitat space is presented graphically. Within a community, the distribution of food captured in the niche space is presented for the blue-grey gnatcatcher (*Polioptila caerulea*) in a Californian oakwood. Between communities the distribution of the chestnut oak (*Quercus prinus*) in the Great Smoky Mountains is presented in terms of altitude on the mountains and a measure of the water relations of the sites occupied, defining a two-dimensional projection of a habitat space.

The original multidimensional concept of the niche, initially elaborated for the plankton, certainly was conceived to extend over more than one community insofar as in a stratified lake several communities are likely to be present one above the other, but with some species common to more than one community, as in the terrestrial example of figure 127, B. It is therefore evident that the original multidimensional niche was not properly distinguished from what would now be called an ecotope. The strength of these concepts lies largely in providing a conceptual scheme for different kinds of diversity. Within the community what is often termed α-diversity is increased by increased species packing; between communities β-diversity may involve having species with essentially identical intracommunity niche requirements, but with rather different tolerance limits so that they are adapted to a different part of a habitat gradient.

What happens in nature?

We have seen that theory suggests that populations dependent on the same resources are unlikely to be able to coexist indefinitely. Moreover, experimental studies indicate that unless the competition is density- or frequency-dependent, coexistence is not observed in homogeneous environments, though it is often possible by introducing a limited amount of heterogeneity to allow two species to live together. The overall theory of competition, ultimately due to Volterra but developed by Lotka and by Gause and Witt, has grown into a quite elaborate conceptual tool for the study of biological associations. The question remains and must now be faced as to how far all these beautiful theoretical developments really apply to the living world.

We will first consider the question as to whether the results of competitive exclusion, namely, that no two species occupy the same niche, are actually observed in nature. It is sometimes argued that since two species must be different in some ways, minute study will always bring to light some divergence in life history, food habits, site preference, or the like, to which mutual coexistence can be attributed. It is therefore supposed that the principle of competitive exclusion is in principle unfalsifiable and so is not a valid scientific proposition. If a very rigid mode of procedure is regarded as mandatory, there may be something to be said for this argument. However, an examination of the history and content of any kind of field biology will probably show to any impartial investigator, whose prime passion is not the falsification of the whole content of a particular part of science, that progress can be made only by not being too rigid. What is needed is a series of investigations from various points of view and a multiplicity of hypotheses at first chosen for their probability, interest, or potential importance more than for their testability. In the present case the way to proceed is to examine several, rather extreme, and quite different types of case, to see what light each of them may throw on the generalization. The more philosophic aspects of such matters will be considered at the end of the book.

The first case to be considered is that of the two Psocoptera, *Mesopsocus immunis* and *M. unipunctatus*, living on the bark of larch and pine trees in England. The two species were studied in Yorkshire by Broadhead and Wapshere[75] in an investigation which is probably more detailed than any other on a pair of closely similar animals. Insofar as the principle of competitive exclusion is testable in nature, Broadhead and Wapshere's work should provide a test. The insects are about 4 to 5 mm long, the adult females are apterous, the males fully winged. Each has one generation a year. Eggs are laid in the late summer, and the first instar nymph overwinters in the eggshell, hatching in March or April; after six molts the adults emerge in June. All stages feed on the green alga, *Pleurococcus*, and on fungal spores.

The insects are parasitized by a minute hymenopterous egg parasite, *Alaptus fasculus*, of the family Myrmaridae, and by another wasp, *Leiophron* cf. *similis*, a braconid, which parasitizes the last two nymphal instars. *M. immunis* is less attacked by the myrmarid but more attacked by the braconid than is *M. unipunctatus*. The adult females of *unipunctatus* have a longer life than those of *M. immunis*, but no such difference is noted in the survivorship of males.

Though the two species live together and feed on the same food, they exhibit different preferences in sites for egg laying. *M. unipunctatus* prefers the tips of twigs for this purpose, while *M. immunis* uses subterminal parts of twigs. The preference is apparently dependent on the diameter of the part of the twig, which is ordinarily less near the tip where *M. unipunctatus* oviposits. Examination of larch from Witton Fell, a more upland station lacking *M. immunis*, and of the same species of tree from Ottery St. Mary in Cornwall, where *M. unipunctatus* does not occur, shows the same types of distribution of eggs as are found when both species are present. The differences are clearly not imposed by competition. Broadhead and Wapshere conclude from their data that owing to this separation in egg-laying sites, which are not of unlimited extent, intraspecific competition for appropriate sites is greater than is interspecific. The conditions for coexistence are therefore fulfilled if the area of suitable egg-laying sites is limiting. It is possible, however, that such coexistence is helped by the steady parasitism on the part of the two Hymenoptera.

Another case in which initially the existence of sufficient niche specificity to permit coexistence seemed doubtful is provided by a group of species of leafhoppers of the genus

75. E. Broadhead and A. J. Wapshere, *Mesopsocus* populations on larch in England: The distribution and dynamics of two closely-related co-existing species of Psocoptera. *Ecol. Monogr.*, 36:327–88, 1966.

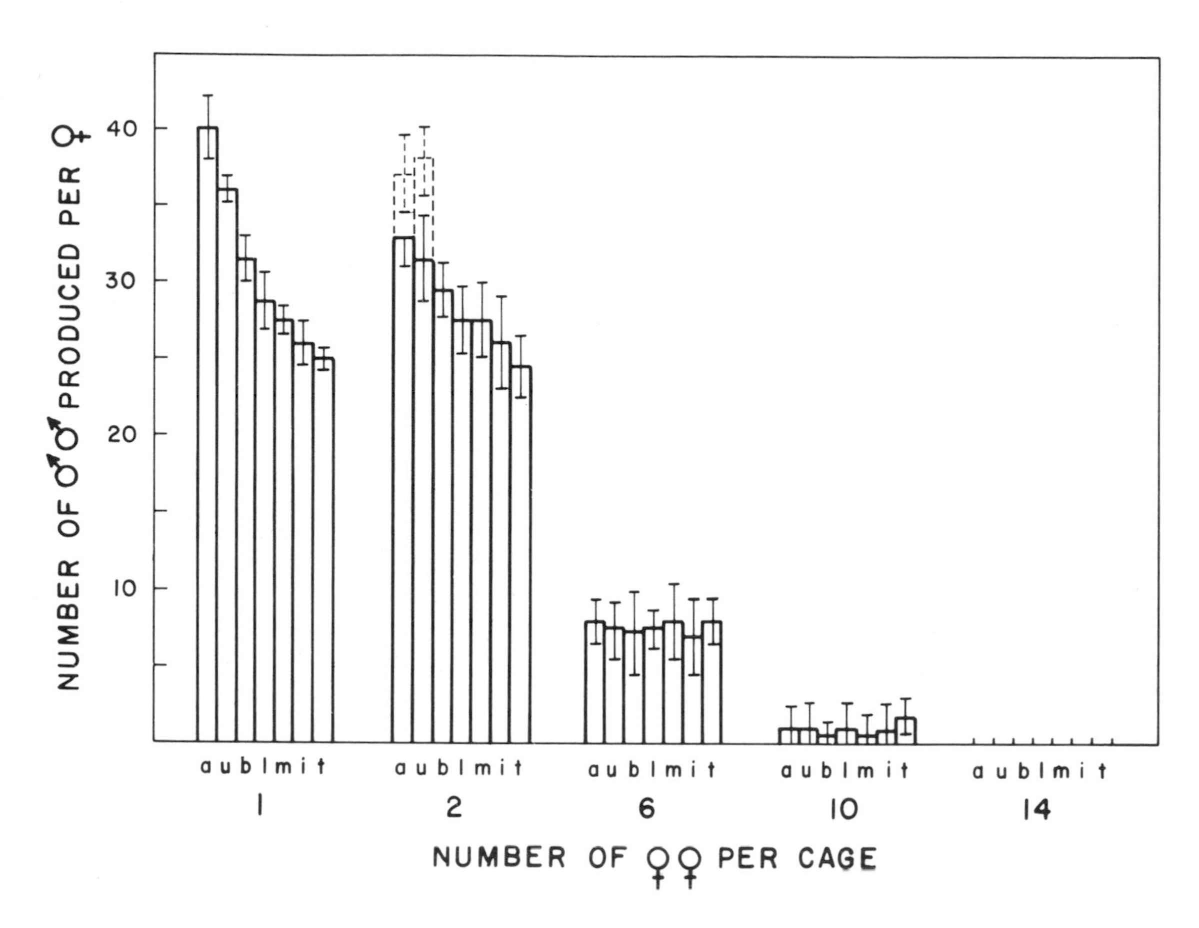

FIGURE 128. Mean number of male progeny, with standard error, in single species rearings of *Erythroneura* with different numbers of females per cage. The broken line columns for the two females per cage group indicate the progeny of *E. arta* and *E. usitata* per female when one female of each species forms the initial population, showing that interspecific is less than intraspecific competition.
a, *E. arta*; u, *E. usitata*; b, *E. bella*; l, *E. lawsoni*; m, *E. morgani*; i, *E. ingrata*; t, *E. torella* (McClure and Price, modified)

Erythroneura. These are small, rather fusiform insects, in color pale greenish with red markings, belonging to the homopterous family Cicadellidae. The family is an enormous and widespread assemblage of sucking insects, sometimes of considerable economic importance, in which many of the genera contain numbers of closely allied species, often distinguishable, at least to the human visual taxonomist, only by the structure of the male genitalia. A group of species of *Erythroneura* living together in various proportions on the American sycamore, *Platanus occidentalis*, has been the subject of study both by Ross and by McClure and Price.[76]

Eight species were present, of which 6 were studied by Ross in his original ecological work, and 7 by McClure and Price in their population experiments. Though at least 5 species can occur together on a single tree, there are some differences in the habitats chosen by the different species, *E. lawsoni*,

76. H. H. Ross, Principles of natural coexistence indicated by leafhopper populations. *Evolution*, 11:113–29, 1957.

H. H. Ross, Further comments on niches and natural coexistence. *Evolution*, 12:112–15, 1958.

M. S. McClure and P. W. Price, Competition among sympatric *Erythroneura* leafhoppers (Homoptera: Cicadellidae) on American sycamore. *Ecology*, 56:1388–97, 1975.

A comparable case is indicated by W. J. Le Quesne, Studies on the coexistence of three species of *Eupteryx* (Hemiptera: Cicadellidae) on nettle. *J. Entom.* (A), 47:37–44, 1972.

the commonest species, being the only consistent inhabitant of trees in dry windswept situations, while *E. morgani* and *E. bella* are usually found in humid valleys.

There are considerable specific differences in the number of progeny that can be reared from a single fertilized female; a mean of 40 male[77] specimens of *E. arta* and of 25 of *E. torella* being produced when each mother was set out alone on a single sycamore leaf. The other 5 species produced intermediate numbers of progeny (figure 128). With progressive increase in the reproductive population and so in damage done to the leaf, smaller numbers of progeny were produced, the proportional decrease being much greater in the initially more productive species. When 6 females were put on each of the single leaves, the mean number of male offspring was about 8 in all species. When the founding population was raised to 14 per leaf, no young survived.

Mixed populations, of which 8 of the possible 21 combinations appear to have been tried, always did slightly better than did pure populations with the same total number of founding females. These results were statistically significant in about half the cases studied, but there was never any evidence of a reverse effect. In some cases the mixed cultures produced surviving progeny when the total founding population consisted of 14 females.

McClure and Price think that there may be some density-dependent competition in the genus, but do not discuss the matter in detail. Populations of *E. morgani* reared with *E. torella* do rather better at low densities and *E. torella* rather better at higher densities, but none of the other data presented suggests a possible difference in the direction of competition in mixed populations with increasing density.

What is very clear about the work is that though density-independent factors, notably the weather, and a considerable amount of migration between trees, also probably dependent on environmental conditions, emphasize irregularities in the distribution, the various syntopic species of *Erythroneura* do differ biologically in such ways that intraspecific competition is regularly more intense than is interspecific competition. This satisfies the formal requirements for coexistence, though without indicating the biological processes that are involved. It is possible that in the development of natural populations, competition for egg-laying sites, which, like the food supply, can be injured by excessive suctorial feeding by the bugs, is of considerable importance.

A very impressive but only partly analyzed case is provided by the water mites of the genus *Arrenurus*, studied by Mitchell,[78] of which genus up to 26 species have been recorded on occasion from the same Michigan pond, in which at least 10 are regularly resident.

The members of the genus are hatched in spring from eggs laid by overwintering females. The six-legged larva becomes a parasite on the winged stage of an insect and feeds on it. The engorged larva then falls off the insect into water and becomes a resting stage or nymphochrysalis. From this emerges a nymph that feeds on microscopic crustacea. About midsummer the nymph forms a second resting stage or teleiochrysalis from which an adult emerges. Like the nymphs, the adults feed on small crustacea. Dispersal takes place very effectively in the parasitic larval stage, when the host insect may fly from one body of water to another.

There is some host specificity. Members of the subgenus *Arrenurus* (s. str) are almost confined to the imagines of the dragonflies

77. Only males are considered in these population studies because when mixed broods are enumerated, there is no practical way of determining the specific identity of the females.

78. R. Mitchell, A study of sympatry in the water mite genus *Arrenurus* (Family Arrenuridae). *Ecology*, 45:546–58, 1964. R. Mitchell, A model accounting for sympatry in water mites. *Amer. Natural.*, 103:331–46, 1969. It is unfortunate that the author in attempting to provide a completely up-to-date model should have dressed his perennially excellent biology in a style that dates so easily.

damselflies as hosts, while members of the subgenera *Megaluracarus* and *Micruracarus* usually parasitize dipterous flies.

The extremely different niches in which the larvae and the active nymphs and adults occur, with a forced migration between them, must often result in species being carried into habitats unsuitable for active nymphal and adult life, or unsuitable as sources of hosts for the larvae. This doubtless explains the frequent casual records, in a particular pond, of species which never become established.

Mitchell has made a study of two species in Japan which indicate in principle the kinds of mechanism that would permit co-occurrence. The two species, *Arrenurus agrionicolus* and *A. mitoensis*, were living in shallow water in Senba Lake, Mito, Japan. Both parasitize the damselfly, *Cercion hieroglyphicum*, and apparently no other species. The naiads of *C. hieroglyphicum* feed in submerged vegetation at a depth of more than 20 cms until two days before emergence. In the locality studied, a narrow zone of littoral vegetation grew marginally in about 2 cms of water. Lakeward of this there was an open zone about 7 m wide, beyond which the plants, among which the *Cercion* naiads had been living, reached the surface of the water. When the naiads prepare for metamorphosis, they either move up to the tops of these plants or migrate marginally to the narrow band of littoral plants.

A. agrionicolus finds practically every potential host, and when the hosts have emerged they carry about 10 times as many of the larvae of that species as of *A. mitoensis*. The larvae of the latter are, however, about 100 times commoner in the plankton samples taken in the submarginal weeds than are those of *A. agrionicolus*. Moreover, the legs are shorter and thicker in the latter species and so better adapted to crawling on the bottom rather than perennially swimming. In this benthic environment *A. agrionicolus* larvae meet the agrionid naiads before those of *A. mitoensis* do so; the latter species must become associated with the naiads as the latter ascend through the weeds, or even after they have emerged and are resting on plants at the surface.

Fixation of the larvae occurs just as ecdysis occurs, the mites tending to attach to the first ventral membrane, between the sternites of the dragonfly exoskeleton, which they meet. *A. mitoensis* starts more quickly attaching than does *A. agrionicolus*, within the first minute of the beginning of emergence, and so most of the larvae become fixed to the thoracic membranes which are the first exposed. *A. agrionicolus* ordinarily starts attachment three minutes after the beginning of ecdysis and usually meets abdominal segment VII, which is just withdrawing from the naiad exuviae. Neither mite has any inherent tendency to settle at a particlar site; if the adult of *C. hieroglyphicum* later lands on floating leaves or projecting stems, it may become infected by *A. mitoensis* anywhere on the venter.

The larvae of the more abundant *A. agrionicolus* are much more evenly spaced than are those of *A. mitoensis*, in which lethal crowding can occur. It will be noted that in the shallow marginal water, naiads arrive accompanied by a full complement of *A. agrionicolus* which attach to their abdomens on ecdysis, but with little parasitism by the planktonic larvae of *A. mitoensis*. If there was a species whose behavior led its larvae to concentrate, lying in wait in the extreme littoral, and then rapidly parasitizing the thorax, a third species might be added to the guild. Mitchell thinks that other ways of settling might be quite feasible and that the possibilities that his example illustrates are unrealistically small in number. It must also be remembered that species could be limited by nymphal or adult food supply as well as by the host relationship of the larva. At least Mitchell's analysis indicates the kind of way in which a complicated water-mite fauna could develop.

Remarkable studies of competition have been made in the sessile organisms of the intertidal zone on various coasts of Europe, North America, and South Africa. Connell[79]

79. J. H. Connell, The influence of interspecific competition and other factors on the distribution of the barnacle *Chthamalus stellatus*. *Ecology*, 42:710–23, 1961.

found that the barnacle, *Chthamalus stellatus*, ordinarily occurs above *Balanus balanoides*, being less sensitive to exposure to air, but easily smothered or displaced by the more rapidly growing *B. balanoides*. He suggested that in general the upper limit of occurrence of littoral sessile species is determined by tolerance to physical factors, while the lower limit is set by biological factors acting in competition. In general, later workers have substantiated this conclusion.

Branch has studied the very complicated interrelations of limpets on the coasts of South Africa, and has elucidated a number of the factors that are operating. Up to ten species of *Patella* and one of *Cellana* are involved, and some of these limpets may enter into competition with barnacles. Thirteen species of *Patella* occur on the coast of South Africa, of these ten species may be sympatric. Not all species are fully investigated. Two ecological and behavioral species groups are recognizable, which Branch initially characterized as migratory and nonmigratory. Only one species, *P. barbara*, seems to be intermediate. The migratory species (*P. granularis*, *P. granatina*, *P. concolor*, and *P. oculus*) settle low on the shore and migrate upward, mainly during the winter; some temporary downward movement may occur in summer. They are generalized in their food habits, nonterritorial, and exhibit rapid growth and high mortality. Usually they have large gonads; *P. oculus*, which has small gonads, is a protandric hermaphrodite spawning twice a year, so that all species have a high natality and may be regarded as having been evolved through *r*-selection. The nonmigratory species (*P. cochlear*, *P. longicosta*, *P. tabularis*, *P. miniata*, *P. argenvillei*, and *P. compressa*) are stationary, taking up a position on a scar and maintaining possession of the area around it. They can sense intruders by contact with the pallial tentacles and then can push them away. They tend to be food specialists; *P. cochlear* and *P. miniata* feed only on lithothamnion, *P. longicosta* and *P. tabularis*, at least as adults, primarily on *Ralfsia*, *P. compressa* on the kelp *Ecklonia*. The nonmigratory species have small gonads, low growth rates, and low mortality and may be regarded as the result of *K*-selection. In both migratory and nonmigratory groups the upper limit of the species seems in general to be set by physical factors. *P. granularis*, a migratory species, ranges highest and lives in areas much frequented by barnacles. Below it *P. concolor* comes into competition with *Cellana capensis*, a limpet of another genus; they have very similar requirements, but differ in *C. capensis* having a greater tolerance of dry conditions on bare rock, while *P. concolor* settles on rock with sand which tends to be wetter. The nonmigratory species tend to occur below the adults of the migratory species. *P. cochlear* feeding on lithothamnion lives at about spring low water, while *P. miniata* remains permanently covered. *P. longicosta* and *P. tabularis* form a similar pair on *Ralfsia*. *P. cochlear* can live only in moderately disturbed water, and on the disturbance axis its niche is included in that of *P. longicosta*. There is usually a remarkable specialization separating the young from the adults, in that the juveniles of *P. cochlear* settle on the shells of the mature individuals feeding on lithothamnion growing there, waiting for a territory to become vacant. *P. miniata* first settles under subtidal rocks and later on their upper surfaces. *P. longicosta* initially lives on *Ralfsia* growing on shells of its own or other species of molluscs, then starts eating lithothamnion, removing it and leaving bare the rock on which *Ralfsia* can grow. The *Ralfsia* garden so established is weeded by the animal eating off invading algae of other species; the main food, however, is *Ralfsia*, which is eaten in strips, leaving uneaten thallus that can regenerate the garden.

A complicated competitive situation is found when the relationship of *P. granularis* to barnacles is investigated. The species occurs higher up the shore than the other limpets,

G. M. Branch, see note 32.

G. M. Branch, Interspecific competition experienced by South African *Patella* species. *J. Anim. Ecol.*, 45:507–29, 1976, and various other papers by this author to which reference is made in these.

in a zone usually dominated by the barnacle, *Chthamalus dentatus*. The population density of the limpets increases with the barnacle cover, but their size decreases. Natality is high when the barnacle cover is low, and falls in two stages to zero when the cover is maximal. The population is thus maintained only by the limpet living slightly deeper than the optimal depth for the barnacles. A number of limpets that have managed to settle must be condemned to sterility.

The four examples that have just been presented were chosen on account of the closeness of the species involved, their number, and the complexity of the environmental circumstances with which they are concerned. Other examples of the same order of complexity, such as Ashmole's seabirds,[80] Diamond's[81] extraordinary studies in New Guinea, Kohn's[82] species of *Conus*, and MacArthur's[83] warblers, already mentioned, could have been used in the same context. In the last two examples, the obvious differences in the food eaten or the way in which it is obtained have become widely known among animal ecologists and so need not now be discussed in great detail. A very large number of cases of two species having differences about as great, both taxonomically and in mode of life, as, for instance, the red-headed and red-bellied woodpeckers, already considered, have been reported in the ecological literature. Such cases concern organisms of many phyla and, on land, of all sizes down from the two African rhinoceroses, locally sympatric in Natal, but one (*Ceratotherium simum*, the so-called white rhinoceros) being a grazer and the other (*Diceros bicornis*, the black rhinoceros) browsing or digging roots.[84] With all this wealth of material one cannot doubt that separation by biologically significant differences in the nature and way of obtaining food and other resources is a very common explanation of coexistence. Enough, however, is known both theoretically and from laboratory experiments to suggest that this is not the whole story.

The most usual, apparent exceptions to the principle of competitive exclusion in nature are those due to predation and quite possibly a class due to failure to achieve equilibrium before conditions in the environment change.

There is also at least one case that gives an appearance of populations of two very similar species being limited solely by the amount of space available for territories. If, as seems to be the case with these species of *Pipilo*, the Abert towhee, *P. aberti*, and the brown towhee, *P. fuscus mesoleucus*,[85] the two, though very much alike, pay no attention to each other, they are formally in two different niches, each requiring and being limited by the supply of its own kind of territory, which happens to occupy the same physical space in the landscape as the territory of the other bird. Some doubt has been expressed about this by various critics in conversation. Ideally such a case is possible; both MacArthur and I wondered if he would find it among his warblers before he started his fieldwork. Clearly, the matter requires further study.

The effects of predation and parasitism

We have already seen that the work of Utida and of Slobodkin[86] on experimental populations and the latter's ingenious development of part of Gause's theoretical arguments show that unselective predation or parasitism of the

80. N. P. Ashmole and M. J. Ashmole, see note 46.

81. See note 4.

82. A. J. Kohn, see note 62; see also G. E. Hutchinson, *The Ecological Theater and the Evolutionary Play* (note 43), pp. 40–42.

A. J. Kohn and T. W. Nybakken, Ecology of *Conus* on Eastern Indian Ocean fringing reefs: Diversity of species and resource utilization. *Marine Biology*, 29:211–34, 1975.

83. R. H. MacArthur, Population ecology of some warblers of northeastern coniferous forests. *Ecology*, 39:599–619, 1958.

84. R. J. G. Atwell: Last stronghold of rhinoceros. *Afr. Wildl.*, 2:35–52, 1948; and information given me and shown me on a film by S. Dillon Ripley. See also *The Ecological Theater and the Evolutionary Play* (note 43), pp. 33–35.

85. J. T. Marshall, Interrelations of Abert and brown towhees. *Condor*, 62: 49–64, 1960. See also *The Ecological Theater and the Evolutionary Play* (note 43), pp. 45–46.

86. S. Utida, Interspecific competition between two species of bean weevil. *Ecology*, 34:301–07, 1953.

L. B. Slobodkin, pp. 146–47.

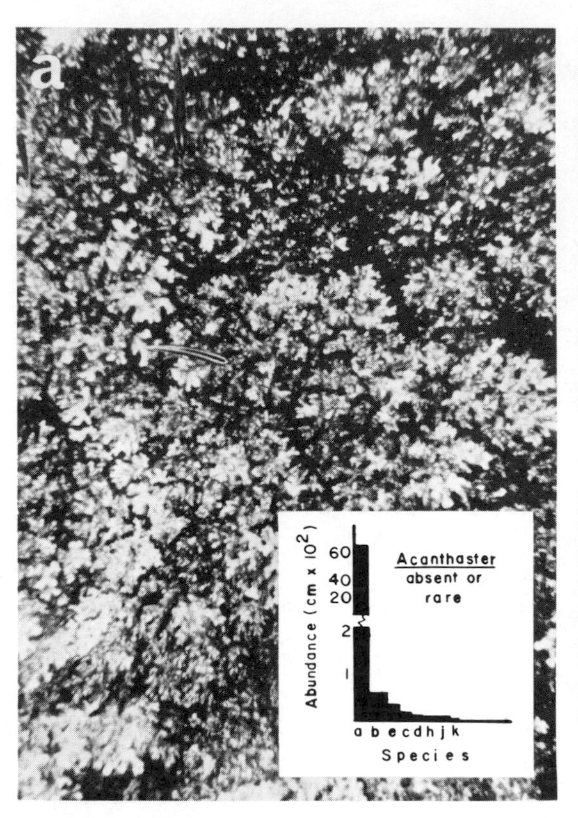

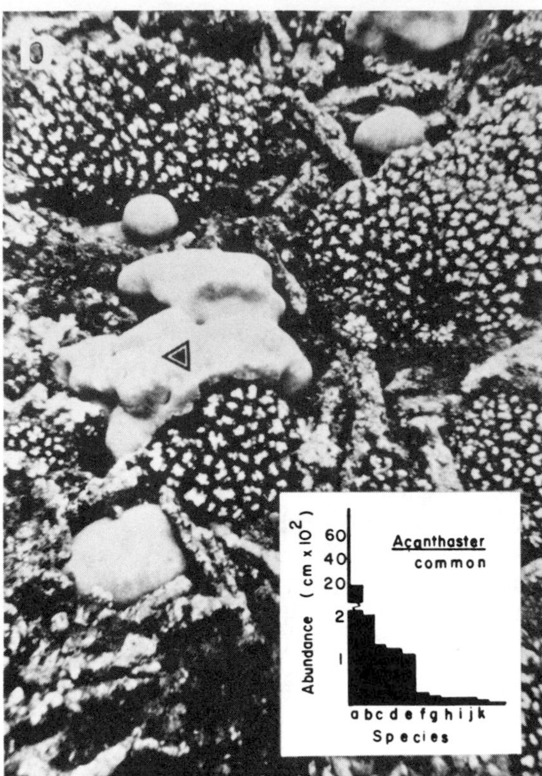

FIGURE 129. Eastern Pacific *Pocillopora* reef crests: (a) an almost monospecific community of *Pocillopora damicornis* without the predatory starfish *Acanthaster*; (b) a partly killed but very diversified reef with *Acanthaster*. In the histograms: a, *P. damicornis*; b, *P. elegans*; c, *Pavona varians*; d, *Millepora* sp.; e, *Porites californica*; f, *Porites panamensis*; g, *Pavona clivosa*; h, *Psammocora brighami*; i, *Pavona ponderosa*; j, *Pavona gigantea*; k, *Tubastrea aurea* (Porter).

right intensity can permit the coexistence of two species where one would be outcompeted in the absence of the predator or parasite. There can be no doubt that this is highly important in nature.

Paine[87] has studied the phenomenon in great detail on the intertidal part of the shoreline at Mukkaw Bay in the state of Washington. Here the removal of the dominant predator, the starfish *Pisaster ochraceus*, though it increased the biomass, reduced the diversity of the biota, in two seasons, from 15 to 8 species, the general tendency being toward a monospecific association of *Mytilus californicus*.

A comparable situation has been studied by Porter[88] in the eastern Pacific, where unpredated coral reefs tend to be composed almost entirely of *Pocillopora damicornis*, the staghorn coral, but when an infestation by the crown-of-thorns starfish *Acanthaster planci* occurs, the specific diversity may become much greater (figure 129).

MacArthur[89] suspected that such rather extreme effects of predation on specific diversity were more likely to occur in communities the animals of which tend to be arranged in two rather than three dimensions. This, however, has neither been established nor refuted. Clearly, the role of predation is often likely to be important in benthic communities and perhaps in grasslands, but it must be kept in mind wherever striking predation is occurring.

87. R. T. Paine, Food web complexity and species diversity. *Amer. Natural.*, 100:65–75, 1966.

88. J. W. Porter, Community structure of coral reefs on opposite sides of the Isthmus of Panama. *Science*, 180:543–45, 1974.

89. R. H. MacArthur, Coexistence of species. In: *Challenging Biological Problems*, ed. J. Behnke. New York, Oxford University Press, 1972, pp. 253–59.

The paradox of the plankton

A peculiar situation has been recognized[90] in the phytoplankton, in which a large number of species of autotrophic algae may coexist, depending on the same nutrients and on a light flux into turbulent water in which it would seem that little spatial heterogeneity is likely to persist for a time of the order of one division of the algae in question.

The nutritive requirements[91] of the algae of both freshwater and the ocean may include the following elements: C, N, P, S, K, Mg, Si, Na, Ca, Fe, Mn, Zn, Cu, B, Mo, Co, and V. In addition, organisms are known that use F, Cl, Br, and I, while some animals need Cr, Sn, and Se in their metabolic machinery and Sr or Ba in their skeletons. Some number between 17 and 26 elements might on occasion conceivably be limiting, as might at least three vitamins, thiamine, cyanocobalamine (B_{12}) and biotin.

The number of species found co-occurring and exhibiting simultaneous growth in their populations is less well known than it should be. In his classical work on the temporal variation of the plankton of the Lunzer Untersee in Austria, Ruttner[92] found on any day up to 8 species abundant enough to enumerate routinely in the surface waters.

In the much more limited number of samples from the shallow lakes of Funen in Denmark, Foged[93] found 40 to 50 species of diatoms and indicates that other algae were present. These lakes are shallow and most of the water sampled was probably freely mixed.

In a study of Castle Lake, California, expressly devoted to a study of the variance of the distribution of phytoplankton in the epilimnetic or freely mixed upper layer of the lake, Rickerson, Armstrong, and Goldman[94] found 48 taxa of phytoplanktonic algae of which 29 were diatoms.

A number of explanations of the diversity encountered have been put forward. Most of them depend on the assumption that these phytoplankton populations are not in equilibrium and that therefore the conclusions of the Volterra model, which are significant only at equilibrium, do not apply. Two possible equilibrium models have also been suggested and may first be considered.

In one equilibrium model,[95] it is suggested that species may coexist in a commensal or symbiotic system, the species that is otherwise the better competitor being dependent on other species for accessory nutrients or vitamins. Only three vitamins are likely to be involved and the possible diversity that could be achieved is obviously not great.

In the second equilibrium model, it is supposed that every species is limited by a different nutrient. Peterson[96] shows that in certain reasonable conditions this could produce stable mixed populations. The three vitamins known to be needed by many algae could obviously also be limiting, but might also be involved in commensalism or symbiosis. Peterson points out that studies by Goldman have shown that a large number of the essential elements may on occasion be limiting and thinks that his scheme might permit 30 coexisting species. Though this may be unduly optimistic, there would be nothing unreasonable in an association of a diatom limited by silicon, a blue-green alga limited by phosphorus but fixing its own nitrogen, another blue-green alga limited by sodium, and one or more chrysophyceans

90. G. E. Hutchinson, The paradox of the plankton. *Amer. Natural.*, 95:137–46, 1961.

91. G. E. Hutchinson, *A Treatise in Limnology*. New York, London, and Sydney, John Wiley, 1967 xi, 1,115 pp.; see especially vol. II, pp. 307–13. Reviews of some of the work involving the more recently discovered roles of minor elements in nutrition may be found in chap. 30 of vol. III of the same work, 1975.

92. R. Ruttner, Das Plankton des Lunzer Untersees: Seine Verteilung in Raum und Zeit während der Jahre 1908–1913. *Int. Rev. ges. Hydrobiol. Hydrogr.*, 23:1–138, 161–287, 1930.

93. N. Foged, see note 30.

94. P. Rickerson, R. Armstrong, and C. R. Goldman, Contemporaneous disequilibrium: A new hypothesis to explain the paradox of the plankton *Proc. Nat. Acad. Sci. U.S.A.*, 67:1710–14, 1970.

95. See note 90.

96. R. Peterson, The paradox of the plankton: An equilibrium hypothesis, *Amer. Natural.*, 109:35–49, 1975.

See also F. M. Stewart and B. R. Levin, Partitioning of resources and the outcome of interspecific competition: A model and some general considerations. *Amer. Natural.*, 107:171–98, 1973.

limited by one or more of the vitamins, a green alga limited by nitrogen and another by carbon dioxide, making at least 6. Peterson points out that since oligotrophic waters are more likely to have all nutrients present in low concentrations, a greater number of co-occurring species of phytoplankton might be expected in them, even though the total biomass is small. Experience in general confirms this.

The nonequilibrium hypotheses are of various kinds. It was initially supposed that since the direction of competition is very generally subject to environmental control and the environment of the phytoplankton is changing in temperature, illumination and chemistry throughout the growing season, continual reversals in the direction of competition may be expected, so that no equilibrium is ever achieved. The hypothesis has been criticized by Klomp,[97] who points out that with sufficient time all but one species is likely to be eliminated. It is, however, extremely unlikely that any body of water constitutes a biologically closed system, so that continued random introductions are always occurring. Species that have been able to inhabit the locality in the past are more likely to undergo successful reintroduction than are species totally alien to the ecological conditions that the water provides, so that some continuity in the flora may be expected.

A more formal study of this kind of disequilibrium has been made by Grenney, Bella, and Curl,[98] who have made simulation models of competition between phytoplanktonic algae under variable nutrient input. Assuming that the various algae in nature have different optimal conditions which are fulfilled by the environment at different times, they conclude that provided the optimum for any species occurs sufficiently often and lasts sufficiently long, species may survive indefinitely in spite of competition from other species with different optima. Merrell's experiments with *Drosophila* provide an interesting analogy.

The field study already mentioned,[99] in Castle Lake, California, showed the rather surprising fact that the distribution of the phytoplankton of the lake is quite patchy. In more than half the taxa present, two statistical tests gave striking evidence of superdispersion. This suggests that small subpopulations can be built up by rapid local reproduction before the water is uniformly mixed by wind-driven currents and the associated turbulence. The idea that the open water of a lake or ocean is a spatially uniform environment thus appears to be false and some of the diversity found may be dependent on this heterogeneity.

Two other possible aspects of the matter must also be mentioned. As with all considerations of competition, the theory says nothing about the time that is needed for one species effectively to displace another. Riley[100] suggested that in the phytoplankton, natural selection has worked to increase the efficiency of all species more or less in the same way, so that although one species might ultimately replace any other, the replacements take so long as to be unobservable in practice. There are hints of this possibility in some of the computer simulations of Grenney, Bella, and Curl.

The second aspect that may bear on the

97. H. Klomp. The concepts of "similar ecology" and "competition" in animal ecology. *Arch. Néerl. Zool.*, 14:90–102, 1961.

98. W. J. Grenney, D. A. Bella, and H. C. Curl, A theoretical approach to interspecific competition in the phytoplankton community. *Amer. Natural.*, 107:405–25, 1973.

D. J. Merrell, see chap. 4, note 23.

99. See note 94.

D. Tilman (Resource competition between planktonic algae: An experimental and theoretical approach. *Ecology*, in press) has examined competition along a resource gradient which may be either in space or in time. Models permitting the coexistence of two species of diatoms can be made in the laboratory, as over a certain range of concentrations one species may be limited by one ion, the other species by the other ion. As has already been pointed out (p. 173), the number of species that can be packed along a gradient depends on its predictability. If the number is considerable and a finite time is needed for competition to run its course, a considerable nonequilibrium flora is possible at any given moment. Tilman's results suggest a combined equilibrium and nonequilibrium solution to the paradox.

100. G. A. Riley, Marine biology: I. *Proc. First Int. Interdiscipl. Congr.* Washington, D.C., Amer. Inst. Biol. Sci., 1963, 286 pp.; see pp. 69–70.

problem is the possibility that important species in the phytoplankton are actually planktonic representatives of benthic species which leave no significant body of descendants, all the competition ecology occurring on a diversified set of solid substrata.[101] There are suggestions of this process in the diatoms of the genus *Tabellaria* and in some of the blue-green algae. In such a case the development of phytoplankton blooms would be rather literally epiphenomenal.

It will be apparent that there are now a good many plausible explanations of the paradox of the plankton. Any or all of them may well be valid in particular cases. There is, however, no satisfactory body of empirical data that permits us to say in any given case which hypotheses are true.

It should be noted that at least one case has been recorded of two very closely allied species living together in a nonequilibrium association. This concerns the two mussels, *Mytilus californicus* and *M. edulis*, near Santa Barbara on the coast of California. Harger[102] has reported that the byssus threads of *M. californicus* are stronger than those of *M. edulis*. However, any *M. edulis* attaching in a clump of *M. californicus* is favored by being able to take up a better position for filtering so that in quiet water *M. edulis* displaces *M. californicus*. In some but not all winters, severe storms occur and when this happens, the clumps that survive are those that have fewest *M. edulis*. The coexistence of the two species therefore seems to depend on the greater strength of attachment of *M. californicus* and on the capacity of *M. edulis* to crowd out the latter. On a more tempestuous coast *M. californicus* presumably would persist, alone, while on a very calm coast only *M. edulis* would be found.

The possibility of symbiosis among closely related species

Since symbiosis or in some cases commensalism may permit two species basically using the same resources to coexist, it is worthwhile to enquire whether in nature any puzzling cases of coexistence may be explained in such a way.

Most conspicuous symbiosis is between very diverse kinds of organisms. One need only mention the various symbiotic algae found in many aquatic animals, the complicated relations between termites and intestinal flagellates, or the curious relationships between hermit crabs and sea anemones. It is probable that there are a number of unrecognized symbiotic associations yet to be discovered.

Some very curious examples involving species much less far apart than hermit crabs and sea anemones have been reported in recent years. The most extraordinary case appears to be that studied by Smith[103] in the colonial icterids, *Zarhynchus wagleri*, *Psarocolius decumanus*, and *Gymnostinops montezuma*, collectively known as oropendulas, and the cacique, *Cacicus cela*, which are all hosts in Panama of the nest parasite, *Scaphidura oryzivora*, the giant cowbird, which also belongs to the Icteridae.

The populations of the oropendulas and caciques appear to be divided into those that nest in close contact with colonies of the wasps, *Protopolybia* and *Stelopolybia*, and of

101. G. E. Hutchinson, The lacustrine microcosm reconsidered. *Amer. Scientist*, 52:334–41, 1964 (reprinted in *The Ecological Theater and the Evolutionary Play* [note 43], pp. 109–20).

102. J. R. E. Harger, The effect of species composition on the survival of mixed populations of the sea mussels *Mytilus californicus* and *Mytilus edulis*. *Veliger*, 13:147–50, 1970.

J. R. E. Harger, Competitive coexistence among intertidal invertebrates. *Amer. Scientist*, 60:600–07, 1972.

103. N. G. Smith, The advantage of being parasitized. *Nature*, 219:690–94, 1968.

Another extraordinary case, involving, however, quite unrelated animals, is recorded by C. E. Valerio, A unique case of mutualism. *Amer. Natural.*, 109:235–38, 1975. A spider *Achaearanea tepidariorum*, living in Costa Rica, is apparently benefited by the parasitism of a small wasp, *Baeus* sp., which develops in the spiders eggs in an egg sac. When the wasps emerge, they stimulate the surviving spiderlings of the brood to intense activity which culminates in cannibalism. Spiderlings which come from uninfested batches of eggs and exhibit no cannabilism have an extremely low expectancy of reaching maturity, the losses imposed by parasitism and cannibalism being much more than compensated by the greater vitality of the survivors.

the stingless, but powerfully mandibulate bees of the genus *Trigona*, and those which have no such association. When the wasps or bees are present, cowbirds are deleterious to the oropendulas and lay mimetic eggs. Where there are no bees, the cowbird eggs are not mimetic but are seemingly tolerated. In such cases a botfly of the genus *Philornis*, which is apparently kept off by the bees and wasps, is apt to lay its eggs on any chicks in the oropendula and cacique nests, so that the nestlings are parasitized. The oropendula and cacique chicks cannot divest themselves of the parasites, but those of the cowbirds can do so, and, moreover, destroy any larvae on a neighboring chick in the nest, whether of their own kind or a nestling of the hosts. As a result, though the host cannot rear a full clutch of its own species in addition to the chick of the cowbird, there is a decided advantage to the host chick in being brought up with a cowbird chick if there are no bees and wasps present to keep off ovipositing *Philornis* females.

A remarkable situation has been studied by Brown,[104] who described a colony of six pairs of eagles, all of different species, nesting quite close together on a single mountain top in Kenya, with two pairs of another species lower down the slope. Other hilltops in the vicinity were without eagles, so the observed distribution cannot have been due to crowding. The most reasonable explanation is that such an arrangement restricts competition to species with but partial overlap in the type of feeding territory preferred and the kind of food eaten. Two pairs of the same species within a hunting territory might compete intraspecifically quite seriously, but several pairs of different species would be able to occupy the area utilizing its resources without serious mutual interference. The mountain would thus appear to be a good place for eagles, the different species attracting each other, but intraspecific territoriality would prevent two pairs of the same species from competing. This arrangement insures maximum utilization of resources, but a minimum of competitive interference.

In the guild of grazing mammals inhabiting the open areas of tropical East Africa, part of the possibility of coexistence appears to depend on the effect of one species in preparing the pasture for another. In the Rukiwa Valley in Tanzania, Vessey-Fitzgerald[105] found that the trampling of the tall and seasonally aquatic grass *Vossia* by elephants and buffalo led to the production of a short sward by the plant which provided food for zebra (*Equus burchelli*), topi (*Damaliscus korrigum*), and in wetter areas puku (*Adenotus vardoni*). Moreover, in other areas, notably Serengeti National Park, it has been observed[106] that zebras lead the grazing succes-

104. L. H. Brown, On the biology of the larger birds of prey of the Embu district, Kenya Colony. *Ibis*, 94:577–620 and 95:74–114, 1952–53. See also *The Ecological Theater and the Evolutionary Play* (note 43), pp. 71–73.

105. D. F. Vesey-Fitzgerald, Grazing succession among East African game animals. *J. Mammal.*, 41:161–72, 1960.

106. M. D. Gwynne and R. H. V. Bell, Selection of vegetation components by grazing ungulates in the Serengeti National Park. *Nature*, 220:390–93, 1968.

The sympatry of zebras and antelopes was noticed in a footnote by Prince Peter Alekseevitch Kropotkin (1842–1921) in his *Mutual Aid: A Factor in Evolution* (London, Heinemann, 1902, xix, 348 pp.), a book written primarily to modify then-current Darwinian ideas about intraspecific competition. The work is in fact an early attempt at constructing a sociobiology, though one with strong political presuppositions. Kropotkin was a distinguished explorer and geomorphologist who became a philosophic anarchist, ultimately finding asylum in England, and late in life protesting, in a letter to Lenin, the conduct of the Russian Revolution. *Mutual Aid* has been reprinted many times; the 1955 edition (Boston, Extending Horizon Books) has an interesting introduction by Ashley Montagu. Kropotkin noted that the quagga and dauw zebra did not associate, but that the former fed with various other ungulates. His dauw zebra is presumably *Equus z. zebra*, the mountain zebra, still surviving in South Africa in a very restricted mountainous area in the Caledon district. Since the word *dauw* appears to be Hottentot, this attribution is doubtless correct, though dauw may have been used by the Hottentots for both the quagga and the mountain zebra (R. Lyddeker, *The Horse and Its Relatives*. London, G. Allen, 1912, xii, 286 pp.) and has also been employed, as in the New Oxford Dictionary, for the allopatric *E. burchelli*. The distributions are given in Austin Roberts, *The Mammals of South Africa*. Central News Agency, S. Africa, 1951, xlviii,

sion and are followed by wildebeest and various gazelles. In such a case the zebras eat more grasses and the ruminants more dicotyledonous plants, which, contrary to expectation, appears to be physiologically optimal for these animals.

A case of very closely allied species is provided by the three 17-year cicadas, *Magicicada septendecim*, *M. septendecula*, and *M. cassini*, and the allopatric vicariant races or species of 13-year cicada *M. tredecim*, *M. tredecula*, and *M. tredecassini*.[107] Within each sympatric group the three species have slight ecological preferences, so that they tend not to occur completely syntopically, but they are snychronized so that the main broods emerge and sing simultaneously, swamping their predators and apparently reinforcing each other's sound to discourage, if not deafen, the latter.

Among plants there is a suggestion of mutualism between co-occurring species of *Eucalyptus* in Australia,[108] one species having and the other lacking a mycorhiza.

In all these cases the value of the symbiosis in the utilization of somewhat different niches in the same sort of way as that in which polymorphs, whether sexual or not, can occupy different niches. Most usually such utilization leads to some exclusion at the competitive niche boundaries; in the symbiotic cases, this may also occur, but it is more than mitigated by the special kinds of benefit each species receives from the other. In the polymorphic cases, interbreeding permits any advantage gained by one parent to increase the fitness of the species. The symbionts use each other in much the same general way as the population of any species takes advantage of what it can find in its environment. When the advantages taken are mutual, the stage is set for at least a certain amount of co-evolution. In general this will not be toward greater similarity of the two species, but rather greater differentiation.

SUMMARY

The concept of the ecological niche seems to have had two origins. In America, Johnson transitorily, and later Grinnell, regarded the niche as a unit to which a single species is adapted, though it was never quite clear what sort of unit this was. In England, Elton saw the niche largely as the functional role of an animal in the environment. Gause adopted the niche from Elton, but interpreted it, quite independently, somewhat in Grinnell's sense. Grinnell's axiom that two species did not occupy the same niche came to be rather generally accepted on account of the mathematics of Volterra, Lotka, and Gause and Witt, and of the experiments of Gause, Park, Crombie, and their successors, without there being a very clear idea as to what the thing was in which two closely allied species did not coexist. An attempt to define a niche more rigorously as a region of a multidimensional niche space, the axes of which represented all the possible environmental variables, stimulated a good deal of work. Though the idea of a many-dimensional niche may at first seem terrifying, division of each axis into 10 steps would provide, if there were three axes, 1,000 orthogonal niche spaces one step wide, and if there were five axes, 100,000 such niches. In most cases, in dealing with a limited part of the biota, two or three dimensions are likely to be sufficient.

The axes clearly are ideally classified as of

---

700 pp. *E. quagga* originally lived in the lowland areas surrounding the mountain habitats of the smaller *E. zebra*. Kropotkin never visited South Africa and seemingly did not know that the quagga was extinct when he wrote. His remarks suggest that Clive Phillips-Wolley's *Big Game Shooting* (2 vols., Badmington Library, London: Longmans, Green, 1894), which I have not been able to consult, may contain early information about coexistence of grazing ungulates in East Africa.

107. R. D. Alexander and T. E. Moore, The evolutionary relationships of 17-year and 13-year cicadas, and three new species (Homoptera, Cicadidae *Magicicada*). *Misc. Publ. Mus. Zool. Univ. Michigan*, 121:1–59, 1962.

M. Lloyd and H. S. Dybas, The periodical cicada problem: I. Population ecology. II. Evolution. *Evolution*, 20:133–49; 20:466–505, 1966.

108. L. D. Pryor, Species distribution and association in Eucalyptus. In: *Biogeography and Ecology in Australia*. Monographiae Biologicae Den Haag, W. June, 1959, 640 pp.; see pp. 461–71.

two kinds, those such as temperature that set the stage but do not involve biological interactions such as competition, and those that involve resources for which there may be competition. This distinction, though valid and useful, is sometimes difficult to apply when what appears to be a *scenopoetic* or stage-setting axis turns out to measure indirectly some nutrient or other *bionomic* variable. When a niche is defined primarily without reference to competitors, but merely in terms of requirements and tolerances, it is *fundamental* or *preinteractive*. When a second competing species has been able to take over a part of the niche, the remainder is called *realized* or *postinteractive*. In some cases the second species takes over a small, but optimal corner of the fundamental niche of the first species. This is described as an *included* niche. In general, the species with the included niche may be regarded as highly *adapted* to the conditions defined by this niche, while the original species with much greater tolerances and so a high *adaptability* can continue to exist suboptimal parts of its now *concave* niche.

If it be assumed that competing species are about equally efficient in their optimal habitats, it is reasonable to suppose that the amount of competition is roughly proportional to the time spent in interaction. It is thus possible to estimate the α_{ij} of the generalized Volterra equation (4.6). For a number of interacting species in a guild, a matrix can be written, formed of all the α_{ij} of the members of the guild. This is usually called the *community matrix*, though it could only refer to a whole community of the simplest kind, such as that of cave crustacea. It is possible to learn from the mathematical properties of the matrix something of the ease or difficulty of invasion by another species. At least in the case of a well-studied cave-crustacean community, it seems likely that no further species could be added without a diversification of resources.

The term *niche breadth* is used to indicate the extent of the occupation by a species along a niche axis. If p_{ih} is the probability that a species i is found over a small length h of the axis, the niche breadth is defined by $B = 1/\Sigma p_{ih}^2$. If there are n sampling points or intervals and the species is equally common in all, the niche breadth will be n. To compare different species it is best to compare the ratio of the observed value of B to this maximal value.

Where there is a great deal of information about the way of life of a particular species, so that one can be fairly certain that observations of time spent in a particular part of the habitat really represent the totality of the activities of the organism, or the niche in Elton's sense, B/n may be regarded as a relative measure of *niche size*.

In the case of the five warblers studied by MacArthur, B/n was found to vary from 0.20 for the Cape May warbler to 0.58 for the bay-breasted.

In many cases all that the measure can give is an indication of what has been called *habitat tolerance* or the range of habitats possible for a species. In the enormously rich bird fauna of the Amazonian rain forest, areas of about 0.1 km^2 may have more than 100 species. Of the species with the greatest habitat tolerance, where B/n is in excess of 0.39, the majority appear to be birds that are fairly specialized in their food habits, but capable of using their specializations in any type of forest. This applies, for instance, to fruit eating or to habitually feeding on prey disturbed by columns of army ants.

Much more elaborate statistical ways of studying niche breadth have been invented, but for the moment the published biological results of such an approach do not seem commensurate with the complexities of the technique.

The phrase *species packing* refers to the number of species that a given niche hyperspace can contain. If the resources are strictly determined, there is no theoretical limit to the fineness of division of the space, but if there is variation in the quantity of resources, the amount of species packing is limited in a way that is very insensitive to the variance of the resources. If one takes some normally distributed function which measures the effectiveness of resource utilization, the permissible niche breadth for each closely packed species is found to be of the same order of magnitude as the standard deviation of this resource utilization function. In a very simple case, if two species habitually eat seeds of different sizes, the mean difference permitting coexistence would be not greatly different from the standard deviations of the seed sizes. What little empirical evidence exists seems to suggest that the theory is likely to be correct. A complication, however, is ordinarily introduced by the fact that it is easier to eat a suboptimally than a superoptimally sized piece of food. The most remarkable case of size being correlated with food niches is found in the fruit-eating pigeons of New Guinea among which eight sympatric species may be so separated. Here the smaller forms have the advantage of being able to perch on small branches, so that the asymmetry just mentioned is largely corrected. Usually allied sympatric species of the same guild differ in size by linear dimensions of 1.2 to 1.4, or roughly $3\sqrt{2}$; in the fruit-eating pigeons, however, the difference is smaller than this.

Though usually one species lives in one niche, there are cases where one species, by sexual or other types of polymorphism, does in fact occupy two or more niches. Cases are known in birds of great sexual dimorphism in beak form and trophic musculature, leading to quite different feeding habits in the males and females. Some of these occur in insular faunas in which few species are present, and the niche breadth is consequently liable to increase. Size differences may permit niche separation in birds of prey and carnivorous mammals. A few cases involving aposematic and cryptic coloration as alternative adaptions in insects are known to involve sexual dimorphism of such a kind that rapid changes in proportion of cryptic and aposematic or apostatic forms would be possible when conditions change. In dry climates male plants may be more xeric and female plants more mesic, differences which probably promote wind dispersal of pollen and a good set of seeds. In some fishes nonsexual polymorphism of a profound kind, putting the various forms of a species in quite different trophic niches, has recently been described.

A quite different geometrical method of treating the competitive aspects of the niche has been introduced by MacArthur and Levins. In general, as many species can coexist as there are discrete kinds of resource, though the various species may utilize mixtures in which different resources are present in different proportions. When competition involves species using similar mixtures, selection favors the specialization of species feeding on the same general resources but living in slightly different habitats. Such species are *fine-grained.* When there is a considerable initial difference, selection favors the accentuation of the differences, so that the species tend to use only their own preferred resource and are *coarse-grained.* Comparison of the animals of higher taxa such as orders or families often indicates fine-grained and coarse-grained component groups.

In some animals, most noticeably birds, but also lizards and probably many insects, the structure of the vegetation cover of an area is clearly an immensely important niche component. MacArthur and his associates pointed out that if we have three measures of the amount of foliage essentially in the herbaceous, shrubby, and arboreal layers, they permit prediction in a well-studied area of the species of birds that will be breeding on any terrain. There is some overlap, probably involving a trophic factor, such as usual height of feeding, but it seems likely that in most cases niches could be defined by the three foliage parameters and one other factor. Such a four-dimensional niche, however, is hardly primary. The foliage diversity may tell the bird, if not the ecologist, that the things that it wants will be there. Moreover, if we measure foliage height diversity and bird species diversity, they are linearly related in deciduous woodland, the slope of the line being the same in eastern North America and Australia. However, it seems likely that the tropical rain forest has a more complex structure both in itself and in its bird fauna. In most cases the species composition of the woodland is not important. This approach has been developed elaborately for grassland, and here again widely separate areas with unrelated birds are very similar in the diversity that they can maintain.

Various statistical methods (multivariate analysis, principal component analysis) can be used to study the structure of an array of points in *n*-dimensional space. These methods permit the minimum number of independent significant orthogonal axes to be recognized and so permit recognition of the minimum number of independent niche dimensions. Methods of this sort have been widely used in plant ecology and are beginning to be of importance in limnology. The axes that can be isolated as significant and independent may not be easily associated with what one would expect to be controlling factors.

Maguire has pointed out that in a multidimensional niche, if the center describes optimal conditions and the periphery conditions in which survival is just possible, as a population develops growing in a more or less logistic manner, the state of the habitat corresponding to any point in the niche will change. The increasing population reduces the favorableness of the habitat until finally all the occupied part corresponds to a point or set of points on the periphery of the niche. Any initially uniform area in the habitat is thus represented by a point describing a trajectory ending at the periphery of the niche. A number of such trajectories would ordinarily start from points representing a number of dissimilar regions in the habitat and might move to different peripheral regions of the niche. Competition can be described by the behavior of such trajectories.

A further development of niche theory has been the separation of the variables, so that the complete hyperspace, termed an *ecotope* by Whittaker, Levin, and Root, is resolved into a *niche*, defined as within a community, and a *habitat*, involving gradients along spatial axes on which numerous communities may have developed.

When we examine the actual situation in nature where closely allied species appear to be co-occurring, it is usual to find that the co-occurrence does involve differences among the species that would be of the kind expected from the theory of competitive exclusion. A very striking case is provided by two species of insects of the genus *Mesopsocus* living on algae and fungal spores on bark, but differing in their preferences for a limited area of egg-laying sites. In a few cases it is possible to go further. In a guild of seven species of leafhoppers of the genus *Erythroneura*, survival is always greater when a population consists initially of two different species than when it consists initially of the same number of conspecifics. The difference is statistically significant only in about half the cases studied, but no cases of the reverse situation were found. Though more work is needed, a distribution of this sort can give valid evidence on the hypothesis of niche separation, for if there were no such separation, the relative survival should either not differ with the initial kinds of population, or there should be some cases of better intraspecific than interspecific survival. Though often small differences of behavior as in *Mesopsocus* or in two well-studied species of water mite may permit coexistence, other cases are known in which apparently closely related species have developed very different types of life history. This is apparent in a quite complicated taxocene of limpets on the South African coast.

There is a good deal of evidence that predation facilitates the coexistence of competitors in nature, as it does in theory and in the laboratory. The possibility of some associations never being in equilibrium still exists, though it has been criticized as inevitably leading to chance extinctions. A number of hypotheses, some involving symbiosis, some limitation by a number of different factors, one for each species, and some dependent on spatial or temporal disequilibrium, have been put forward to explain the paradox of many species of phytoplankton living in the same way in the same volume of water. Such hypotheses are hard to test even if, between them, they explain the observations.

Though chemical mutualism, with two species each synthesizing a vitamin that the other cannot make but needs, might explain some coexistence in the phytoplankton, the closely related sympatric symbiotic species that have been described live in different niches and in effect contribute to the breadth of each other's niche by the symbiotic relationship.

Chapter Six

How Is Living Nature Put Together?

WE have so far studied individual populations, and the interaction of populations when they are in competition, or more rarely, in symbiosis. These possibilities obviously do not exhaust what we are likely to find in nature.

The concept of the food chain

Any animal population must have food, and this food must come either directly from a population of plants, in a living or sometimes dead and decomposing form, or from other organisms that have lived on such plants. The interrelations of the various trophic levels, the autotrophic plants feeding on simple chemical compounds and sunlight, the primary consumers or herbivores, the secondary, tertiary, and higher levels of consumers or carnivores, the parasites, and the decomposers living on dead organisms larger than themselves, returning material to be used again by autotrophic plants, form a complicated food web. This food web can be treated generally in several ways.

Alfred Russel Wallace,[1] in his part of the joint communication on natural selection to the Linnaean Society of London in 1858, wrote: "The general proportion that must obtain between certain groups of animals is readily seen. Large animals cannot be so abundant as small ones; the Carnivora must be less numerous than the Herbivora; eagles and lions can never be so plentiful as pigeons and antelopes."

The same idea was put forward independently and much later by Elton.[2] His arguments were very simple and so fundamental. It is first usually apparent that small animals are commoner than large. The small animals are able to reproduce more rapidly than the large and produce more descendants than are needed to replace them. Ordinarily a predator must be stronger, if not larger, than its prey. The predators in an equilibrium community, subsisting on the surplus production of the herbivores, will therefore be fewer in number than the latter and generally larger in size. The second level of predators will be rarer and larger than the first and so on until a point is reached at which the carnivore is too rare to support any further predator. Such an arrangement was called by Elton a *food chain*. Elton quotes a Chinese proverb that "the large fish eat the small fish; the

1. A. R. Wallace, On the tendency of varieties to depart indefinitely from the original type. In: *On the tendency of species to form varieties; and on the perpetuation of varieties and species by natural means of selection,* by Charles Darwin . . . and Alfred Wallace. . . . *Proc. Linn. Soc. Lond.*, 3:53–62, 1858.

2. Charles Elton, *Animal Ecology*. London, Sidgwick and Jackson, 1927, xx, 207 pp.

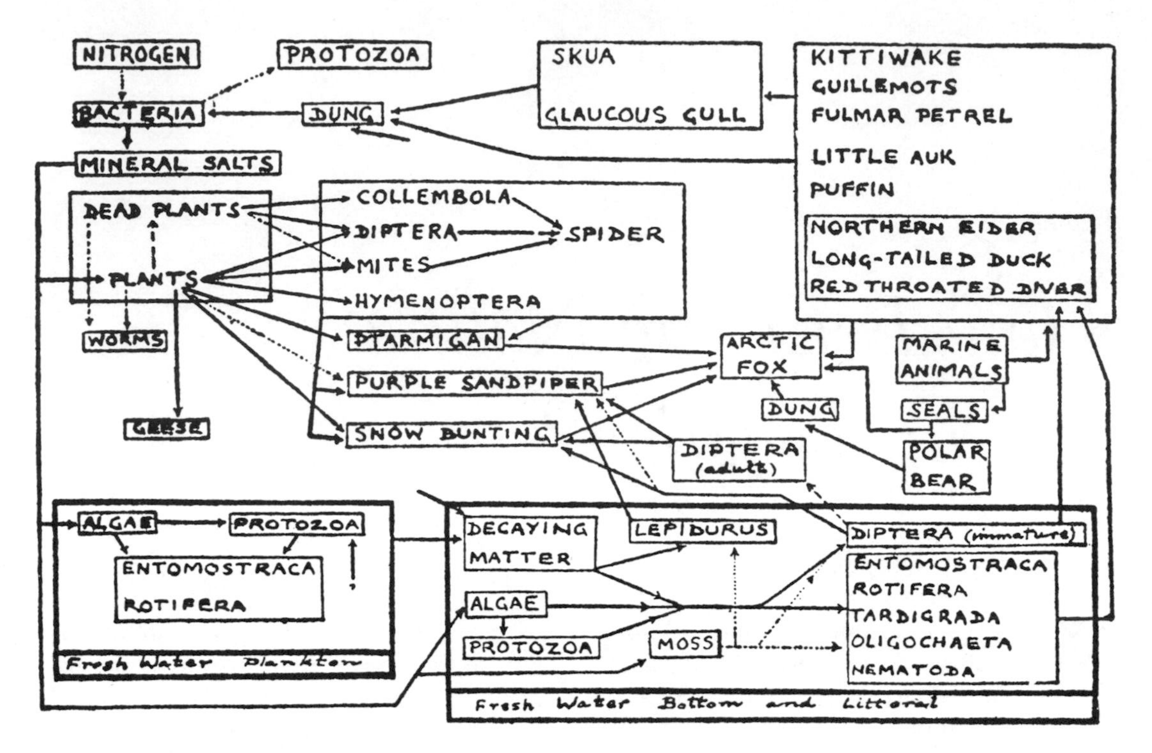

FIGURE 130. Elton's diagram of the food cycle on Bear Island in the Barents Sea.

small fish eat the water insects; the water insects eat plants and mud." He also points out that a second kind of food chain with its pyramid can also be observed among parasites, corresponding at least to the jingle, "Big fleas have little fleas upon their backs to bite 'em, and little fleas have lesser fleas, and so *ad infinitum*."

Elton used the term *food cycle* to indicate the sum of all the food chains in a community. It is clear that he had been thinking about the nitrogen cycle, which was becoming used in textbooks, in a rather stereotyped form, a little before he wrote. The diagram (figure 130) for the known and suspected food relations on Bear Island, south of Spitsbergen, which is labeled in *Animal Ecology* as "Food cycle among the animals on Bear Island," appears in Summerhayes and Elton's original paper[3] of 1923 as a nitrogen cycle, which in a way it is, even though no nitrogen compound other than atmospheric nitrogen is specifically mentioned.

Clements and Shelford[4] rather characteristically replaced Elton's use of the food cycle by *food nexe*, which I have never read or heard anyone else use. The contemporary term in English is *food web*.

Elton and very many other investigators have given detailed diagrams of the food webs in particular habitats. In order to investigate certain properties of such complex systems further, some degree of simplification may be imposed on them intellectually, though the basic complexity must always be kept in mind whatever we do.

The system may be thought of as composed of a series of levels. Each level will usually consist of many species, mutually competing in the ways studied in the two previous chapters. All the green plants, as photosynthetic producers, are included in the first level (Λ_1), all the herbivores in the second level (Λ_2), all the carnivores feeding on her-

3. V. S. Summerhayes and C. S. Elton, Contributions to the ecology of Spitsbergen and Bear Island. *J. Ecol.*, 11:214–86, 1923. The diagram of the "Nitrogen Cycle" is figure 2 on p. 232. There is a statement of the size pyramids in food chains on p. 233.

4. F. E. Clements, and V. E. Shelford, *Bio-Ecology*. New York, John Wiley, 1939, vi, 425 pp.

bivores in a third level (Λ_3), and so on. We can now leave, for a short time, the realm of population ecology, which is *merological*, dealing with collections of distinct parts or individuals whose properties largely determine those of the collections, and enter that of biogeochemistry, which is *holological*, and in which we measure the transfer of matter and energy through the various levels of a food web without considering the individual organisms that compose it.

If we now look at Elton's pyramid we can see that though ordinarily the first or producer level Λ_1 will consist of a greater quantity of matter than the herbivores of the next level Λ_2, this is not really necessarily true. If Λ_1 is composed of sufficiently small and rapidly growing organisms that can capture energy fast enough, the first level might maintain fairly slow-growing and not too rapidly metabolizing herbivores, the total mass of the latter Λ_2 being greater than the total mass of the producers. This has been claimed to be the case in various relatively unproductive freshwater lakes in which the ratio of zooplankton volume to phytoplankton volume seems to be perennially considerably greater than unity. We can, however, in such a case be quite sure that in an equilibrium community the mean rate at which energy is transferred from the first level to the second is less than the rate at which it enters the first level as solar radiation, and greater than the rate at which it enters the third level, as food derived from the second level and so on.

We may write

$$\frac{d\Lambda_n}{dt} = \lambda_n + \lambda'_n \qquad (6.1)$$

where λ_n is the rate at which energy enters Λ_n from Λ_{n-1}, while λ'_n, which is negative, is the rate at which energy is dissipated metabolically or returned to the environment as dead organisms, feces, and the like, or passed onto Λ_{n+1} as food.

At equilibrium,

$$\lambda_n = -\lambda'_n \qquad (6.2)$$

and

$$\ldots \lambda_{n-1} > \lambda_n > \lambda_{n+1} \ldots \qquad (6.3)$$

The last expression is in fact a generalization of the Eltonian pyramid, which thus appears as a commonsense extension from the second law of thermodynamics.

The energetics of ecology were developed by Lindeman[5] in this form in a classical paper published in 1942. Lindeman realized that ordinarily the efficiency of the first step in the process, which can be symbolized as λ_1/λ_0 where λ_0 is the flux of solar radiation, is very small, of the order of 0.*n* percent, while in the transfer of material from the producer plants to the herbivores and onto the various levels of carnivores, i.e., λ_n/λ_{n-1}, values of 5–15 percent would be reasonable. Lindeman suspected the efficiencies to rise as *n* increased, but this now seems doubtful. Slobodkin gives reasons for thinking that the upper limit of such efficiencies is likely to be about 15 percent.

We might picture an ideal food chain by supposing that the members of each level were on an average 10 times as heavy, or about 2.15 times as long as the level below, that the efficiency of transfer was 10 percent, and that the energy flux through each level was proportional to biomass. If we started with a soupy suspension of 1,000,000 phytoplankton cells per milliliter of water in a very eutrophic lake, we could have 10,000 herbivores (Λ_2) but only 1 second-level carnivore (Λ_4) per milliliter. There would be 1 fifth-level carnivore (Λ_7) per cubic meter

5. The notation used was introduced in a mimeographed set of notes: G. E. Hutchinson, *Lecture Notes on Limnology*. New Haven, Osborn Zoological Laboratory of Yale University, 1941. This work had a small distribution beyond the classes for which it was prepared. It is referred to by Lindeman as "Recent Advances in Limnology (in manuscript)."

R. L. Lindeman, The trophic-dynamic aspect of ecology. *Ecology*, 23:399–418, 1942. Lindeman had used similar ideas in his treatment of the Eltonian pyramid. The genesis of Lindeman's classic paper has been considered by R. E. Cook, Raymond Lindeman and the trophic-dynamic concept in ecology. *Science*, 198:22–26, 1977.

L. B. Slobodkin, Ecological Energy Relationships at the Population Level. *Amer. Natural.*, 94:213–36, 1960.

and 1 tenth-level carnivore (Λ_{12}) in 10 cubic kilometers. The entire hydrosphere with a volume of about 1.4×10^{24} cm^3 would accommodate with a little room to spare a single individual of the fifteenth-level carnivores (Λ_{17}).

Actually in this example the size factor between the levels is probably too small. The Worthingtons[6] give an interesting case of a food chain in Lake Albert. Here detritus formed from dead plankton is eaten by molluscs, which are the food of a small fish, *Haplochromis albertianus*, not more than 10 centimeters long. These fish are caught by the local fisherman by putting out bundles of brushwood or grass in which they hide. They are then used as bait to capture tiger fish, *Hydrocyon*, which may be up to $\frac{1}{2}$ meter in length, and this is used as bait for the very large *Lates albertianus*, which may be when mature from 1 to 2 meters in length. A five-link food chain of this sort is likely to be present in well-developed freshwater communities and slightly longer chains may occur in the sea.

In all cases it must be remembered that food chains are not discrete entities. Species may alter their position as they grow, so blurring the chain, a process that has been called *metaphoetesis*.[7] There is evidence that cannibalism on the young by larger adults may be an adaptive mechanism by which surplus young are set to work collecting food items too small to be taken by the adults.

In the aquatic systems that we have just noted, one would find the herbivores of the free water consisting of small microphagous animals, ciliates, rotifers, cladocera, small copepods, and in the ocean also innumerable larvae of larger organisms. The first carnivore level would contain many species belonging to the same higher taxa. Some plant material would always be falling to the bottom, there to be consumed in a more or less decomposed state by a great variety of animals of very diverse sizes feeding on detritus, while in the sea a great variety of benthic microphagous animals feed by filtering or sedimenting the plankton of the water that passes through or over them.

On land the situation is quite different. Though damp soil can support a rich algal flora, it inevitably forms an essentially two-dimensional community; anything that can grow up into the air, whether a few centimeters or 100 meters, can shade and so outcompete such an algal flora. The land plants on which nearly every herbivore depends are macroscopic. There is no need for the herbivores to be larger than their food, for such plants seldom have active mechanical means of flight or of fighting off a predator, but rather rely on passive structural or chemical means of defense. Many herbivores, such as the ungulates, are in fact large, but vastly more, such as the insects, are small, though still macroscopic. Only in the detritus layer, which on land we speak of as soil, is there a great variety of microscopic species.

The relative large size of most terrestrial herbivores, when compared to the planktonic herbivores of lakes and oceans, tends to shorten terrestrial food chains at their lower ends. At the upper end the development of homoiothermy, which is expensive, has a comparable effect. There are of course homoiotherms, notably seals, whales, and porpoises, in the higher levels in the ocean, but they are accompanied by many more large poikilotherms, particularly fishes but also giant squids, in the sea. In both ocean and on land, there is a marked tendency for the food chains to become shortened when they have warm-blooded species at their upper ends. The largest terrestrial animals, whether sauropod dinosaurs, or mammals such as *Baluchotherium* and modern rhinoceroses, or the fossil and living elephants were or are all herbivores. In the ocean the largest forms are not herbivores, but cases such as that of a whalebone whale feeding on *Calanus* probably belong in the third level of the web. The Sirenia or seacows of course supply a moderate-sized marine ana-

6. S. and E. B. Worthington, *Inland Waters of Africa*. London, Macmillan, 1933, xix, 259 pp.

7. G. E. Hutchinson, Homage to Santa Rosalia; or, Why are there so many kinds of animals? *Amer. Natural.*, 93:145–59, 1959.

logy to the large terrestrial herbivorous mammals.

Loop analysis
Although the holological approach developed by Lindeman threw much light on the problems of efficiency in nature, it must never be forgotten that each level is complicated, consisting of many species, which may often be competitors, coexisting by niche separation, frequency-dependent competition, or varying predator pressure. In any level the properties of its component parts may be expected to determine the properties of the whole, but not necessarily in obvious ways. Moreover, the levels are not really discrete. Any carnivore is likely to grow from one level to another. Large predators, though they may reach down into a lower level less efficiently than do small, nevertheless usually have a potentially wider selection of food. No very simple theory of any complicated food web therefore can be expected.

The questions posed by food webs and food chains are essentially those of complexity and of stability. Complexity can be increased in at least two ways, by the increase in species packing along a resource axis of niche space, and by an increase in the length of the food chains of which a web is composed.

At any stage in the evolution of any guild under consideration, species packing is, as we have seen, primarily determined, in a variable environment, by the variance of the resource-utilization function in formal terms, or more simply by the range of possible kinds of resource that the morphology and physiology of an animal permit it to utilize.

The lengths of food chains are presumably determined partly by the interaction of two opposing tendencies. The species at the top is by definition predator-free, but relative to the primary production in any habitat it is the least efficient. There is therefore, as we have noted, a counterselection toward the shortening of food chains, notably if homoiothermal species are involved. Though a great deal can be done by simulating actual food webs, composed of species that have ascertained properties, in computer models, the only convenient formal general theory which has been proposed for the study of food webs is that of loop analysis, introduced into population genetics many years ago by Sewall Wright[8] and developed very recently by Levins[9] as a powerful method in population ecology.

If we take the simplest possible situation with one species S_2 feeding on another S_1, we may write

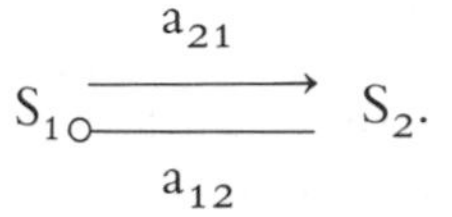

$$S_1 \underset{a_{12}}{\overset{a_{21}}{\rightleftarrows}} S_2.$$

Here the upper arrow indicates that the interaction of the two species, as measured by the community matrix coefficient a_{21}, promotes an increase in S_2, by its feeding on S_1, while the line ending with a circle implies that the opposite interaction promotes a decrease of S_1. In this example no provision is made for self-limitation of either organism. Ordinarily at least the lower-level species will be self-limiting by intraspecific competition; thus we write as

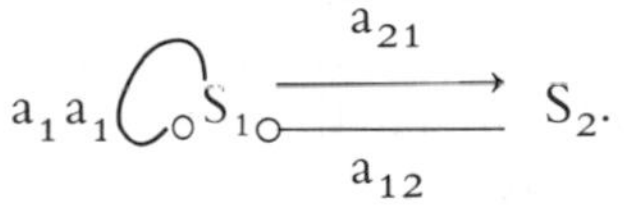

$$a_1 a_1 \circlearrowleft S_1 \underset{a_{12}}{\overset{a_{21}}{\rightleftarrows}} S_2.$$

If we now add a third species simply feeding on S_2, maintaining the self-limiting nature of the herbivore S_1, we have

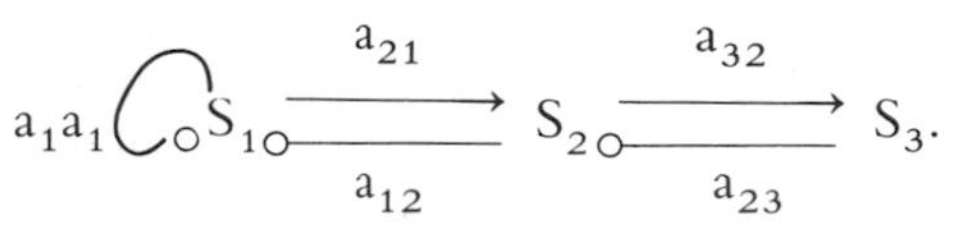

$$a_1 a_1 \circlearrowleft S_1 \underset{a_{12}}{\overset{a_{21}}{\rightleftarrows}} S_2 \underset{a_{23}}{\overset{a_{32}}{\rightleftarrows}} S_3.$$

All the interactions can be expressed as a community matrix **A**: In the first case, with no self-interaction

8. S. Wright, Correlation and causation. *J. Agric. Res.*, 20:557–85, 1921.

9. R. Levins, Evolution in communities near equilibrium. In: *Ecology and Evolution of Communities*, ed. M. L. Cody and J. M. Diamond. Cambridge, Mass. Belknap Press of Harvard University Press, 1975, pp. 16–50.

$$\mathbf{A} = \begin{bmatrix} 0 & -a_{12} \\ a_{21} & 0 \end{bmatrix}. \qquad (6.4)$$

In the second

$$\mathbf{A} = \begin{bmatrix} -a_{11} & -a_{12} \\ a_{21} & 0 \end{bmatrix}. \qquad (6.5)$$

In the third

$$\mathbf{A} = \begin{bmatrix} -a_{11} & -a_{12} & 0 \\ a_{21} & 0 & -a_{23} \\ 0 & a_{32} & 0 \end{bmatrix}. \qquad (6.6)$$

Any real system is of course likely to be much more elaborate; some examples of the kinds of elaboration to be expected are given below. The mathematical treatment is initiated in the following way.

Each species or resource occupies a point or vertex in a diagram of the kind given in figure 131, in which three competing species, all self-limited by intraspecific competition and all competing with each other, are shown. The matrix is

$$\mathbf{A} = \begin{bmatrix} a_{11} & a_{12} & a_{13} \\ a_{21} & a_{22} & a_{23} \\ a_{31} & a_{32} & a_{33} \end{bmatrix}. \qquad (6.7)$$

The determinant of this matrix is $a_{11}a_{22}a_{33} - a_{11}a_{23}a_{32} - a_{12}a_{21}a_{33} - a_{13}a_{22}a_{31} + a_{12}a_{23}a_{31} + a_{13}a_{21}a_{32}$. This quantity is derived by multiplying the terms on the diagonal, which represent self-loops of length 1; then permuting the second subscript of a pair of terms, at the same time changing the sign, giving the three negative terms; and then performing a double permutation, again changing the sign twice, which gives the last two positive terms; all $3! = 6$ possible permutations are thus represented. This standard procedure is seen to produce the product of three loops of length 1, the products of each self-loop with its disjunct loop of length 2, and the two loops of length 3 going round the triangle clockwise and counterclockwise.

The determinant may be written

$$D_n = \Sigma(-1)^{n-m}\, L(m,n), \qquad (6.8)$$

where $L(m,n)$ is the product of n links of the form a_{ij} which constitute m disjunct loops, or loops not sharing a vertex.

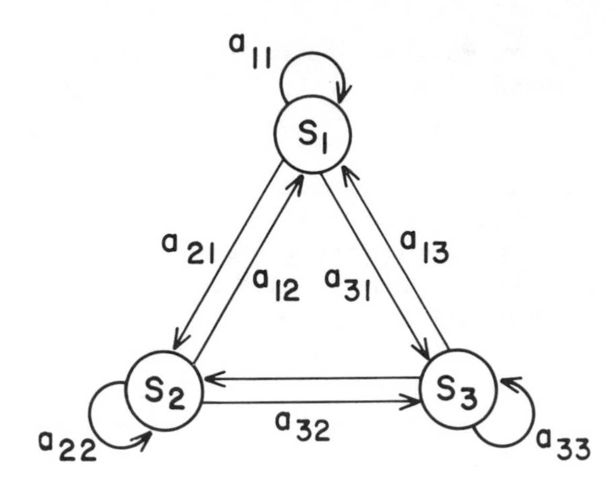

FIGURE 131. Loop diagram of three self-inhibiting interacting species (Levins, modified).

For all possible products involving k vertices we may define

$$F_k = \Sigma(-1)^{m+1}\, L(m,k). \qquad (6.9)$$

This quantity is termed the *feedback at level k*. The term $(-1)^{m+1}$ assures that if m is odd, the feedback will be positive; if even, negative.

The feedback at level 1, which consists solely of the self-loops is

$$F_1 = \Sigma a_{ii}. \qquad (6.10)$$

The feedback at $k = 2$ is

$$F_2 = \Sigma a_{ij}a_{ji} - \Sigma a_{ii}a_{ji}. \qquad (6.11)$$

The feedback at $k = 3$ is

$$F_3 = \Sigma a_{ii}a_{ji}a_{kk} - \Sigma a_{ii}a_{jk}a_{kj} + \Sigma a_{ij}a_{jk}a_{ki}. \qquad (6.12)$$

It can be shown that all the feedback terms in a stable system are negative; furthermore, that negative feedback from the long loops can promote instability if too great. For three vertices, or species, this second condition is given by

$$F_1F_2 + F_3 > 0, \qquad (6.13)$$

or that the positive product of the feedback from the short loops must be algebraically greater than the negative feedback from the loops of length 3.

In the case given in equation (6.6), for a predator, herbivore, and self-limited plant,

$$\begin{aligned} F_1 &= -a_{11} \\ F_2 &= -a_{12}a_{21} - a_{32}a_{23} \\ F_3 &= -a_{11}a_{23}a_{32} \\ F_1F_2 + F_3 &= a_{11}a_{12}a_{21} > 0, \end{aligned}$$

so that both criteria for stability are fulfilled.

It is possible to go very much further than this, as Levins has shown, by considering what happens as changes are introduced into the values of the coefficients a_{ij} by natural selection. In the particular case with a self-limiting plant, a single herbivore, and a carnivore, this approach leads to the surprising result that selection for viability or fecundity of the herbivore has no effect on its population but merely produces more food for the carnivore population, which increases. A comparable evolutionary change in the plant is found to increase its population; the herbivore has more food, but again its population does not increase, the abundance being handed on to the predator. Evolution in the carnivore reduces the population of the herbivore, and this increases the population of the plant. The only way for the prey herbivore to increase is for it to develop adaptations that make predation upon it more difficult. An even more surprising result, less easily explained in nonmathematical terms, is that a predator which produces diversity by eliminating a dominant prey species can in some circumstances select itself to extinction.

It must be remembered that all this argument ultimately depends on the use of a community matrix which assumes linearity in the operation of its component coefficients. It is therefore really applicable only when the population is near equilibrium. Extreme extrapolation of any of the more surprising conclusions may be inappropriate, but it is clear that situations can arise that seem to run contrary to initial intuition.

Experience has shown that the very simple prey-predator community is in fact often quite unstable. This is dramatically indicated by the introduction of rabbits onto the island of Laysan in the Hawaiian Leeward Chain,[10] which destroyed almost the whole terrestrial vegetation and practically all the associated endemic terrestrial fauna. Elton,[11] moreover, quotes from W. C. Tate an account of rabbits being introduced onto Berlenga Island, off the coast of Portugal, and the subsequent establishment of a population of cats to control the rabbits. The cats exterminated the latter and then died of starvation for lack of any other prey. This last case would seem to correspond to the plant-herbivore-carnivore situation just analyzed, which should result in a stable equilibrium.

Levins gives an analysis of a special situation in such a simple plant-prey-predator chain which shows the kind of complexity that may arise. If the prey is a small mammal, perhaps a vole, which is more vulnerable to predation when feeding than when resting, anything that increases the lushness of the grass acts as a check on the efficiency of the predator by decreasing the time that the prey needs to obtain a given amount of food. This has the result of introducing a second-order term since what determines the death rate of the prey is the product of the predator and the plant populations. This in fact provides a case of the efficacy of predation being reduced to the benefit of the prey, though the reduction may not have resulted from any selection pressure on the latter.

The stability of natural communities

It is common experience that some years are more favorable than others for certain crops. Moreover, the favorableness may be expressed in more than one way. A very large harvest may imply a small or otherwise low-quality fruit. The finest wines are unfortunately apt to be those of years when the vintage is not great.

We ordinarily think of these variations as determined primarily by the weather, but it

10. The tragic history of Laysan is summarized in: G. E. Hutchinson, The biogeochemistry of vertebrate excretion. *Bull. Amer. Mus. Natur. Hist.*, 96:1–554, 1950. S. O. Schlanger, and G. W. Gillett, A geological perspective of the upland biota of Laysan atoll (Hawaiian Islands). *Biol. J. Linn. Soc.*, 8:205–16, 1976.

11. *Animal Ecology*; see note 2.

is evident from the facts that not every crop reacts in the same way, that certain years produce great quantities of acorns or a particular species of insect, but not of other comparable products, and that not merely the weather but the biological response to it is involved. We may therefore suppose that every species, while tending to build up a population according to its rate of natural increase, will be limited in an individual way. This is in fact implied by the belief that each species occupies its own niche. The varying responses to varying environmental changes are likely to produce extremely complicated changes in any community. However, since an increase is finally checked when the population of any species S_i reaches K_{it}, its saturation population at the state of the whole community at time t, and since any decrease must ultimately either be reversed or lead to extinction, any species that is actually present in a community, unless it is a transitory newcomer, must have exhibited some tendency to persistence in a population varying around an average number. We have seen that there may be evidence of such a tendency in cases where it can be shown that any increase is more likely to be followed by a decrease than by a further increase. Even the most opportunistic species will behave in this way over a long enough period. The question naturally arises as to the magnitude of this type of variation, which, when a whole community is considered over long enough time, may be regarded as an ecological equivalent of thermal noise. There is, moreover, in many cases an internal tendency to fluctuate quite apart from changes in the environment. We have seen that this is so in the case of time delays. Moreover, when there are well-marked breeding seasons, which usually involve interaction of an external periodicity with physiologically determined time lags such as pregnancy, so that finite difference equations are appropriate, not merely can fluctuations be set up, but they may be of so complicated a form that, though strictly determinate, they appear chaotic (see p. 33–35).

When a community is mature and complicated enough for all the niches to be filled, invasion will be difficult. The possibility of an invasion will depend on the niche of the invader being represented in the habitat, and by the invader being able to displace any previous occupant in at least part of this niche. In an old and undisturbed community in a habitat set geographically in a region with a large pool of species within migrating distance, not only will the niches be filled, but they will be efficiently filled, by species molded through natural selection to both the physical or scenopoetic features of the habitat and the biological features of the community. That a random invader should be better adapted than a resident species is unlikely.

If, however, a habitat has been disturbed, either naturally or artificially, the available niches after the disturbance may well be different from those originally present, and adaptable invaders may be able to establish themselves in them. Any experienced naturalist will be able to think of cases of this phenomenon. Very striking examples are provided by the earthworms, in which group there is a small number of species, usually termed peregrines, that have been artificially spread and have colonized disturbed soils throughout the temperate parts of the world. Most are lumbricids of Palaearctic origin, but there are a few species of *Pheretima* from the Oriental region and a couple of members of the genus *Microscolex* that may be austral.

In the humid coastal region of South Africa, in undisturbed localities having at least 500 mm of rainfall annually, the earthworm fauna consists of endemic species, mostly of the megascolecid subfamily Acanthodrilinae, which has differentiated on an extraordinary scale, producing about 80 localized species.[12] The invasion of undisturbed situations by introduced earthworms is not unknown, but is extremely rare. When an area is cleared of native vegetation and cultivated, this native earthworm fauna declines drastically before any invaders arrive. The locality then may be colonized by 4 or 5 species of lumbricine peregrines and an odd species of *Pheretima* in some places.

In contrast to this, we may note that Elton, who has given great attention to this matter, believes that in Wytham Woods near Oxford,

12. G. E. Pickford, *The Acanthodriline Earthworms of South Africa*. Cambridge, Heffer, 1937, 612 pp.

in which there are at least 2,500 and perhaps as many as 5,000 species of animals, only 4 are clearly recent invaders, the most conspicuous being the grey squirrel, *Sciurus carolinensis*, of American origin.[13]

In any system of species, however it is organized, natural selection will presumably be more rigorous and more effective when a species is observed to be declining than when it is increasing. This situation, which produces character release in a rapidly increasing population, presumably increases fitness during the decline of a population, though only relative to the conditions experienced at the time of the decrease. Since genetic variance is lost in the process, recovery under other conditions may be difficult, but the process, which is analogous to the prudent predator effect of Slobodkin,[14] may provide some sort of stability to species in a fluctuating system. If they have survived at all, they are somewhat adapted against going all the way to perdition at the times of future vicissitudes. There is some evidence that this may happen in small mammal populations in which selection favors aggression at high, but reduces aggression at low, densities.

The extent to which populations fluctuate seems to vary greatly, but really good comparative data are sparse. Leigh,[15] who has devoted much thought to the matter, believes that in general, fluctuation is not less in tropical regions than in temperate, though from this generalization he excludes the most extraordinary cases, namely, the mammals of the arctic and subarctic parts of the northern hemisphere.

Lotka-Volterra oscillations

One type of population behavior which has been widely discussed and has stimulated much work, though it may actually be seldom, if ever, realized in nature, is the cyclical type of change deduced by Lotka[16] and independently and in greater detail by Volterra.[17]

If we suppose that the number (N_y) of a prey species growing exponentially with a rate of increase r is controlled solely by predation, and that the number of predators (N_r) depends on the number of prey captured and on a constant mortality m, we may write, assuming the predation rate depends solely on the encounter rate which is proportional to N_yN_r,

$$\frac{dN_y}{dt} = rN_y - \gamma_y N_y N_r$$

(6.14a, b)

$$\frac{dN_r}{dt} = \gamma_r N_y N_r - mN_r.$$

Here γ_r/γ_y gives the efficiency with which prey eaten is converted into predator and will ordinarily be ≈ 0.1.

The solution of this pair of equations is

$$\left(\frac{N_y\gamma_r}{m \exp N_y\gamma_r/m}\right)^m \left(\frac{N_r\gamma_y}{r \exp N_r\gamma_y/r}\right)^r = C, \tag{6.15}$$

where C is a constant of integration. This equation represents a family of closed curves around a singular point (inset, figure 132) $N_y = m/\gamma_r$, $N_r = r/\gamma_y$. It will be noted that the larger the mortality of the predator, the larger is the population of the prey; while the larger the rate of increase of the latter, the larger is the population of the predator. Both populations are decreased by increases in the coefficients γ_y and γ_r, which indicate the efficiency of predation. In particular, if these coefficients are very high, the point

13. Elton has considered these problems in great detail in:

The Ecology of Invasions by Animals and Plants. London, Methuen, 1955, 181 pp.

The Pattern of Animal Communities. London, Methuen; New York, John Wiley, 1966, 432 pp.

14. L. B. Slobodkin, How to be a predator. *Amer. Zool.*, 8:43–51, 1968.

L. B. Slobodkin, Prudent predation does not require group selection. *Amer. Natural.*, 108:665–78, 1974.

15. E. G. Leigh, Jr., Population fluctuations, community stability, and environmental variability. In: *Ecology and Evolution of Communities*, ed. M. L. Cody and J. M. Diamond. Cambridge, Mass., Belknap Press of Harvard University Press, 1975, pp. 51–80.

16. A. J. Lotka, *Elements of Physical Biology.* Baltimore, Williams and Wilkins, 1925 (reissued as *Elements of Mathematical Biology.* New York, Dover, 1956).

17. V. Volterra, Varizioni e fluttuazioni del numero d'individui in specie animali conviventi. *Mem. R. Acad. Naz. dei Lincei* (ser. 6), 2:31–113, 1926.

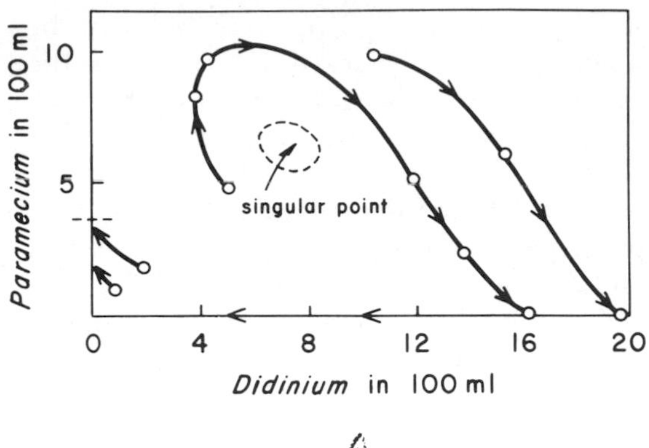

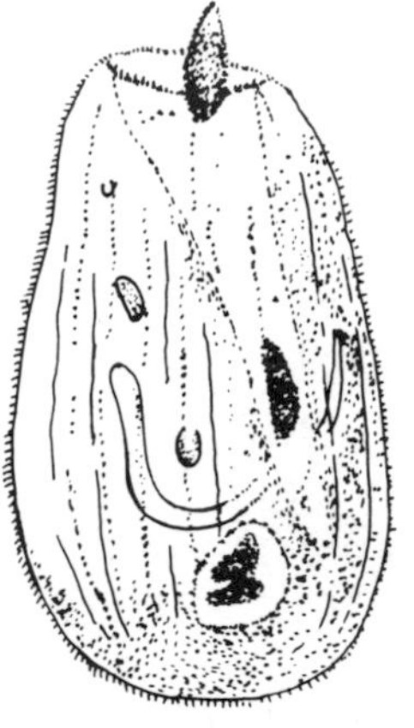

FIGURE 132. Trajectories, in the neighborhood of the singular point, of *Didinium nasutum* feeding on *Paramecium caudatum*, with a figure of the former species (Gause, modified).

around which the population oscillates is very close to the origin. This means that in a small population any random variations can easily cause extinction when the two populations are near their minima. Comparable oscillations in finite difference models of parasitism or predation were early developed by Nicholson and Bailey[18] and probably stimulated as much interest as did the continuous analytic treatment of Volterra.

Oscillations can be set up at least under some conditions when the prey population has an upper limit independent of predation, and when time lags are introduced.[19]

In experimental studies it is often possible by setting up rather artificial circumstances to produce what seem to be noisy Lotka-Volterra cycles. The limitation imposed by the equilibrium point lying near the origin may easily vitiate the experiments, and so may changes in the behavior of the predator as the prey becomes scarcer. Thus when *Didinium nasutum* was used as a predator by Gause,[20] *Paramecium* used as prey became scarce, the *Didinium* continued dividing, so producing an increased number of smaller individuals. This means that the number of cytostomata or cell mouths per unit mass of predator increased; the predation became increasingly more effective until all prey were removed and both populations died out. In general, the experimental verification seems to proceed best in a mechanically dense medium, such as in experiments by Gause[21] on flour mites living in flour (figure 133), or when spatial heterogeneity is introduced, as in experiments by Huffaker[22] on mites living on oranges. Here usually the predators almost exterminated their prey and then died off, leaving a few undiscovered prey to produce a new unpredated population. When extreme environmental heterogeneity was introduced, by covering most of the surface of the oranges with vaseline, an oscillatory population might develop for a time. Such treatment has the effect of reducing the predation coefficients γ_y and γ_r when averaged over time. It is just possible that the variations in protozoal and bacterial populations[23] that have been observed in soil, through which movement of the protozoa cannot be very fast, are in part Lotka-Volterra oscillations.

18. A. J. Nicholson, and V. A. Bailey, The balance of animal populations. *Proc. Zool. Soc. Lond.*, 3:551–98, 1935.

19. P. Wangersky and W. J. Cunningham, Time lag in population models. *Cold Spring Harboz Symp. Quant. Biol.* 22:329–77, 1957.

20. See Chap. 4, note 7.

21. See Chap. 4, note 8.

22. C. B. Huffaker, Experimental studies on predation: Dispersion factors and predator-prey oscillations. *Hilgardia*, 27:343–83, 1958.

23. W. D. Cutler, L. M. Crump, and H. Sandon, A quantitative investigation of the bacterial and protozoan population of the soil, with an account of the protozoan fauna. *Phil. Trans. Roy. Soc. Lond.*, 211B:317–50, 1923.

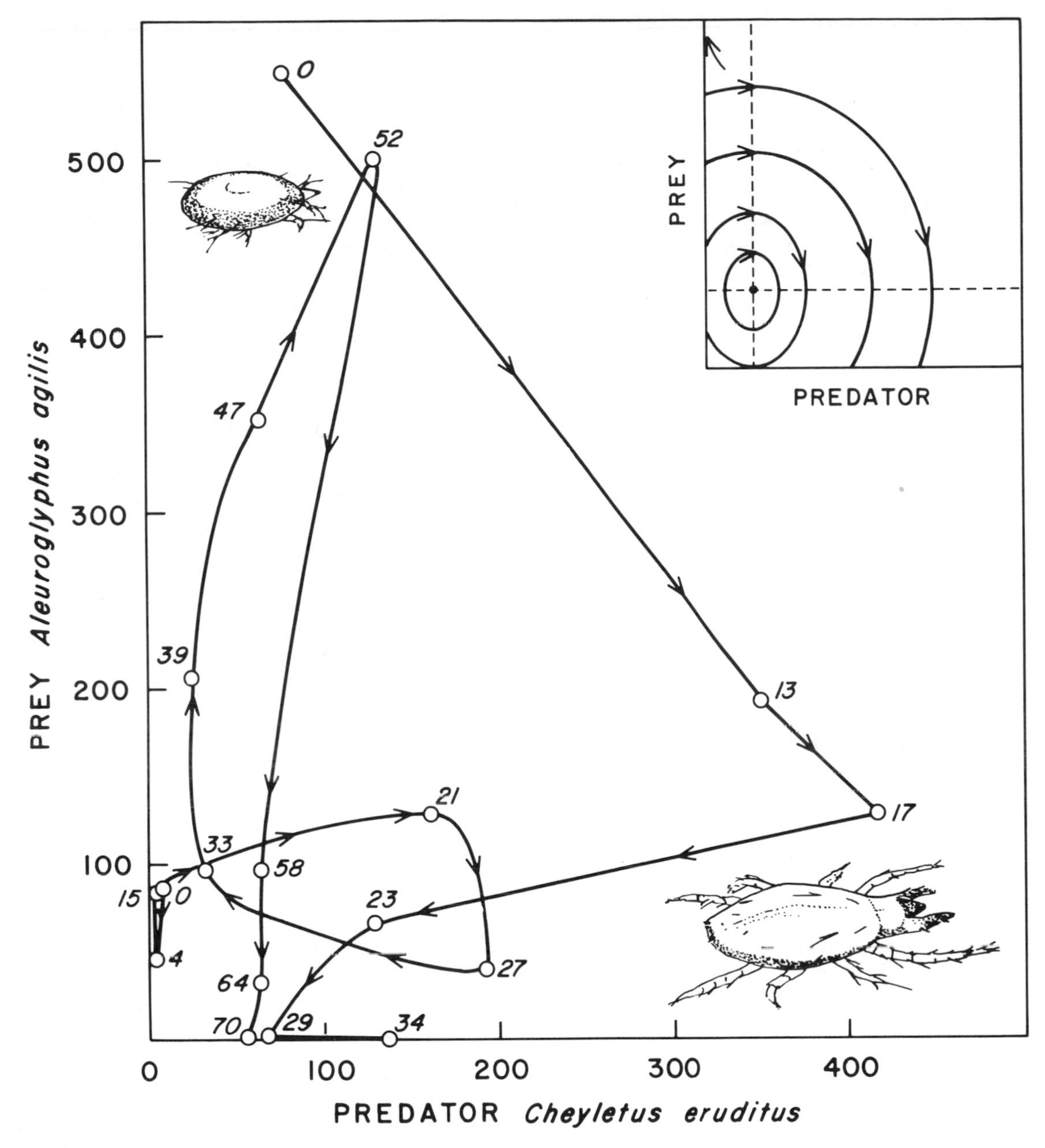

FIGURE 133. Trajectories of numbers of prey plotted against predators in two experiments with the flour mite, *Aleuroglyphus agilis*, fed on by the carnivorous mite, *Cheyletus eruditus*. The mites are figured in insets, as is an ideal Lotka-Volterra type of model. At very high densities of *A. agilis* the predator is inhibited by metabolic products of the prey and disappears. Time in days is indicated in italics (Gause, modified).

Very regular oscillations were produced by Utida[24] in the bean weevil, *Callosobruchus chinensis*, and its parasite, *Heterospilus prosopides*. These seem to continue for a time and then become damped; later some unidentified disturbance may increase the variation again. Possibly here we are dealing with variations in *C* imposed on a population in which natural selection is producing some degree of damping (figure 134).

24. S. Utida, Population fluctuation: An experimental and the oretical approach. In. *Cold Spring Harbor Sympo. Quant. Biol.*, 22:139–51. 1957.

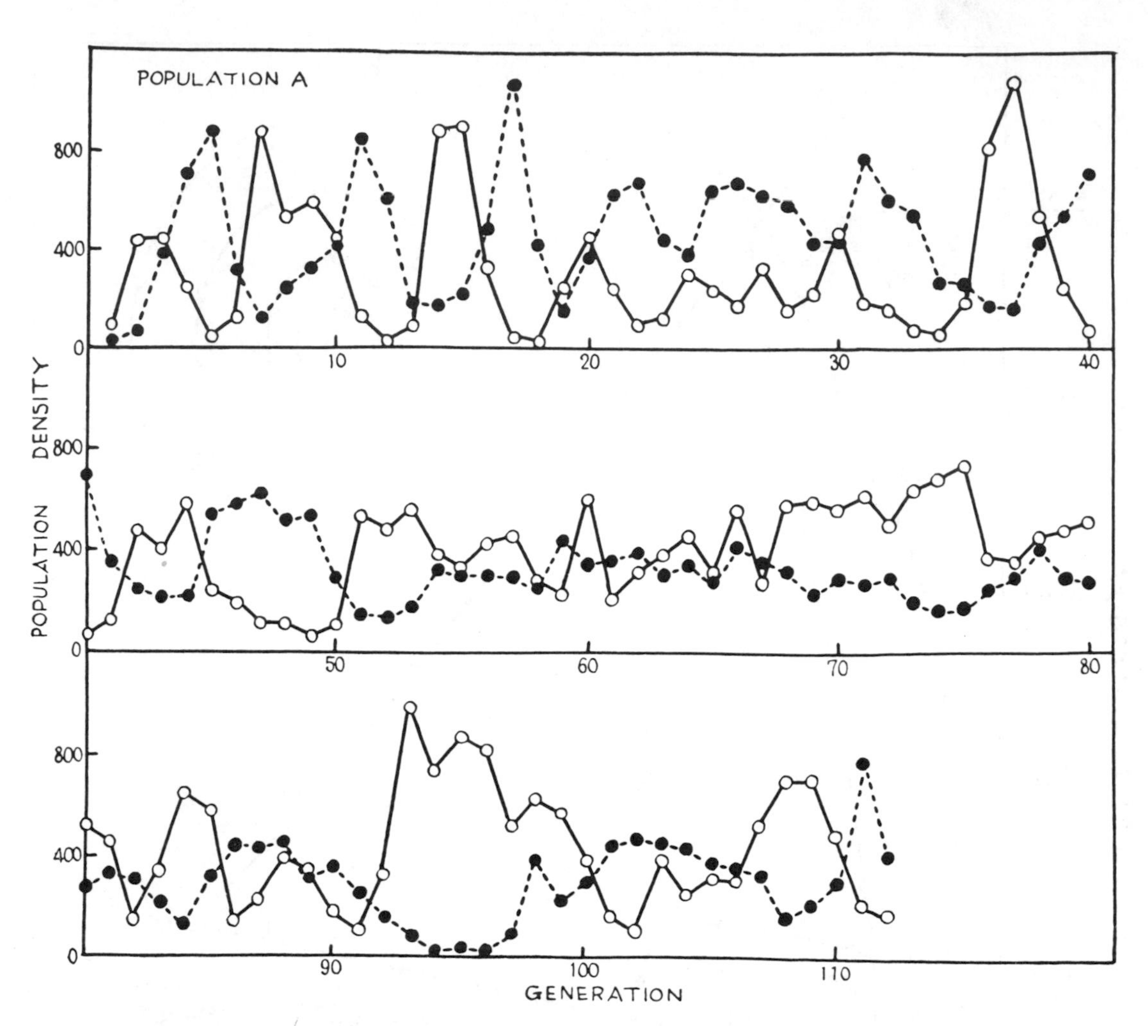

FIGURE 134. Oscillation of an experimental population of the azuki bean weevil, *Callosobruchus chinensis* (open circles), and its parasite, the braconid wasp, *Heterospilus prosopidis* (solid circles). Note that after the 54th generation the oscillations are damped out for about 10 generations, but then appear again in a very irregular manner (Utida).

The regular oscillations of boreal mammals

The very extraordinary oscillating populations of mammals, notably lemmings and the foxes (figure 135, c, d), that feed on them, or snowshoe hares (*Lepus americanus*) and the lynxes (*Lynx canadensis*) which are their characteristic predators (figure 135, a, b), have often been supposed to be either Lotka-Volterra cycles generated by the carnivores interacting with the rodents, or by the latter interacting with the vegetation. Since these cycles are at present being treated exhaustively by J. Patrick Finerty in a book[25] which should appear only a little later than this one, and which adopts much the same theoretical attitude, they will not be treated here at length. They are certainly real and not due to accidental properties of random number sequences, as was for a time often believed.[26] Their most extraordinary property is their persistence, the lynx and snowshoe rabbit having apparently shown cyclical

25. J. P. Finerty *Cyclic Fluctuations in Biological Systems: A Revaluation.* Ph.D. Thesis, Yale University, 1971.

26. L. C. Cole, Some features of random population cycles. *J. Wildl. Mgt*, 18:1–24, 1954. Cole pointed out that the mean distance between maxima in any series of random numbers converges on 3 as the series becomes long. Some natural populations may show this effect. It can be distinguished from a real periodicity by the use of serial correlation. The data are correlated

variation for over 200 years. May[27] points out that the variations do not look the least like Lotka-Volterra oscillations. They may run for many years varying only by about a factor of 2 in amplitude, but when changes in amplitude do occur, they seem to be random. There is no hint of an environmentally produced change in C, which then persists for a time.

May himself believes that the 4-year cycle depends on r being above a critical value when reproduction can be expressed by a finite difference equation (see p. 35) and at present nothing is known that contradicts such a belief. The genesis of the 10-year period is more puzzling. If, as seems possible,[28] there are actually several independent 10-year cycles in Canada, with comparable periods, but not in phase, the possibility of an external control, to which the populations react with different lag periods, cannot be summarily dismissed. It was suggested[29] some time ago that the cycle might be produced by some mechanism in the population resonating to an appropriate frequency in a wide spectrum of nonbiological disturbances. This idea has very recently been put forward independently in a more sophisticated form by Nisbet and Gurney.[30]

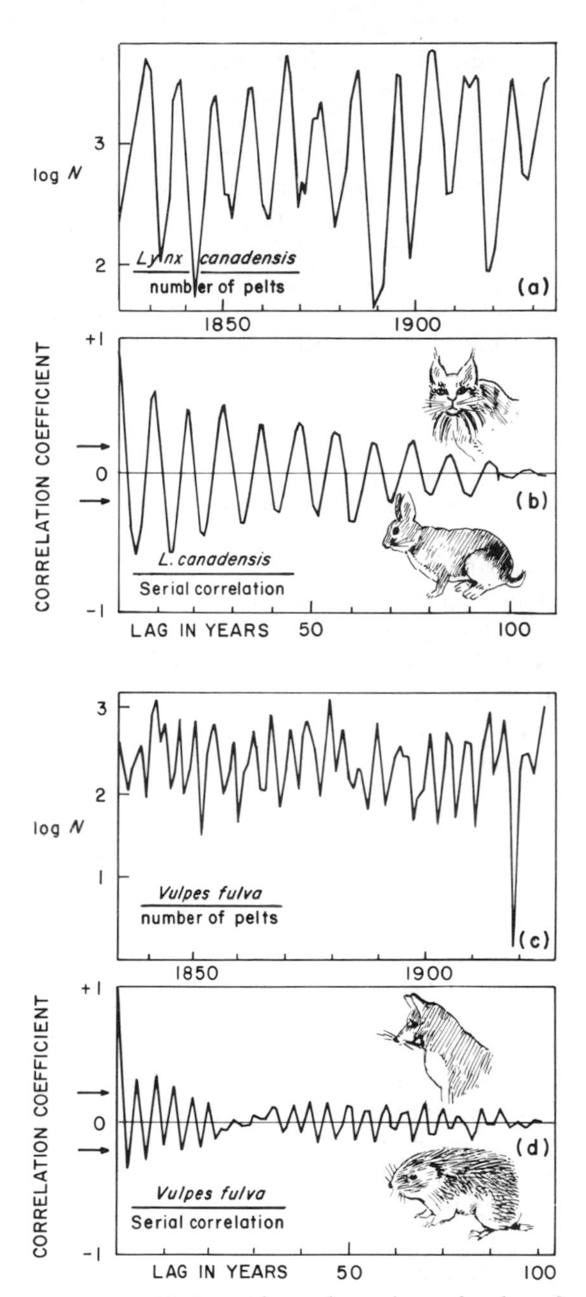

FIGURE 135. (a) Logarithm of number of pelts of *Lynx candensis* traded by the Hudson Bay Company between 1821–34, indicating the abundance of this species and indirectly of its prey, the snowshoe hare, *Lepus americanus.* (b) Serial correlation obtained by calculating the correlation coefficient between the data of (a) and the same data displaced 1, 2, . . . , n years. This gives the most easily appreciated visual evidence of the existence of a real cycle, of about 10 years. (c) Logarithm of number of pelts of the fox, *Vulpes fulva*, feeding primarily on lemmings, reported as traded by the Moravian Missions in Labrador. (d) Serial correlation showing the 4-year cycle. Arrows indicate 95 percent confidence limits (Finerty).

with themselves displaced by 1, 2, 3, . . . , time units. It is evident that if a completely regular periodicity existed, as in a sine curve, the correlation would be −1 when the displacement were by 0.5, 1.5, . . . , periods, and +1 when it were by 1.0, 2.0, . . . , periods. This technique which gives a visually very convincing demonstration of the periodicity, as in figure 135, was introduced into ecology by P. A. P. Moran, The statistical analysis of the sunspot and lynx cycles. *J. Anim. Ecol.*, 18:115–16, 1949.

27. R. M. May, *Theoretical Ecology: Principles and Applications.* Philadelphia, Toronto, W. B. Saunders, 1976, viii, 317 pp.

28. M. G. Bulmer, Phase relations in the ten-year cycle. *J. Anim. Ecol.*, 44:609–21, 1975.

29. G. E. Hutchinson, Theoretical notes on oscillatory populations. *J. Wildl. Mgt.* 18:107–09, 1954.

30. R. M. Nisbet and W. S. C. Gurney, A simple mechanism for population cycles. *Nature*, 263:319–20, 1976.

The mechanism is mathematically simple, but there is no indication as to what, physiologically or environmentally, it really is.

Leigh[31] believes the regular cycles of boreal mammals to be adaptive, preventing the predator from becoming so abundant that it completely destroys its prey. In the same way[32] it has been supposed that variation in the set of seeds by trees at high latitudes, which appears to exhibit an almost biennial oscillation, may prevent the excessive development of birds, squirrels, and other animals feeding on nuts and other large seeds and fruits. These explanations do not elucidate the mechanisms of the cycles involved; they may, however, throw light on why the cycle is preserved or possibly accentuated by natural selection, rather than being damped out.

Diversity

The complexity introduced into animal nature by the elaboration of the food web is probably the most obvious cause of biological diversity. Since the successive levels in an Eltonian pyramid decrease in size quite rapidly, the diversification introduced by trophic relations in their simplest form is limited when qualitative differences in food are involved between the species of any level, as they always are when at least two species are present; the diversity also involves not only the length of the food chain but the closeness of species packing. This is limited, as we have seen, by the variance in the feeding habits of the individual species. A third major factor in producing diversification is clearly the evolution of biological substrata, notably plant cover on land and any kind of sessile benthic community, plant or animal, under water; by their existence these not merely add species in a primary way to the community but produce a great variety of sites and foods for secondary species, which constitute what marine biologists call the epifauna. From one point of view biological communities can be divided into those in which this does not happen, notably the plankton and the benthos or infauna on muddy bottoms, as contrasted to the communities of all vegetated areas on land as well as the higher vegetation and all kinds of zoogenous reefs under water. The enormous preponderance of species of insects on land must mainly be due to this kind of diversification.

At the present moment the study of diversity is an extremely active part of ecology. A full account of the subject would extend beyond the limits of population ecology adopted in this book. Moreover, several good accounts have recently been published, by Whittaker,[33] by May,[34] and by Pielou,[35] authors whose works have the virtue of clear and elegant expression unusual in scientific writing today. A short final section indicating how the problem stands is all that is needed to complete the design of this book.

The simplest measure of diversity is merely the number S of species present. This can in some cases be counted directly, as with birds holding territory in a woodland or field. More usually S must be determined by sampling. The statistic S, however obtained, is obviously an inadequate measure of diversity. If we have two samples in one of which every species except one is represented by single specimens and one by a very large number, we feel in some way justified in believing that the sample is less diverse than if all the species in it were present in equal numbers. It is usual to measure diversity by one of two statistics H and H', which were introduced into ecology from information theory by MacArthur[36] and by Margalef.[37] For an indefinitely large assemblage that can be sampled without affecting significantly the number of the different species present we

31. See note 15.

32. C. E. Bock and L. W. Lepthien, Synchronous eruptions of boreal seed-eating birds. *Amer. Natural.*, 110:559–71, 1976.

33. R. H. Whittaker, Dominance and diversity in land plant communities. *Science*, 147:250–60, 1965.

R. H. Whittaker, *Communities and Ecosystems*. New York, Macmillan, 1970.

R. H. Whittaker, Evolution and measurement of species diversity. *Taxon*, 21:213–51, 1972.

34. R. M. May, Patterns of species abundance and diversity. In: *Ecology and Evolution of Communities*, ed. M. L. Cody and J. M. Diamond. Cambridge, Mass, Belknap Press of Harvard University Press, 1975, pp. 81–120.

35. E. C. Pielou, *Ecological Diversity*. New York, Wiley–Interscience, 1975, viii, 165 pp.

36. R. H. MacArthur, Fluctuations of animal

may write

$$H' = -\Sigma p_i \log_n p_i \qquad (6.16)$$

where p_i is the fraction of the whole sample composed of species i and n is the base of the logarithm used. If one wishes to express H' in the units (*bits or binits*) of information theory, $n = 2$; if one prefers $n = 10$ the units of diversity are called *decits*, while if $n = e$ the unit is a *nat*. There is no standard procedure, but the investigator should not emulate those of his predecessors who do not indicate clearly what base is being used.

It will be evident that if a single species only is present, $p_i = 1$ and $\log_n p_i = 0$ so that $H' = 0$. If all species are equally common, for any value of $S_n H'$ will be maximal, but will of course increase if S_n is increased. Equation (6.16) was introduced by Shannon;[38] it was originally used in information theory. If we have all specimens in a population of the same species, when one has been identified no further information is added by examining another specimen. If every specimen is different, corresponding to H'_{max}, examining a new one will always add further information.

In a completely censused community the proper procedure is to determine diversity as

$$H = \frac{1}{N} \log \frac{N!}{\Pi N_i!} \qquad (6.17)$$

where N is the total number of individuals, $N!$ is $N(N-1)(N-2)\ldots$ or factorial N and $\Pi\, N_i!$ is the product of all the values of factorial $N_1 \ldots N_i \ldots N_s$.

The numbers involved in calculating H, even if we had only 1,000 individuals divided among 10 species, are obviously very great, but shortcuts to their calculation exist.

H is always rather less than H', which in fact is the measure of H as the numbers of individuals approach infinity. It is often convenient, when H' has been calculated from the data, to compute H'_{max} for the number of species encountered. The ratio $H'{:}H'_{max}$ gives a measure of the unevenness of the representation of the different species of which the sample is composed, though it is not independent of sample size.

The various pitfalls encountered in using these measures of diversity are admirably discussed by Pielou in her very recent book.

The treatment of diversity derived from information theory does not depend on any specific type of distribution; H' can be calculated whatever the actual distribution may be. It has, however, been found that in nature the relationship of number of species to number of individuals takes a limited number of forms, which seem to imply known statistical distributions. In general, four such forms or variants thereon have been found to be important.

Perhaps the first significant treatment of the problem was that of Motomura,[39] put forward in Japanese in 1932, and little appreciated in the Western World until Whittaker's highly important publications called attention to it over 30 years later. Motomura's model is constructed by assuming that the first species to enter a habitat will take a certain proportion of the available resources, the second species some proportion of what is left and so on. In the most general case, the available resource will be solar energy.[40] The

populations, and a measure of community stability. *Ecology*, 36:533–36, 1955. MacArthur used Shannon's expression for H'.

37. R. Margalef, Información y diversidad especifica en las communidades de organismos. *Investigacion pesq.*, 3:99–106, 1956.

R. Margalef, La teoriá de la información en ecologia. *Mem. R. Acad. Cien. Artes, Barcelona* (3 v.), 32, no. 13, 1957, 79 pp.

Margalef used H, due originally to Brillouin.

38. C. E. Shannon and W. Weaver, *The Mathematical Theory of Communication*. Urbana, University of Illinois Press, 1949.

39. I. Motomura, A statistical treatment of associations [in Japanese]. *Jap. J. Zool.*, 44:379–83, 1932. [Not seen, ref. Whittaker, 1965, 1972; see note 33.]

40. H. T. Odum, J. E. Cantlon, and L. S. Kornicker, An organizational hierarchy postulate for the interpretation of species-individual distributions, species entropy, ecosystem evolution, and the meaning of a species-variety index. *Ecology*, 41:395–99, 1960.

The authors prefer a cumulative or integral curve, which is, however, based on the geometrical postulate. They trace the idea back to H. A. Gleason, On the relation between species and area (*Ecology*, 3:158–62, 1922), which is itself a criticism of O. Arrhenius, Species and area (*J. Ecol.*, 9:95–99, 1921).

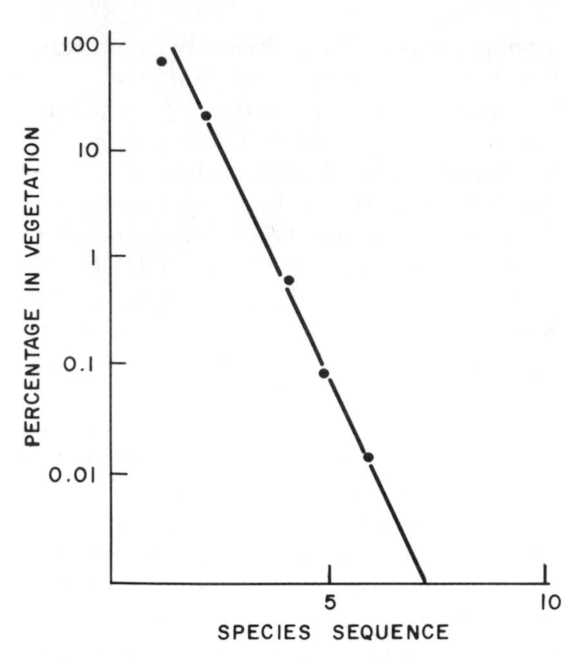

FIGURE 136. Logarithmic decline in importance of species in a subalpine community (Whittaker, modified).

resulting distribution, if the species are of approximately the same size and general biological properties, is a declining geometric series as one passes from the population of the most abundant species to the next and so on. If the species are ordered in decreasing abundance arithmetically, the logarithms of their populations will fall off linearly (figure 136). The distribution is of some significance in plant ecology, particularly under harsh conditions in which but a small number of species are present.

Of greater historical significance was the observation made by Corbet[41] when studying the distribution of numbers of individuals per species in a collection of butterflies made in the Malay Peninsula.

Initially Corbet had concluded that for the rarer species in the collection, if the number represented by a single specimen be n_1, the number of species represented by two specimens will be about $n_1/2$, by three specimens about $n_1/3$, and so on. When common species are reached, the agreement with this harmonic series is progressively less good. In the highly important later paper[42] in which Corbet was joined by R. A. Fisher to provide a statistical analysis and by C. B. Williams, who had a wealth of further data relating to British moths taken at light traps, it became apparent that the number of species having N individuals in any collection is approximately given by $\frac{\alpha}{N}x^N$ where α is a constant independent of the size of the collection, and x, which depends on the size of the sample,

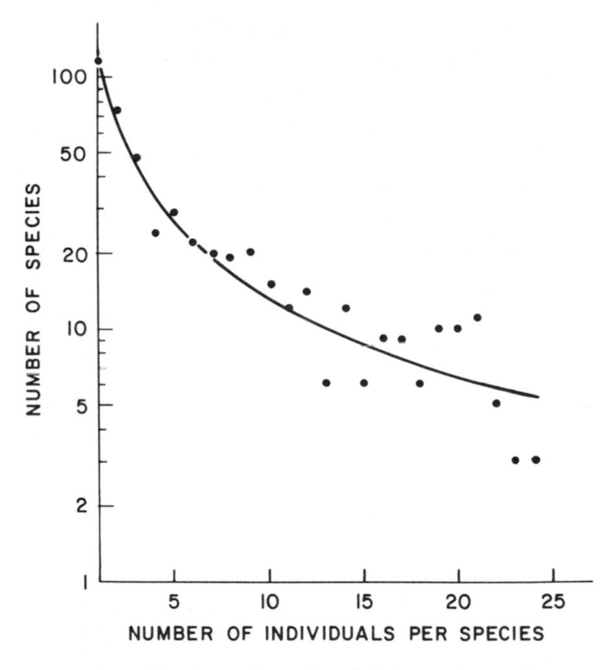

FIGURE 137. Number of species of Malayan butterflies represented by 1, 2, 3, . . . , specimens. The corresponding value of the index of diversity α is 135.5, while $x = 0.997$. (Data of Corbet in Fisher, Corbet, and Williams.)

41. A. S. Corbet, The distribution of butterflies in the Malay Peninsula (Lepid.). *Proc. Roy. Entom. Soc.*, ser. A., 16:101–16, 1941.

Corbet mentions J. C. Willis, *Age and Area* (Cambridge University Press, 1922, x, 259 pp). In this work the main quantitative treatment concerns the number of species in a genus. However, Willis had clearly observed the kind of distribution studied by Corbet, as is indicated in J. C. Willis, The endemic flora of Ceylon, with reference to geographical distribution and evolution in general. *Phil. Trans. Roy. Soc. Lond.*, 206 B:307–42, 1915.

42. R. A. Fisher, A. S. Corbet, and C. B. Williams, The relation between the number of species and the number of individuals in a random sample of an animal population. *J. Anim. Ecol.*, 12:42–58, 1943.

is a number very slightly less than unity. The number of species with only one specimen will therefore be αx. Since x is almost unity, α, the *index of diversity*, will be just a little greater than the number of species known from single individuals. The distribution of Corbet's original sample is shown in figure 137.

It has been suggested that the log-series distribution, as the distribution of Fisher, Corbet, and Williams is usually called, is characteristic of taxocenes in which niche preemption occurs, as in the geometrical distribution, but at random times. In view of the fact that it is most frequently found in insects, which are largely oligophagous, if not monophagous, this explanation seems unlikely.

Preston,[43] in 1948, in a paper clearly influenced by that of Fisher, Corbet, and Williams, put forward a quite different distribution. Using bird censuses, and counts of moths at light traps, he grouped the species, by the number of individuals represented, arranged in octaves. Thus in the first octave 0–1 half the species with one individual was entered; in the second octave 1–2, half those with one and half those with two; in the third octave, 2–4, half with two, all with three, and half with four, and so on. The numbers per octave were then plotted against the ordering of the octaves. When this is done, a truncated normal curve is obtained (figures 138 and 139). As censusing continues,

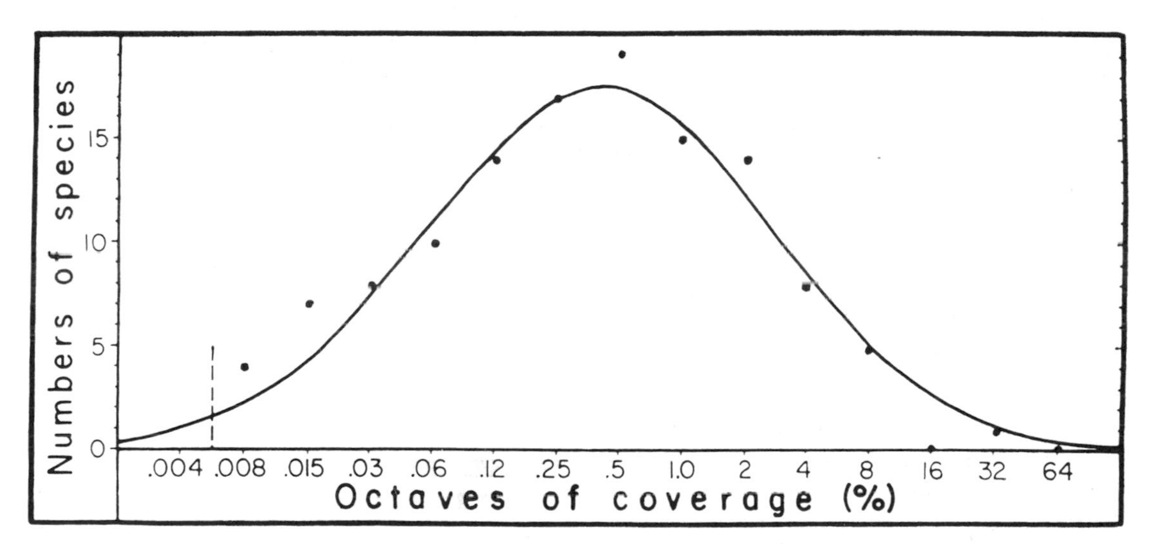

FIGURE 138. An almost completely unveiled log-normal curve for a rich north-slope shrub phase in the Sonoran desert communities of the Santa Catalina Mountains of Arizona. The veil line is the broken line on the left. The species pool consists of 126 species (Whittaker).

43. F. W. Preston, The commonness, and rarity, of species. *Ecology* 29:254–83, 1948.

F. W. Preston, The canonical distribution of commonness and rarity. *Ecology*, 43:185–215, 410–32, 1962.

In his second paper Preston shows that not only the number of species in an octave is log-normally distributed, but that another log-normal curve, to the right of that for the number of species, is obtained if the total number of specimens is plotted. If the mode of this second log-normal curve corresponds to the octave containing the species of greatest abundance, as is usually the case, the distribution is said to be *canonical*. The distribution implied by the canonical log-normal curve falls approximately midway between the log-series and the broken-stick distributions. The canonical log-normal distribution plays a considerable role in the theory of how species accumulate and become extinct on islands, a theory developed by R. H. MacArthur and E. O. Wilson. The truncated log-normal curve has been extensively used by Patrick and her associates in studies on the ecology of diatoms, the area under the unveiled curve given in (6.19) providing an estimate of the species pool which is a useful index of diversity. It is, however, clear that at least in some diatom communities, continued counting while moving the curve to the right very slowly also moves it upward so that S_T increases, as appears in figure 139, given me by Ruth Patrick, for my *A Treatise on Limnology*, vol. II. New York and London, John. Wiley, 1967, xi, 1,115 pp.

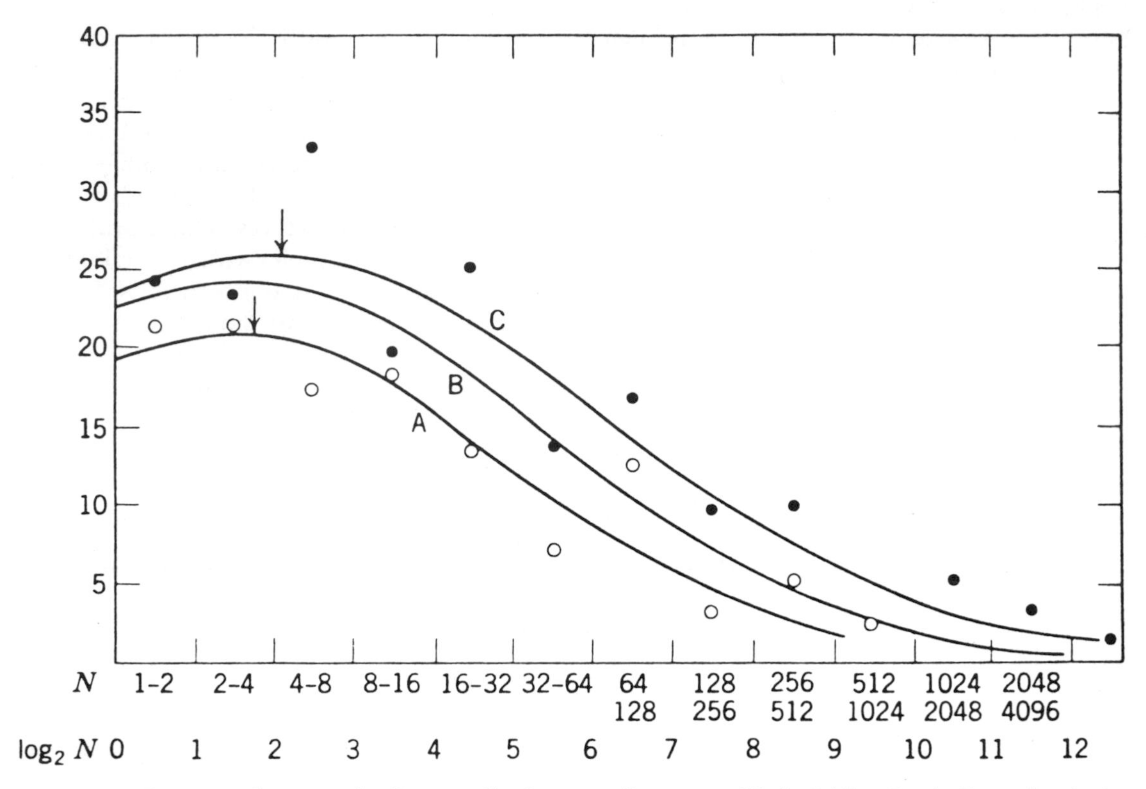

FIGURE 139. Log-normal curves of a diatom collection on a diatometer slide in Ridley Creek, Pennsylvania, in the process of unveiling as more and more individuals are counted. A based on 5896 specimens (open circles), with a species pool estimated at 188 species; B based on 15,732 specimens; C based on 28,831 specimens, (filled circles) species pool 263 species (Patrick in Hutchinson).

the mode of the normal curve moves to the right, while more and more of the curve is unveiled on the left.

The curve as plotted is a log-normal curve to the base 2. The equation of the curve may be written

$$S_x = S_M e^{-(x_M - x)^2 / 2\sigma^2}, \qquad (6.18)$$

where S_x is the number of species in any octave x, S_M the number in the modal octave, and $(x_M - x)$ the distance of the octave under consideration from the mode. The total species pool is given by

$$S_T = S_M \sqrt{2\pi\sigma^2}. \qquad (6.19)$$

Ordinarily $\sqrt{2\sigma^2}$ is not greatly different from 5, so that S_T is usually about 9 times as great as the number of species in the modal octave.

MacArthur[44] had pointed out that for any population, if the rate of increase at any time t is $r(t)$,

$$r(t) = \frac{1}{N(t)} \frac{dN(t)}{dt}, \qquad (6.20)$$

whatever law of growth be operating.

Integrating we may write

$$\ln N(t) = \ln N(0) + \int_0^E r(t)\,dt \qquad (6.21)$$

where the species is a strict equilibrium species in $N(t) \simeq \ln N(0)$ and the integral is too small to be considered. Where the species is opportunistic, $\ln N(t)$ is equal to the accumulated integral up to t. This integral is made up of random elements of history and is likely to be normally distributed. N is therefore log-normally distributed. If, however, we are dealing with equilibrium species, their abundance may be determined by the

44. R. H. MacArthur, On the relative abundance of species. *Amer. Natural.*, 94:25–36, 1960.

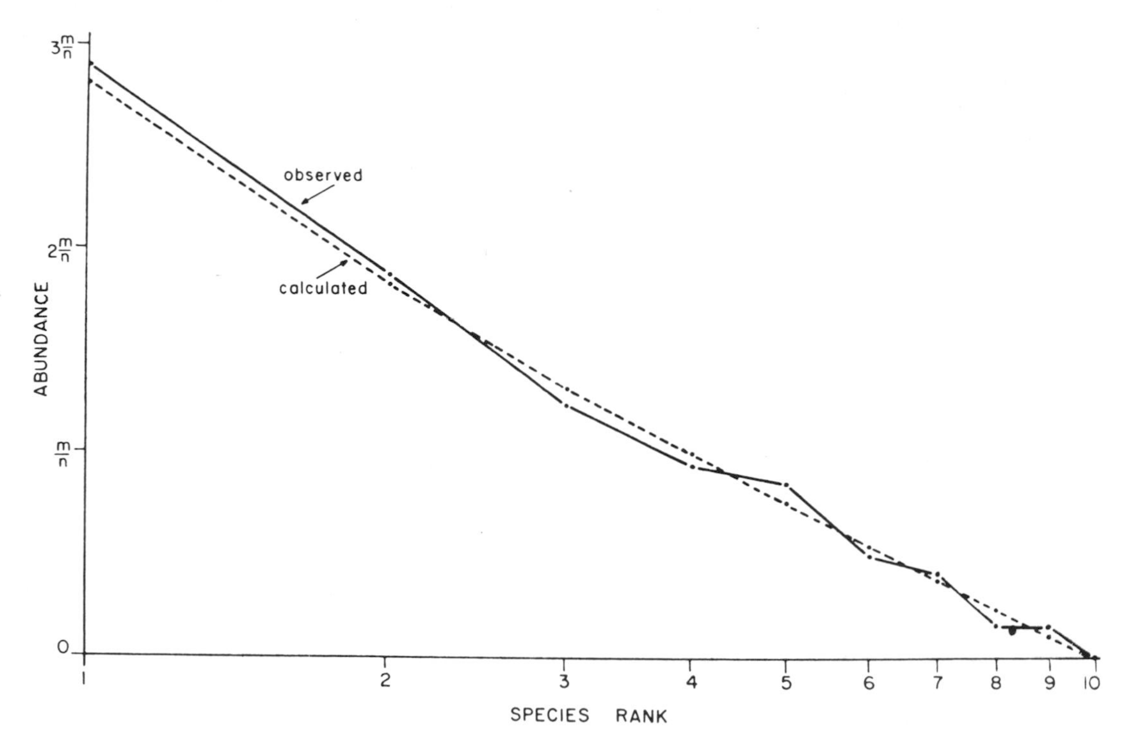

FIGURE 140. Rank order of abundance of species of *Conus* on a Hawaiian beach with the calculated value for MacArthur's "broken-stick" distribution (Kohn).

product of many small environmental events, or by the role of evolution depending on the abundance of beneficial genes fixed at a rate proportional to population size. Such processes would tend to give log-normal distributions. In general, the Preston distribution is therefore the one most likely to be found, and this is in fact apparently the case.

It may be noted that any variable quantity which cannot be zero, and of which the mode is nearer to zero than to the maximum recorded value, is apt to show a log-normal distribution. This is exemplified by the usual kinds of distributions of the less common elements in rocks.

In 1957, MacArthur,[45] stimulated by verbal discussion of some of Preston's results, discovered another distribution of significance, which is generally called the "broken-stick" distribution. Here the numbers of individuals in any one of n species tend to correspond to the lengths of a line divided at random into n segments by $n - 1$ points, designated at random as breaking points along the line. The distribution may in fact be imagined by supposing all the skins of birds collected in a certain homogeneous area to be arranged head to tail, by species on a very long bench. The broken-stick distribution implies that any segment, representing a certain species,

45. R. H. MacArthur, On the relative abundance of bird species. *Proc. Nat. Acad. Sci. U.S.A.*, 45:293–95, 1957. The idea of using the broken-stick distribution occurred to MacArthur as the result of a critical discussion of the end paragraphs of G. E. Hutchinson, The concept of pattern in ecology (*Proc. Acad. Natur. Sci. Phila.*, 105:1–12, 1953), where the significance of the variance of the log-normal curve gave rise to some, as it now seems, unjustifiable speculation.

It is likely that there is an influence of Willis on Corbet; Fisher, Corbet, and Williams certainly influenced Preston, and Preston likewise had an important influence on MacArthur. It is also interesting that of the four distributions that are empirically significant, two were discovered by scientists working professionally in areas very far from population ecology. Corbet, as an agricultural biochemist, had earlier established the production of N_2O by denitrifying soil bacteria, a finding of great importance to atmospheric chemistry. Preston is a glass technologist.

has a length corresponding to a random non-overlapping division of the line. The distribution expected is that the rth rarest species S_r in a population of S_s species and N_s individuals has an abundance given by

$$N_r = \frac{N_s}{S_s} \sum_{i=1}^{r} \frac{1}{S_s - i + 1}. \qquad (6.22)$$

If the rank order is plotted logarithmically and N_r arithmetically, an almost straight line is obtained (figure 140). Since in general any two habitats will not contain the same number or kind of species even in collections consisting of the same number of individuals, two such collections if pooled will not produce an approach to a straight line. The broken-stick distribution is often found in assemblages of species of birds provided they live in a *homogeneously* diverse habitat;[46] by this is meant a habitat in which the grain size defining its structure is small compared with the mean free path of the animals under consideration. The broken-stick distribution was discovered as the result of an effort to find out the most probable division of a biotope into individual habitats, each corresponding to a niche and so containing a single kind of organism. The distribution is now known[47] to arise in a number of different ways. Its biological meaning is not obvious; empirically it is most likely to occur in taxocenes of somewhat mobile animals living in a homogeneously diverse environment. It has not been encountered in plant communities. Though the various kinds of distribution are clearly interesting, so far they lack a really full and convincing interpretation. One study shows that a further investigation can be very fruitful.

If a plot be made of the species number S and the information-theoretic diversity H', it is possible to indicate for any S what value of H' will correspond to the broken-stick distribution. This is often called the most equitable distribution. Studying the fossil cladocera in lake sediments, Goulden[48] found that for any S the H' was never significantly greater than it would be on the broken-stick hypothesis, so that the line relating H' to S on that hypothesis (figure 141) constituted the envelope of empirically determined points. Moreover, there is some evidence that, after disturbance, H' may be depressed and then rise until the envelope is reached. This indicates that the type of distribution implied by the broken-stick model has a significant stability.

At this point the study of diversity becomes not a study of populations but rather of species and so outside the somewhat idiosyncratic limits of this book. The reader who has studied Pielou's admirable little volume will realize that the issues that now arise concern the relative importance of area, productivity, and environmental stability in promoting diversity. None of the statistical models just discussed really bears on this matter, though they do clearly indicate that real and quite deep differences may be expected when different communities are compared.

One final point should, however, be mentioned, as it has long been the subject of some misunderstanding. It was suggested long ago[49] that a very simple community consisting of two species of plants and one herbivore might well be more stable than one consisting of a single species of plant and a herbivore, because if the two species had rather different seasons of maturation, palatability, and ease of discovery, the less easily and less palatable species would be left

46. G. E. Hutchinson, Concluding remarks. *Cold Spring Harbor Symp. Quant. Biol.* 22:415–27, 1957. See also A. J. Kohn, The ecology of Conus in Hawaii. *Ecol. Monogr.*, 29:47–90, 1959.

47. A large literature has grown up around the broken-stick model, most of it listed in May's and in Pielou's bibliographies. Of particular importance are: J. E. Cohen, *A Model of Simple Competition*. Cambridge, Mass., Harvard University Press, 1966, 138 pp. J. E. Cohen, Alternate derivations of a species-abundance relation (*Amer. Natural.*, 102:165–72, 1968), in which various other ways of generating the distribution are described.

48. C. E. Goulden, Developmental phases of the biocoenosis. *Proc. Nat. Acad. Sci. U.S.A.*, 62:1066–73, 1969. Goulden's paper did much to restore the broken-stick model to a respectable place in ecology after MacArthur himself had become somewhat disillusioned about it.

49. An argument of this sort was perhaps first

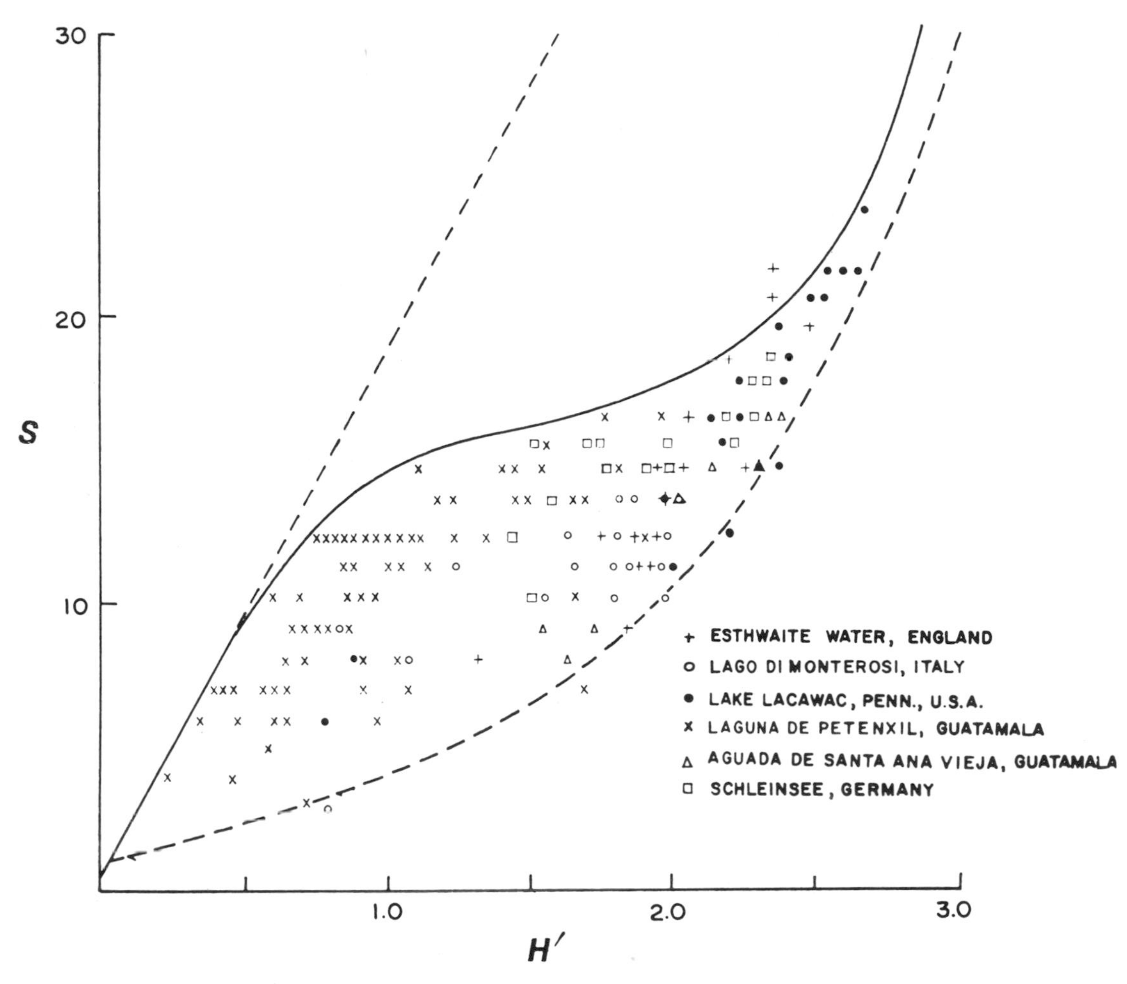

FIGURE 141. Number of species of Cladoceran faunas in six lake cores from Europe, North and Central America, plotted against H'. The lower broken line gives the number expected according to the broken-stick hypothesis. The upper broken line corresponds to an extreme situation with all species but one corresponding to 1 percent of the fauna (Goulden).

until the first was eaten down and might provide a reserve food supply as the first species regenerated. Moreover, if the two species varied in their resistance to extremes of environmental variation, one being greatly reduced by summer drought but more cold-resistant than the other, in exceptional years there might always be something to eat. Variations on this theme are easily imagined.

More elaborate analysis suggested that as a general ecological principle the idea that diversity promoted stability is erroneous. In its simplest form such analysis indicates that if each species responds to any disturbance in its own individual way, adding a species to the web increases the number of ways in which the web can be disturbed without much chance of the new way tending to cancel out one of the established ways. The condition for adding a new species is that the system be stable enough to take it. Diverse systems must be stable, and since there are

put forward in E. P. Odum, *Fundamentals of Ecology*. Philadelphia, London, W. B. Saunders, 1953, xii, 384 pp., see p. 40.

Ramifications which still may yield highly significant results are exemplified by C. E. King, Resource specialization and equilibrium population size in patchy environments. *Proc, Nat. Acad. Sci. U.S.A.*, 68:2634–37, 1971.

usually potential invaders trying to gain a foothold, few stable systems will not be diverse. This argument, fundamentally due to May, has recently been modified somewhat by Leigh who concludes, after a fairly elaborate mathematical analysis, that if we have a food web in which the top predator eats all species save its own, the next predators all save their own and the top predator, and so forth, adding new links increases the stability to disturbances that affect all species in the same way. It may be possible to recognize elements of this kind of situation in real food webs, so that to some extent the old view that stability promotes diversity may have a little validity.

SUMMARY

Any biological community that is self-sustaining must be based on a population of photosynthetic plants. Herbivores feed on these, primary carnivores feed on the herbivores, secondary carnivores on the primary, and so on. This organization into a series of levels $\Lambda_1, \Lambda_2, \Lambda_3, \ldots$, leads to a pyramidal structure. The rate λ_1 at which energy enters the plants as primary producers, Λ_1, is small, under 1 percent of the rate at which solar radiation falls on the earth, but at higher levels the rate is usually such that λ_n is 5 to 15 percent of λ_{n-1}. In general, predators are larger and stronger than their prey; they are also almost inevitably less abundant, not only in numbers but in biomass. Any animal community thus tends to have a pyramidal form with large and rare higher-order carnivores at the top and small and common herbivores at the bottom. A less conspicuous pyramid of parasites and hyperparasites also exists. The most general form of the pyramid is

$$\ldots > \lambda_{n-1} > \lambda_n > \lambda_{n+1} > \ldots$$

It seems likely that the number of levels is greater in aquatic than in terrestrial communities. If homoiotherm or warm-blooded animals are at the top, they are more expensive to maintain, and so there can be fewer of them. The food chains of which the systems are composed must tend to shorten as high-level predators tend to reach down to smaller and smaller forms, so outcompeting low-level predators, while herbivores become increasingly hard to eat by increasing their size and by other specializations. There is an alternative tendency to attempt, by predators, to prey on the highest-level predator and so lengthen the chain. There are probably, even in water, very rarely more than six levels.

Insight into the dynamics of food webs can be obtained from the use of loop analysis, in which the trophic relations of a community can ideally be represented as the determinant of a community matrix. It is possible to consider evolutionary changes in the system. Some of these may at first sight prove surprising. It can be shown that natural selection operating on a community near equilibrium can raise the natality or decrease the mortality of an herbivorous prey species without increasing its numbers, the advantage being merely passed onto its predator. Selection increasing the size of the plant population will again also benefit the predator rather than its prey, while selection benefiting the predator reduces the populations of both prey and its plant food. Only by the prey becoming harder to capture or eat can its population be increased.

Complicated food webs are hard to analyze, but in any but the most stable climates, seasonal and sporadic meteorological changes will be continually benefiting some and damaging other species, even when the resulting fluctuations continue around a persistent set of means. When the fluctuations are long compared with the time of a generation, there might be considerable *r*-selection, for rapid recovery when the population is falling, and *K*-selection, for efficient utilization when it is maximal. Such selection will tend to promote the stability of the system.

In addition to fluctuations directly caused by variations in the environment, there are likely to be effects due to time lags, which can be very complicated and may appear random. Such fluctuations, if their periods are longer than those of single generations, will have similar selective effects as do externally forced fluctuations.

It has long been known that internally generated oscillations can in theory be set up by the interaction of predator and prey. Such Lotka-Volterra oscillations involve in their simplest case a relation between the two species that implies that relative to each other the species (N_y, N_r) fluctuate around a point

$$N_y = \frac{m}{\gamma_r} \; N_r = \frac{r}{\gamma_y}.$$

Any increase in the predation coefficients γ_r and γ_y brings this point nearer to the origin, and so increases the chance of extermination of one or other species by random fluctuations when the numbers are low. A comparable theory for discrete generations was independently developed by Nicholson and Bailey.

The constant of integration which sets the amplitude of the fluctuations presumably depends on the initial conditions. It may be susceptible to environmental change, but one would expect that after a change in amplitude, the new value would persist over several cycles. This is perhaps suggested in experiments with bean weevils and their hymenopterous parasites, but has not been observed in nature.

The most extraordinary cases of regular oscillation in animal populations, namely, the cycle of about 10 years in the lynx and snowshoe hare in boreal Canada, and that of between 3 and 4 years in the lemmings and foxes of arctic Canada and the mountains of Norway, show some variation in amplitude, but in a very irregular manner. The first of these cycles has certainly persisted for over 200 and the second probably for over 100 years.

Though such cycles were often supposed to be Lotka-Volterra oscillations, involving either the mammalian prey and predator, or the plant cover and the prey, this now seems rather unlikely. The 3- to 4-year cycles of the lemmings and the foxes feeding on them are most plausibly attributable to the rate of natural increase being high enough to generate oscillations in a population breeding very discontinuously. The nature of the 10-year cycle is more obscure. It has been suggested that the oscillations control the predator population, and so are adaptive, just as shorter cyclical variation in the seed production of northern trees is believed to prevent development of excessively high populations of seed-eating mammals and birds.

The diversity of any community clearly must depend on the length of the food chains in its food web and more importantly on the species packing possible in any of the lower levels of the chains. This clearly depends not only on the specialization of food habits, but also on the development of vegetation, or animal-built reefs under water, which increase the opportunities for shelter and food in highly diversified ways.

The measurement of diversity has been greatly developed in recent years. The simplest measure is the number of species present, but this gives no indication of the rarity or commonness as contributing to the diversity as it would be experienced by any organism living in the community. The most commonly used measure which corrects this defect is the Shannon information-theoretic index

$$H' = -\Sigma p_i \log_n p_i$$

where p_i is the fraction of any sample that consists of species S_i, n being any convenient base, in information theory usually 2. Four known distributions have been recognized in data relating number of species to number of individuals.

The geometrical distribution assumes that a certain fraction of available space, energy, or other competitive variable is preempted by the first species, the same fraction of what remains by the second species, and so on. It appears to characterize some plant associations, especially under rigorous climatic conditions.

The log-series distribution is nearly harmonic when only rare species are considered. If n_1 is the number of species in a collection represented by one specimen, then n_2 is nearly $n_1/2$, n_3 nearly $n_1/3$, and so on. The most satisfactory expression is

$$S_N = \frac{\alpha x^N}{N}$$

where S_N is the number of species represented by N specimens, x a number just under unity, and α an index of diversity. This distribution appears to fit many assemblages of insects, but the explanations that have been given are not fully satisfactory.

A third type of distribution is seen in many cases if the species are arranged in order of increasing abundance, and then grouped in octaves 0–1, 1–2, 2–4, etc. The number of species in each octave is seen to be normally distributed so that the whole assemblage is log-normal. This type of distribution can arise quite easily and may be regarded as the most general case.

A fourth kind, the broken-stick distribution, is found in collections of organisms from within a fairly homogeneous environment. In it the proportion of each species corresponds to the most probable lengths of the segments of a stick broken into n parts by the random designation of $n - 1$ breaking points. There is evidence that in homogeneous environments such as the bottom deposits of lakes, animal communities tend toward this distribution as the most stable state.

Though it is possible to say something about the meaning of these different types of distribution, they do not go very far. Most explanations of diversity, if they go beyond the three factors of food-chain length, species packing, and environmental complexity, are based on a reasonable, but not yet fully substantiated belief that in an area of given environmental complexity and given access to a species pool, the diversity depends directly on area, on productivity, and on the stability of the physical conditions. While it has often been supposed that diversity promotes the stability of a community, the conditions under which this may happen appear to be very restricted and more generally it is likely that the opposite is true.

Aria da Capo and Quodlibet

WITTGENSTEIN wrote, on the last page[1] of the *Tractatus Logico-philosophicus*, which, whatever else it is, may be one of the greatest poems of this century, "My propositions are elucidatory in this way: he who understands me finally recognizes them as senseless."

When one is writing science, neither Wittgenstein nor anyone else would want such a statement to be a true ending to a book. There is, however, a contemporary school[2] that is very skeptical of the validity of the kind of theoretical argument that I have used, finding it to be tautological and claiming that when tautologies are suitably mixed with empiricisms, which seem to be what most of us broadly call observations, the result of the summation is metaphysics, to be avoided like poison. Such criticism is healthy in that it continually makes us keep track of what we are doing. The criticism can, however, easily become unhealthy if it discourages certain kinds of work by too rigid an adherence to dogmas about the proper scientific procedure. Very great men[3] indeed have believed that the entire area of evolutionary biology is outside the legitimate domain of science as it cannot be approached experimentally in a direct way. Such ideas, expressed nearly half a century ago, have not prevented us from acquiring, during that half century, a vastly increased understanding of evolutionary processes.

Logico-mathematical theories are of course tautological in the sense that they are derived analytically from a set of axioms. If we are studying science, certain axioms, here called postulates, are the formal statements of something believed to be possible about the external world. Theory derived from correct logico-mathematical manipulation will produce propositions which will be true when the postulates are true statements about the external world, but may be either true or false if a postulate is a false statement. It is not possible, as is well known, to verify in an absolute sense the truth of any statement used as a postulate in this sense, but it is possible to falsify it, for all practical purposes, though even then one always might make a

1. L. Wittgenstein, *Tractatus Logico-philosophicus*. London, Routledge Kegan Paul, 1932, 189 pp. The quotation is from 6.54 on p. 189. Though this or the later English version are usually quoted, perhaps the full quality of the work is more apparent from "am Ende als unsinnig erkennt" than from "finally recognizes them as senseless."

2. References may be found in R. H. Peters, Tautology in evolution and ecology. *Amer. Natural.*, 110:1–12, 1976.

3. I am particularly thinking of my revered and much-loved former chief, Ross Granville Harrison, the greatest biologist with whom I have had a close day-by-day relationship over a number of years. I think, however, that this point of view was quite widely held in the 1920s and 1930s. Harrison, though looking at science in ways that could be very different from mine, always supported me as a junior colleague. He had a marked, if indirect, effect on ecology through his interest in allometric growth, which interest developed partly in response to Joseph Needham's work. It affected E. S. Deevey, R. Lindeman, and myself, as we all lunched with Harrison on the fourth floor, under the roof, of the Osborn Zoological Laboratory.

mistake in the process of comparison of the result deduced with the observations, and so reject an agreement that is actually adequate.[4] In a satisfactory case in which we have fairly good prior reason to think that a certain hypothesis may be true, we may find either that we are wrong, the hypothesis becoming too improbable to worry about, or that we are right, meaning that the hypothesis is a bit more probable than we previously had thought. The process is qualitatively the same whether we are increasing the probability from 0.900 to 0.999 in physics or from 0.501 to 0.510 in evolutionary biology.

Since we would not have made the deductions and comparisons if we had not initially had an intuition of being on the right track, the most practical thing to do when we have falsified a hypothesis is to modify one of the postulates used and try again. Over time, as any set of concepts develops, it will be found to have embraced a number of alternative hypotheses, some proving to be almost certainly false, others proving to be very probably true in a certain set of circumstances. The set of all these possibilities, some falsified under all, some verified under some, circumstances, would, if the subject matter were important enough, constitute what Kuhn calls a paradigm.[5] When all the possibilities of such a set that are readily conceived have been tried, before the ingenious investigator attempts to turn to something radically new, the results are filed away for future reference in papers and books. A few such results which immediately appear as important become part of the intellectual currency of science and are taught in classes and stated in textbooks; others remain nostalgically in the minds of their discoverers, on whose death they pass into the unread older literature, well summarized in the reviews of N or M.

If the work was worth doing, concepts such as the logistic may be rediscovered, after a rest of 80 years in limbo, and then may undergo a new development. This process has been amply documented in the preceding pages. More usually the best immortality that an idea can hope for is to be disinterred by a historian of science. Though historians of science probably look upon themselves as students of the development of the human mind, their value as quarrymen or miners of good forgotten ideas should not be overlooked.

All the theoretical arguments in this book are obviously based in part on postulates derived from biology. As logico-mathematical statements, they are tautological; their biological interpretation, however, is not tautological and can be falsified. Their implications can be compared with biological reality and in every case the simple initial statement is found to be wrong in at least some important circumstances. Most significantly, the logistic is wrong when the resources that determine K are not constant, as in man, or when K-selection is continually improving the efficiency with which these resources are used. The Volterra competition equations are wrong whenever the outcome of competition depends on the frequency of one competitor. This is presumably due to the competition functions being of too simple a form. As Cardinal Newman[6] put it, "In scientific researches error may be said, without a paradox, to be in some instances the way to truth, and the only way."

The concept of r- and K-selection, which is widely believed to be useful in understanding much of the behavior of populations, and the concept of frequency-dependent competition, which is likely to become a very important explanation of specific coexistence, would never have been discovered without the development of the simple theories. There are indeed also some cases where the simple theories are reasonably well confirmed by experiments.

A further criticism is that in the supposed confirmation of theory in the laboratory or

4. This specific point, as well as much else, I owe to Harold Jeffreys.

5. T. Kuhn, *The Structure of Scientific Revolutions*. Chicago University Press, 1962, xv, 172 pp.

6. Quoted from Henry Chadwick, presumably from Newman's *Idea of A University*, in *Times Lit. Suppl.*, August 13, 1976, p. 1003. col. 2.

in nature, the confirmation turns out merely to be entry of cases in a classification obtained by a tautological process and so covering every possible contingency. The example of frequency-dependent competition, not implied by the original Volterra equations, but recognized by the use of some of their properties, shows that such classifications may initially be deceptive, and need to be examined in terms of their own internal structure. Any classification which proves to be sound, however tautological, can be used in an important way merely by enquiring what are the empirical frequencies of occurrence of events in the different categories. The explanation and finally the prediction of such probabilities may require quite different kinds of theory from that used in making the classification. What the classification does is to tell us what sorts of phenomena we can encounter; further work may tell us which of these we will encounter, and how often we are likely to do so.

In the case of the various kinds of competition, it is reasonably clear that in a great many cases, fairly similar species cannot live together; but that when they can, it may be due to niche separation, or to frequency-dependent competition, or to the effects of predation. We have at the moment no idea which alternative is the most probable. Fifteen years ago niche specificity would have seemed the obvious answer, but now Connell[7] can write: "Predation should be regarded as being of primary importance, whether directly determining species composition or in preventing competitive exclusion, except where the effect of predation is reduced for some reason." Connell goes on to consider the two more important of such reasons, namely, refuges and defenses. They often apply, but very qualitatively the quoted sentence, though not free from tautology, implies a by no means negligible probability for the importance of predation. The volume in which the quotation occurs, presents the best general evaluation of the state of knowledge, in 1975, of this type of problem, but it apparently contains only one mention of determination of the direction of competition by frequency dependence (p. 495–96) and part of that discussion is probably wrong. Since poppies and pomace flies are not closely related organisms and since the effect also seems to occur in daphnid Cladocera, the probability of it occurring in many other groups and having some general importance can hardly be negligible. Though in this case we cannot really get near to assessing numerical values of the probabilities, our informed guesses, such as they are, certainly have changed and hopefully have improved in a decade or so.

The present aim of the part of ecology with which this book is concerned is therefore largely to uncover possibilities, by any kind of theoretical analysis that proves helpful, and then to see how many of these possibilities are indeed realized in nature. This activity may well take the form of mapping onto any suitable classification of nature, systematic, ecological, geographic, or temporal, the frequency of occurrence of particular kinds of demographic or ecological phenomena. This is what we do when we say, for instance, that long food chains more often occur in aquatic than terrestrial communities, or make any statements about the rules that seem to govern clutch size in birds. As such maps develop and become more precise, more and more hypothetic-deductive analysis of their contents should become possible. In the case of food chains and egg numbers, this of course has largely been done, if in a rather informal way. What is most likely to impede the growth of this part of science, as of others where the approach is extensive, is not heresy in scientific method but ignorance and lack of imagination.

There is, moreover, another pressing danger. Long before we have reached even an elementary knowledge of the distribution of

7. J. H. Connell, Some mechanisms producing structure in natural communities: A model and evidence from field experiments. In: *Ecology and Evolution of Communities*, ed. M. L. Cody and J. M. Diamond. Cambridge, Mass., Belknap Press of Harvard University Press, 1975, pp. 460–90; see specifically pp. 475–76.

FIGURE 142. Leguat's figure of the solitaire, *Pezophaps solitarius*.

the kinds of ecological phenomena, they may have disappeared, owing to the continual erosion of nature that is characteristic of our era.

This book has partly been written to show that there is a considerable amount of available knowledge set in an irregular way in a vast area of ignorance. We have already noted that in the case of the now extinct huia, the observations recorded suggest, but do not quite establish, the existence of a unique type of mutual feeding behavior between the sexes.

An even more tantalizing story is told of another extinct species, the solitaire of Rodriguez in the Indian Ocean. The island was colonized temporarily by a group of Huguenots in 1691. François Leguat,[8] who was the leader of the group, left a fascinating account of their adventures on the island. The book first appeared in French, but was published in London in 1707. In it Leguat gives a detailed account of the behavior of a flightless bird, a little taller than a turkey, which he called the solitaire, a name probably derived from another less well-known extinct bird, living on Réunion in the seventeenth century. The solitaire of Rodriguez was a highly territorial bird mating for life. Each territory had a radius of about 200 yards and was defended during the incubation of the single egg and during a period of several months while the young bird was not capable of independent life.

The male had a bony mass under the wing feathers that was used in defense, though Leguat says intruding females were driven off by the female, males by the male. The general form of the bird (figure 142), the moderate reduction of the wing, and the bony mass were confirmed in the nineteenth century when much skeletal material was recovered from the island. Leguat also describes the female as having a cleavage in the arrangement of the feathers on the breast, which delighted him as representing *merveilleusement un beau sein de femme*. This was doubtless an incubation patch. Except for the specific roles of the sexes in defense, part of his account is confirmed by the subfossil evidence, the rest so far seems reasonable. We now, however, reach the extraordinary part of the story. Leguat says that when the chick had left the nest for some days, a company of 30 or 40 adults brought another chick and, being joined by the parents, the two young birds were escorted to some unoccupied territory. This extraordinary ritual he describes as the marriage of the solitaire. Leguat seems so reliable on most other aspects of the bird's biology, and the marriage, if it occurred, must have been so conspicuous, that it is hard to doubt its reality. Now the bird is extinct and we shall never know the truth.

Of course, we can learn little of the ecology and almost nothing of the behavior of the several million extinct species that have existed in geological time. Nevertheless, the evidence that we have as to the limits of what evolution can do at any level of organization is precious. Moreover, it is the highly specialized, slow-breeding, *K*-selected species which are most vulnerable, but which give most information about such possible limits. Think what we might have learned if Gerald Durrell had been able to establish a colony of *Pezophaps solitarius* on the island of Jersey. Some well-meaning people will doubtless wonder what the preservation of odd organisms merely for their oddness does for suffering humanity. I can only reply that experience shows that the most unpromising knowledge is always proving useful and that some of us have a duty to foster and make it available. It is well to bear in mind that the niches illustrated in figure 101 are totally inadequate for man.[9] We also need[10] all the wondrous things under heaven:

> Their leaues that differed both in shape
> and showe
> (Though all were greene) yet difference
> such in greene
> Like to the checkered bent of Iris bowe.

8. See chap. 3, note 32.
9. Cf. Matth. 4:4.
10. From a poem by Christopher Marlowe, on p. 480, in *England's Parnassus; or, The choysest flowers of our moderne poets*. London, for N.L. C.B. and T.H., 1600, x, 510 pp.

Appendix

Ratiocinator Infantium

or The Modicum of Infinitesimal Calculus Required for Ecological Principles

(1) Meaning of dy/dx:

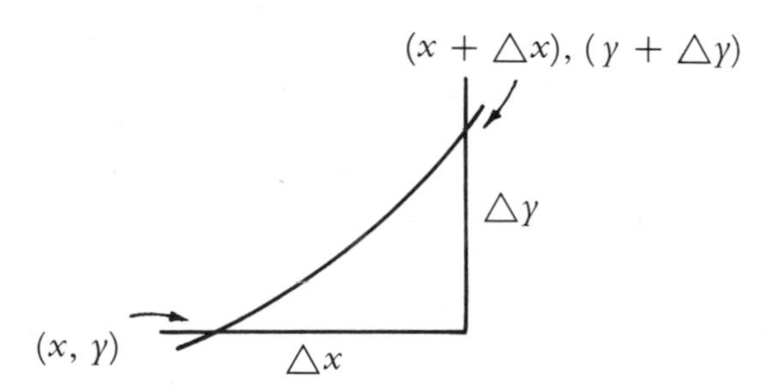

Consider two points on a curve (which might be a curve of a growing population or anything else that interests you). Let the coordinates of one point be (x, y), of the other $(x + \triangle x), (y + \triangle y)$. The ratio of $\triangle y/\triangle x$ gives an approximation to the slope of the curve between the two points. What we want to know is how fast y changes with respect to x, not approximately between the two points, but accurately at one of them. Take $(x + \triangle x, y + \triangle y)$ nearer and nearer to (x, y). Since the curve becomes shorter and shorter and can do less wobbling, the approximation to a true slope at a point becomes better and better. The small length, $\triangle x$, can be made as small as one likes. Ordinarily, for any value of $\triangle x$ there will be a definite value of $\triangle y$ which is also made smaller and smaller. The ratio $\triangle y/\triangle x$ will thus approach closer and closer to what we want. Let us call the ratio at x and y, which we are alooking for, dy/dx. Then we may say, as $\triangle x$ approaches 0, the limit of $\triangle y/\triangle x$ is dy/dx. It is theoretically important to note that since we must not make x and $y = 0$ (otherwise we get the meaningless ratio 0/0), so we cannot "really" get to dy/dx. This difficulty was the basis of Bishop Berkeley's attack on the calculus. For scientific as opposed to logico-mathematical purposes, this difficulty appears not to matter; it continues to worry some mathematicians.

(2) The process of obtaining the slope of a curve from its equation is called *differentiation*. It is easy to show that this process can be applied to any equation piecemeal, differentiating each individual term separately. Constant terms, since they do not change, disappear. It is also easy to show that if

$$y = bx^n$$
$$dy/dx = bnx^{n-1}$$

so that for any point on the graph of a linear equation:

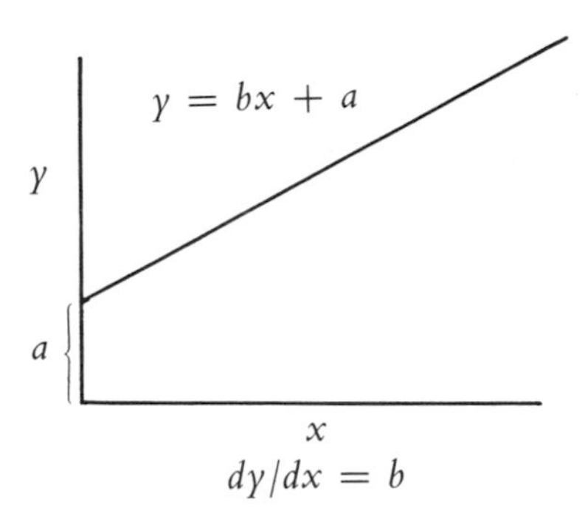

$$dy/dx = b$$

The slope of the line is b, which will be obvious.

Or in the quadratic equation:

$$y = cx^2 + bx + a$$

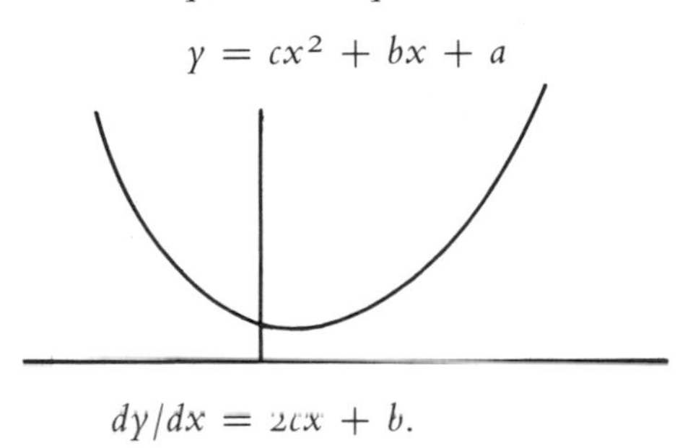

$$dy/dx = 2cx + b.$$

This means that if we want to know what the slope is at any value of x, we just write that value of x into the equation for dy/dx. For $x = 0$ the slope is b (a negative quantity in the illustration since the curve is going upward to the left), for $x = 1$ the slope is $2c + b$, etc.

(3) Where we have a minimum or a maximum

$$dy/dx = 0$$

momentarily y is not changing with respect to x. In the above example

$$dy/dx = 2cx + b.$$

Set this equal to zero

$$2cx + b = 0$$
$$x = -b/2c.$$

This tells us where the minimum is at once.

(4) If we want to study the change of dy/dx with x, we can draw a new graph evaluating and plotting dy/dx against x. The usual way of indicating this second derivative is by the symbol d^2y/dx^2, while further derivatives are d^3y/dy^3, d^4y/dy^4, etc.

In the example above

$$y = cx^2 + bx + a$$
$$dy/dx = 2cx + b$$
$$d^2y/dx^2 = 2c.$$

Since the second derivative is a constant $2c$,

$$d^3y/dy^3 = 0$$

and the process stops.

(5) Two theorems, here not proved, are convenient:

(a) $dy/dx = dy/dt \, . \, dt/dx$ where t is any new continuous variable.
(b) $dyt/dx = t \, dy/dx + y \, dt/dx$.

(6) There is a very important special case of great biological significance, namely, the case in which

$$dy/dx = y.$$

It is found for this case that

$$y = e^x$$

where e is the base of natural logarithms or 2.718. It is, furthermore, evident from (5a) above that when

$$y = e^{ax} + k$$

where k is a constant

$$dy/dx = ae^{ax}.$$

This last expression is of great biological importance.

(7) Let us assume a population of N unicellular individuals, where N is large and so the growth of the population reasonably continuous. Under ideally favorable conditions, the growth rate of the population will depend on the division rate b (divisions per hour or day per individual) and on the number of individual N, or symbolically:

$$dN/dt = bN$$

whence from the considerations above

$$N = e^{bt} + k.$$

The constant, which depends on initial conditions of the population, is called the *constant of integration*. In this case$(1 + k)$ is the number of individuals when $t = 0$, so if we start with one organism multiplying by binary fission, $k = 0$. Most of this book is about the fact that no organism can keep this up indefinitely.

(8) The process of going from the derivative (i.e., dN/dt) to the original equation (i.e., $N = e^{bt} + k$) is called *integration*. Symbolically we denote this process in the case of the example by

$$\int bN\,.\,dt = e^{bt} + k.$$

(9) If we have an equation

$$\frac{dy}{dx} = f'(x)$$

where $f'(x)$ is some function of x, we write in general,

$$y = \int f'(x)\,.\,dx$$

The only general way of finding out what the expression y is, is to differentiate all the possible kinds of expression containing x, and then pick out the one which gives $f'(x)$ as its derivative. It is obvious that this is an infinitely laborious procedure. We do know, however, that

$$\int 2c\,.\,dx = 2cx + b$$

$$\int (2cx + b)\,.\,dx = cx^2 + bx + a$$

as well as the exponential case already given in (8). Integration is therefore not quite hopeless. Lots of tables of integrals exist.

(10) In general, the integral represents the growth of an area below a curve as we move along the curve. In the simplest case when

$$\frac{dy}{dx} = b,$$

$$y = \int b\,.\,dx = bx + a$$

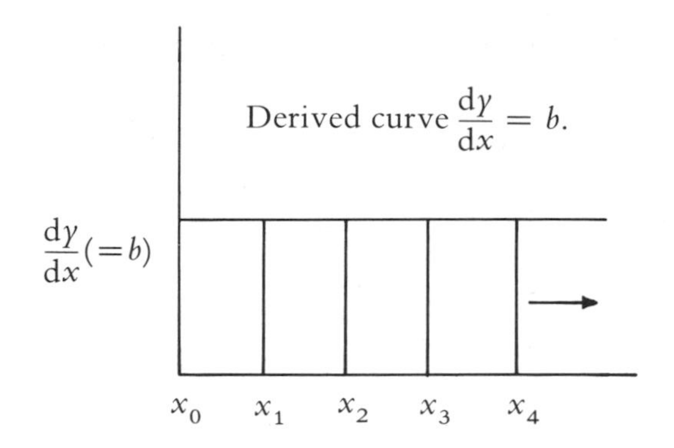

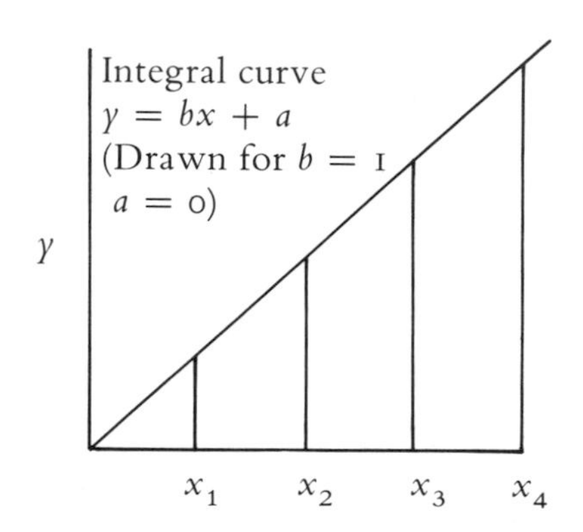

The area under the line between x_0 and x_2 in the example $(b = 1, a = 0)$ is twice that between x_0 and x_1. In the *integral curve* on the right we find y at x_2 is twice y at x_1.

The numerical value of the integral between two values of x, say x_1 and x_2, is written

$$\int_{x_1}^{x_2} f'(x)\,.\,dx.$$

The easiest way to get it is to draw the curve $y = f'(x)$. Draw perpendiculars at x_1 and x_2, and measure with a planimeter.

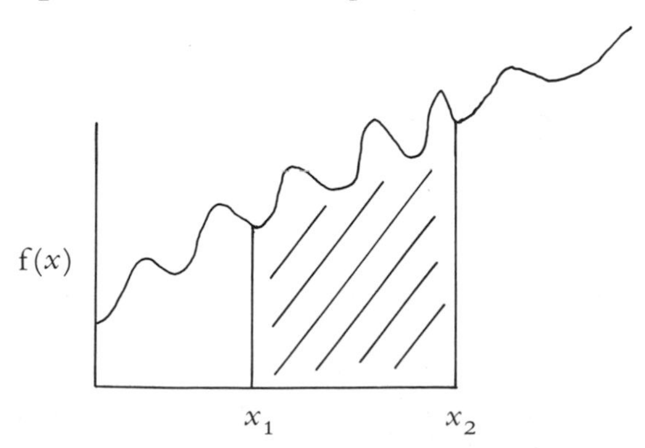

This permits one to get values of the integral even if one does not know how to write it down in symbols. Mathematicians think this is not really fair, even though they might not be able to treat the example given in the last graph in any rigorous way. There are anyhow some quite respectable-looking functions that can not be integrated. One of these transcendental integrals is $\int e^{-x^2}dx$, which is apt to turn up in unexpected places, such as in oxygenation of water and in statistical theory.

(11) The following may be found useful

$$y = \log_e x,\ dy/dx = \frac{1}{x}.$$

Index of Authors

Index of Genera and Species of Organisms

Botanical conventions are used in citing authors' names of plants, and zoological conventions in citing those of animals. Vernacular names with initial capitals are accepted names for the species listed, those that are uncapitalized merely indicate the general nature of the organism.

General Index

Inclusion of proper names implies more than reference to the contents of a published paper or book.